An Anatomy of Risk

Wiley Series on Systems Engineering and Analysis
HAROLD CHESTNUT, Editor

Chestnut
Systems Engineering Tools

Hahn and Shapiro
Statistical Models in Engineering

Chestnut
Systems Engineering Methods

Rudwick
Systems Analysis for Effective Planning: Principles and Cases

Wilson and Wilson
From Idea to Working Model

Rodgers
Introduction to System Safety Engineering

Reitman
Computer Simulation Applications

Miles
Systems Concepts

Chase
Management of Systems Engineering

Weinberg
An Introduction to General Systems Thinking

Dorny
A Vector Space Approach to Models and Optimization

Warfield
Societal Systems: Planning, Policy, and Complexity

Gibson
Designing the New City: A Systemic Approach

Coutinho
Advanced Systems Development Management

House and McLeod
Large-Scale Models for Policy Evaluation

Rowe
An Anatomy of Risk

An Anatomy of Risk

William D. Rowe, Ph.D.

A WILEY-INTERSCIENCE PUBLICATION
JOHN WILEY & SONS, New York • London • Sydney • Toronto

Library of Congress Cataloging in Publication Data:

Rowe, William D 1930–
An anatomy of risk.

(Wiley series on systems engineering and analysis)
"A Wiley-Interscience publication."
Bibliography: p.
Includes index.
1. Technology assessment. 2. Risk. 3. Decision-making. I. Title.

T174.5.R68 614.8 77-5048
ISBN 0-471-01994-1

Printed in the United States of America

10 9 8 7 6 5 4 3 2 1

To our children—in the hope that an
understanding of risk will clarify their options.
Perhaps their choices will be better than ours.

A reasonable estimate of economic organization must allow for the fact that, unless industry is to be paralyzed by recurrent revolts on the part of outraged human nature, it must satisfy criteria, which are not purely economic.

R. H. Tawney
Religion and the Rise of Capitalism

SYSTEMS ENGINEERING AND ANALYSIS SERIES

In a society which is producing more people, more materials, more things, and more information than ever before, systems engineering is indispensable in meeting the challenge of complexity. This series of books is an attempt to bring together in a complementary as well as unified fashion the many specialties of the subject, such as modeling and simulation, computing, control, probability and statistics, optimization, reliability, and economics, and to emphasize the interrelationship among them.

The aim is to make the series as comprehensive as possible without dwelling on the myraid details of each specialty and at the same time to provide a broad basic framework on which to build these details. The design of these books will be fundamental in nature to meet the needs of students and engineers and to insure they remain of lasting interest and importance.

Preface

This book examines the means by which our society, or its agents in Congress and regulatory agencies, can set levels of acceptable risk for technological systems and programs. Can formal acceptable levels of risk be set? Why should they be set? How can they be set? These questions, recently only barely perceived, are quickly taking on considerable importance.

A few years ago, as a member of a federal regulatory agency, I became aware of the degree to which many regulatory decisions are based on assessments of risks, and of the inability of bureaucratic organizations to deal with such assessments. My work forced me to take risks into consideration in regulatory decisions, and I undertook to determine what we meant by "risk" and what means were available to analyze it. I found a surprising diversity in definitions of risk and risk acceptance and a singular lack of understanding of the means to assess risks. Unfortunately, the concept of risk was so little understood that it was often misapplied, resulting in erroneous use of risk concepts. Thus clarification and better understanding of the concepts involved in risk assessment appeared necessary.

As a result, I attempted to define "risk" for the program at hand, and to state how the program could deal with it. It was to be a three- or four-page paper. I immediately discovered that the problem of risk assessment was so complex that for every question I addressed, 10 new questions arose. The brief paper never emerged, but over 18 months a 200-page monograph evolved instead. Three hundred copies of this initial version of *An Anatomy of Risk* were distributed for peer review in March 1975. The response was far greater than expected. Not only were voluminous comments received, but there were more than 2300 requests for copies. The interest and helpful comments persuaded me to take the next step and formalize *An Anatomy of Risk* into a book.

Although the readers of the monograph will recognize the underlying theme, little remains of the original effort. The text has been completely revised, and more than half of the material is entirely new.

This study on risk was undertaken at my own time and expense. It is a personal work that in no way represents any policy of any agency or organization with which I have been associated. However the objective of the book, though by no means the last word on risk assessment, is to provide further insight into risk and risk assessment, and to stimulate further discussion and effort in the field.

As you read this book, keep in mind that I feel I have asked more questions than I have answered, and that I often have taken stances to provoke discussion. There are no "rights" or "wrongs" in this area, only paths to better understanding and methods whose validity can only be tested by their usefulness in resolving issues. In this light, the assessments in Chapter 19 are for illustrative purposes only and must be refined considerably before being used in practical applications. Nevertheless, some inevitably will use the evaluations in a manner different from that intended. I know of no other way to illustrate the method than by example, and I exhort readers not to use my results as conclusive.

I acknowledge the encouragement of Harold Chestnut, editor of the Systems Engineering and Analysis series, and commend his desire to expand the scope of the series. I also thank those who have commented on the initial monograph, and I hope that I have addressed their concerns here.

WILLIAM D. ROWE, PH.D.

Falls Church, Virginia
September 1976

Contents

An Anatomy of Risk

1

Introduction

The only certainty in life is death; uncertainty lies in when and how death occurs, and whether it is final. Man strives to delay its onset and extend the quality of life in the interim. Threats to these objectives involve risks, some natural, some man-made, some beyond our control, and some controllable.

As the length and quality of life have increased, and thereby its value, society has become increasingly concerned with avoiding risks, particularly those imposed without offsetting benefits to the risk taker. Although every activity involves some risk, there are some kinds of risk and some levels of risk that members of society are unwilling to assume.

Simultaneously, the increasing complexity of technologically based industrial societies has resulted in growing concern about risks imposed by man-made technology. In the 30 years since the end of World War II, the *scale* of man's activities has increased sharply. There has been a unique quantitative jump in industrial society's ability to produce goods and services, and to create new demands for these goods and services. There has also been a qualitative change as technology has made breakthroughs into new areas, such as genetics, atomic energy, and the development of new materials. For example, man has learned to compound substances unknown to nature, which nature has no ability to counteract.[1] As a result, man is faced with a bewildering array of risks that must be reconciled continually. Increased awareness of new technological hazards, together with increased attention to the impact of these systems on our safety, health, and environment, has resulted in new anxieties toward man-made risk. These anxieties are exacerbated by rapid communication and dissemination through the news media, which have major impact on the public's perception of risks. Many of these imposed technological risks are perceived by different groups as unacceptable and result in major confrontations in the courts, in referenda, and in regulatory proceedings.

The need to address these problems in a regular and consistent manner has focused attention on the formal analysis of risk, as well as on the need to address the more subjective problems of how risks are deemed to be acceptable or unacceptable. A great deal of attention has been devoted to the means to estimate risk. The scientific establishment has developed new analytical techniques for dealing with technological and, to some extent, societal problems. Systems analysis, operations research, and decision theory are all examples of techniques that have been applied to problems involving risks. However these methods have been comparatively unsuccessful in the more subjective areas of risk acceptance and interaction of estimates of risk and levels of societal risk acceptance. Certainly the analysis of a seemingly simple concept, with which man has always dealt pragmatically, has not resulted in significant solutions. The subject of risk is in reality very complex. Since risks involve man, the analysis of risks will be as complex as the individual, group, and societal behavior at hand. Oversimplification of complex issues is a temptation that often results in masking of important considerations and may lead to misrepresentation of results.

The purpose of *An Anatomy of Risk* is to investigate this complexity of concept, to provide dimensions and definitions that encompass and describe the subject of risk, and to address a variety of methods for dealing with the analysis of risk. In accomplishing this purpose, definitions must be explicit, and we should revisit previous work in the field. The objective is to clarify the subject of risk and, where possible, to provide techniques useful for addressing risk problems. There are no final answers; only some clues and paths to follow to where and how such answers might be obtained.

In reaching this objective, attention is focused on the total scope of risk analysis and the provision of a framework for the subject. The use of this framework is directed at the development of means to arrive at formal decisions on the acceptability of risk. Robert Anthony has provided a *raison d'être* for such a framework:

> To be useful, material dealing with any broad subject needs to be organized within a framework of topics and subtopics. If the topics and subtopics are well chosen, the available material can be so arranged as to make it possible for one to reach conclusions generally applicable to each classification but not applicable to other classifications. Such conclusions or generalizations, furthermore, will have validity and significance impossible in the absence of such a framework.[2]

Anthony goes on to use biological systems as an example of the usefulness of frameworks:

> Within a given subject area, different frameworks can be useful for different purposes. Thus, in biology, the classification of plants and animals by phylum, class, order, family, genus, and species is useful for some purposes. Comparative anatomy, which cuts across these classifications, provides a useful framework for other purposes.[2]

This analogy has provided the rationale for the title, *An Anatomy of Risk*. Not only are risks and components of risk to be classified, but an approach that cuts across the classifications to develop a methodology for risk assessment is sought. This involves comparison of risks of different types.

In developing a framework, it is important not to reinvent the wheel. The work of other investigators is used extensively; indeed this effort, relies heavily on existing material.

In this light the study attempts to answer three major questions:

1. What is risk, and how many different aspects of risk can be categorized? Included is consideration of such topics as why it is important to understand risk, and if and when exposure to risks should be regulated by government.
2. How may different types of risk be compared for making risk decisions? Consideration of different methods and data bases for establishing comparative risk levels is involved.
3. How can acceptable levels of risk for specific activities be determined? Included is consideration of individual, group, and societal risks *in toto,* alternate methodologies for risk assessment, and the meaningfulness of such judgments based on the need for implied subjective value judgments.

Two major underlying concepts make this study different from earlier work in the field. First is the idea that the assessment of risk is as important as the quantification of risk; and second, that the subjective perception of risk is the basis for risk acceptance regardless of the objective or quantified evaluation. In a recent treatise on risk by Lowrance,[3] a major contribution to the estimation of risk that summarizes the ways an industrial society may methodically appraise risk, only 11 pages deal with acceptability of risk, and that only qualitatively. In his review of Lowrance's book, Kai N. Lee states:

> . . . Yet the book is disappointingly unprovocative: safety analysis, it appears, is little more than common sense routinized.
>
> The fundamental distinction in this subject is between risk and safety, Lowrance says. Risk is here taken to mean the (objectively ascertainable) probability of harm. "Safety" denotes the social and legal judgment that that probability is appropriately low; "a thing is safe if its risks are judged to be acceptable." The statement has a peculiar ring to it.
>
> What is peculiar is its flatness. Lowrance omits a crucial signpost that would help to orient the reader in the conceptual marshes of hazard assessment. The long-running dispute over the interpretation of probability measures is of central, if unacknowledged, importance for the definition of risk. One school holds that probabilities are primarily reflections of the actual frequency of occurrence of events; probabilities are therefore objective, as Lowrance postulates. Another school holds that the assignment of probabilities primarily reflects the assigner's belief or

confidence that the events in question will occur. This subjective view would, if adopted, blur the difference between risk and safety that Lowrance seeks to discern: probability and judgment are intertwined, not separate.[4]

The assessment of risk is now becoming a specific discipline of interest as evidenced by considerable recent activity on this subject. An Engineering Foundation workshop on risk-benefit methodology and application held in Asilomar, California, in September 1975, brought together a spectrum of practitioners. The International Institute for Applied Systems Analysis (IIASA) in Laxenburg, Austria, has a major interdisciplinary effort underway in the area of risk assessment. Kates[5] has addressed the problem of risk assessment directly, albeit in a qualitative manner. His many insights are referred to often in this book.

On the other hand, risk comparisons are often made erroneously. Voluntary risks such as those involving motor vehicles are used to justify smaller risks imposed involuntarily on a risk taker. Of particular concern here is the treatment of involuntary risks imposed inequitably on groups in society who either knowingly or unknowingly fail to reap the benefits of the activity causing the risk. It is the responsibility of representative government to assure that such inequities are minimized. Although Congress and the judiciary are involved in ameliorating inequities, the major role is played by regulatory agencies in the executive branch. Public utility laws, the National Environmental Policy Act of 1969, the Federal Trade Act of 1913, and the Consumer Protection Act of 1975* are all examples of this process in action. The question may be narrowed to a determination of how a regulatory agency fairly assesses acceptable levels of risk in a visible, reportable manner in the best interest of the public.

The difficulty of addressing this type of problem lies in the imprecision of intrinsic human values on which decisions are based. The resulting decisions are often expressed in explicit, objective terms that belie the subjectivity of the value judgments employed in arriving at them.

A regulatory agency involved in setting standards of acceptability for society is responsible for making the required value judgments for society, but it has the added responsibility of employing procedures ensuring that all affected parties are heard from, that the best available expert testimony and technical information are used, and that the value judgments are made in a visible, reportable manner for all to question and revisit if necessary.

Such value judgments fall into three classes: (1) technical, (2) societal, and (3) managerial. Technical value judgments are made by experts, in the absence of hard technical information, when faced with the difficulties in

* Magnusson-Moss Act.

obtaining further information. Scientific bodies, such as the National Academy of Sciences, usually make technical judgments on the basis of consensus. The 1972 "BEIR Report"[6] on low-level ionizing radiation and the recent report on ambient air pollution hazards[7] are examples.

The societal value judgment involves balancing benefits, costs, and risks, and attempting to minimize inequities in benefit-cost imbalances. It is important to realize that scientists or technical experts have no more expertise in this area than any other well-informed* interested citizen. Uninterested citizens who choose not to participate disenfranchise themselves.†

The managerial value judgment involves interpretation and modification of the societal value judgment in order that the societal expression may be implemented and enforced. The State Implementation Plans required by the Clean Air Act are a reasonable example of managerial value judgments. Here, technicians once again are involved; however lawyers, law enforcement personnel, scientists, engineers, and many others are involved also.‡

This book deals primarily with the first two types of value judgment—technical and societal. First, a conceptual framework for the total field of risk assessment is presented, using parts of several different risk models merged into a single qualitative approach. The structure is by no means complete, but it provides insight to the many accompanying problems. Second, an attempt is made to quantify and compare different types of risk and the levels society seems willing to accept. Finally, these are used to develop a methodology for establishing acceptable levels of risk on a regulatory basis. This methodology is not an end in itself but serves to demonstrate that the development of such methods is feasible, desirable, and useful. Before proceeding, it is worthwhile to briefly review some previous work on risk evaluation.

A great deal has been published on decision theory formulations of risk

* This assumes that the required knowledge is easily available to the citizens who want it.

† A glaring example of intrusion of the technician into the societal area is in the field of economics. Economists propounding their own theories of economics, attempt to influence the economic decisions of the government and the nation. This can result in devastating oscillations of the economy as each makes his "profound" judgment to the press. Individuals seem more concerned with setting forth pet theories than with the health of the economy. More properly, a body of economic experts should examine and review the theories of all proponents, expose the pros and cons of each method, synthesize new approaches as required, and present this information to the appropriate governmental agency to make the systemic value judgments for implementation.

‡ Paul C. Tompkins in a personal communication has pointed out that "much of the time, the societal or managerial value judgments of the technical experts have a major influence on the technical judgments." Unfortunately, he is quite correct and such influence is exerted all too often.

that differentiate between decisions involving risk (probabilities of all alternatives are known and given) and decisions under uncertainty (probabilities not known). These statistical decision theory approaches, which are well documented by Schlaiffer[8] and Luce and Raiffa[9], depend on subjective probability functions in the form of expected value or expected utility of various outcomes. The process, once utility values are assigned to outcomes, is mechanistic.

Approaches that depend more on psychological processes follow the works of Edwards,[10] Coombs and Pruitt,[11] Pruitt,[12] Lichtenstein,[13] Kogan and Wallach,[14] and Van der Meer,[15] where decision making involving risk is seen as a process in which an individual or a group maximizes a combination of subjective probability and utility, where alternatives exhibiting higher potential gain or loss (more variance) are deemed as more risky. These decisions always involve the probability of loss.[16]

Finally, there are the approaches to risk that develop monetary equivalents of premature death, and so on, by examining what people actually seem to do in society. This effort has been led by Starr,[17] Libby,[18] Siccama,[19] and Sagan.[20]

The study has attempted to cover all these areas, as well as many other areas not heretofore considered. The effort is a total multidisciplinary approach that incorporates philosophy, anthropology, psychology, statistics and decision theory, systems analysis, social and political sciences, economics, engineering, medicine, and public health. The author does not claim to be an expert in all these fields, and he has drawn on the work of others more skilled for use in this synthesis.

Broad studies of this type raise more questions than are answered. However this study may serve as a baseline for further efforts in the field, or it may show that some areas of endeavor are dead ends. Its value probably lies between these two extremes; the reader will have to judge for himself.

The study is divided into five major parts, A through E. The introduction to each part lays out its contents and advises the reader on what material he might cover, depending on his interest. In this light this study is aimed simultaneously at two types of reader. First is the reader who is not directly involved as a professional in evaluating risk situations from a purely analytical and, perhaps, academic point of view. This noninitiated reader may be involved in making decisions on risk acceptance in which risk parameters have to be weighed, yet he does not need to be bogged down with all the technical details as long as the main points are made evident. At the same time, the serious professional will want to understand all the analytical aspects that have been studied and used to reach the conclusions given in this treatise, and if he disagrees, he may wish to take issue. The initiation of critical discussion in the area of risk assessment is the prime motivation for this study.

Parts A and B make up the basic descriptive framework for risk analysis. Part A deals with the nature of risk, addresses the question of why it is important to examine risk analytically, especially from a regulatory point of view, and provides a formal structure for risk analysis. The difficulties involved in analyzing risk are identified, some examples of societal risk analysis are provided, and a formal structure for estimating risks is developed. Part B expands the framework by laying out and qualitatively addressing the many factors that affect the assessment of risk and valuation of risk consequences.

Parts C and D treat the quantification of risk and risk assessment in order to develop means to compare different types of risk for decision-making use. Part C takes up the problems and methods involved in quantification of risks, recognizing the subjective, often intangible, nature of values associated with these risks. Part D attempts to quantify a variety of different kinds of risk based on society's historic behavior toward specific risk situations.

Finally, Part E deals with the problem of developing formal methods for risk acceptance on a societal basis. A particular methodology is developed and minimally tested to provide a "straw man" for further discussion of the feasibility of such methodologies, their value and limitations, and the acceptance of the results of the application of these methods by society. To demonstrate its utility, the methodology is used to evaluate four fuel systems for generating electrical energy. Subsequently, areas for future effort are discussed, to lead to a wider discussion of risk analysis problems in society.

It should be recognized at the outset that a wide variety of ideas is presented. The object is to be complete, not right. In this light, neither the author or the reader can accept the content without challenge. Many premises are laid out to identify problems and evoke response, not necessarily to establish final solutions. It is too early in the development of risk analysis to expect such solutions. The total scope of risk analysis must be considered, however, if results are to be valid. Expansion of scope represents enlightenment.

REFERENCES

1. E. F. Schumacher, *Small Is Beautiful* (New York: Harper Colophon Books, Harper & Row, 1973, p. 17), gives a more complete and convincing argument on this subject.
2. Robert N. Anthony, *Planning and Control Systems, A Framework for Analysis*. Boston: Harvard University, Graduate School of Business Administration, Division of Research, 1965.
3. William W. Lowrance, *Of Acceptable Risk: Science and the Determination of Safety*. Los Altos, Calif.: Kaufman, 1976.

4. Kai N. Lee, "Assessing Danger" Book Review, *Science,* Vol. 193, July 9, 1976, p. 139.
5. Robert W. Kates, "Risk Assessment of Environmental Hazard," SCOPE Report No. 8, International Council of Scientific Unions, Scientific Committee on Problems of the Environment, Paris, 1976.
6. "The Effects on Populations of Exposure to Low Levels of Ionizing Radiation," Report of the Advisory Committee on the Biological Effects of Ionizing Radiation (BEIR), Division of Medical Sciences, National Academy of Sciences, National Research Council, Washington, D.C., November 1972.
7. "Air Quality and Automobile Emission Control." Washington, D.C.: National Academy of Sciences, September 1974.
8. Robert Schlaifer, *Analysis of Decisions Under Uncertainty.* New York: McGraw-Hill, 1969.
9. R. Duncan Luce and Howard Raiffa, *Games and Decisions: Introduction and Critical Survey.* New York: Wiley, 1971.
10. Ward Edwards, "Subjective Probability in Decision Theories," *Psychological Review,* No. 79, 1969, pp. 109–135.
11. C. H. Coombs and D. C. Pruitt, "Components of Risk in Decision Making: Probability and Variance Preferences," *Journal of Experimental Psychology,* No. 60, 1960, pp. 265–277.
12. D. C. Pruitt, "Pattern and Level of Risk Taking in Gambling Decisions," *Psychological Review,* No. 69, 1962, pp. 187–201.
13. S. Lichtenstein, "Bases for Preferences Among Three Outcome Bets," *Journal of Experimental Psychology,* No. 69, 1965, pp. 162–169.
14. N. Kogan and M. A. Wallach, *Risk Taking: A Study in Cognition and Personality.* New York: Holt, Rinehart & Winston, 1967.
15. H. C. Van der Meer, "Decision Making: The Influence of Probability Preference, Variance Preferences, and Expected Value on Strategy in Gambling," *Acta Psychologica,* No. 21, 1963, pp. 231–259.
16. Siegfried Streufort and Eugene A. Taylor, "Objective Risk Levels and Subjective Risk Perception," Purdue University Technical Report No. 40, Lafayette, Ind., August 1921.
17. Chauncey Starr, "Social Benefit Versus Technological Risk," *Science,* Vol. 165, No. 169, September 19, 1969, pp. 1232–1238.
18. L. M. Libby, *Technological Risk Versus Natural Catastrophe.* Santa Monica, Calif.: The RAND Corporation, March 1971, p. 4602.
19. E. H. Siccama, "The Environmental Risk Arising from the Bulk Storage of Dangerous Chemicals." Paper presented at the Conference on Hazard Evaluation and Risk Analysis, Houston, August 18–19, 1971.
20. L. A. Sagan, "The Human Costs of Nuclear Power," *Science,* Vol. 177, No. 4048, August 11, 1972, pp. 487–493.

Part A

THE NATURE OF RISK

This part is an introduction to the basic concepts involved in understanding risk. Chapters 2 to 5, aimed at the general reader; cover basic questions in understanding risk. What risk is and how risk analysis may be structured are the major concerns of Chapters 2 and 3, respectively. Chapter 4 investigates problems and considers the manner in which individuals and society determine what risks are acceptable or unacceptable, and how action to ameliorate risk is undertaken. Chapter 5 examines the basic requirements of methods that might be applied formally to implement decisions on risk acceptance and risk aversion.

These four qualitative chapters furnish background. Chapter 6 attempts to cap this background with examples of risk assessment to provide some understanding of its application and problems associated with implementation. Building on this, Chapter 7 sets forth a formal structure for risk estimation. It is quantitative for structural purposes, not for manipulative implementation. As such it is a formal characterization of risk estimation. The casual reader may want to omit Chapter 7.

2

What Is Risk?

Everyone is constantly subjected to an array of risks, both as an individual and as a member of various societal groups. Generally these risks are accepted qualitatively, even questioned and deliberated in this manner, rather than analyzed quantitatively. As a rule risks are quantitatively assessed only in classic gambling games (e.g., playing the odds at craps), in business and insurance decisions, and in some governmental regulatory actions. Reams of data on risks are published annually by institutions throughout the world. However these data are usually descriptive of risks experienced by segments of the population, rather than an analysis of risk decisions. The manner in which society adapts to risks tends to make the concept of risk appear simple, although in reality it is very complex. As a result, analysis of risk data for evaluation of risk decisions is not easily accomplished.

Nevertheless, risk has been a subject of interest and study in the recent past, especially in the insurance industry. As early as 1901 Willett defined risk as "the objectified uncertainty regarding the occurrence of an undesirable event."[1] Knight defines risk as "measurable uncertainty."[2] Denenburg et al.[1] call risk an "uncertainty of loss." All these definitions involve some aspect of uncertainty. Thus an understanding of risk must begin with a consideration of uncertainty, which in turn begins with some basic philosophic considerations.

2.1 PHILOSOPHICAL CONSIDERATIONS OF UNCERTAINTY IN RISK

2.1.1 Mortality as the Root Source of Uncertainty

The only certainty in life is death. Man and biota are mortal. However the time and manner of death are uncertain, and man does not know with certainty whether death is final.* The history of our species reveals that

* An immortal might be concerned with the uncertainty of unexpected death by accident, purpose, or disease, as opposed to death by aging.

until the last few centuries, people did not consciously consider direct control of the time and manner of an individual's death as a possibility. Certainly, various cultures have found dignity in the manner of death important. The Norsemen believed death in battle to be the ultimate goal, rewarded by an afterlife in Valhalla. The Japanese devised *hara kiri* as a culturally acceptable way to maintain dignity in death by controlling the means and the manner.

The "value of life" in many cultures is held low by and for the masses, but high for rulers. The ancient Egyptians used to erect monuments to sustain the afterlife of a ruler—but we know that the pyramids cost the lives of innumerable slaves—a good example of a society's dichotomous attitude toward a significant value.

2.1.2 Religion as a Means of Dealing with the Uncertainty of Mortality

Most religions are based in part on man's fear of the unknown,* particularly the time and mode of death, and the possibility that death might not be final and that life might continue on other planes (e.g., the immortality of the soul in Judeo-Christian metaphysics).† Thus the occurrence of death in a particular time or manner was understood to be an act of God(s),‡ and favor of God(s) sought by personal and group endeavor. Ritual developments evolved into ethical structures that might please or not please God(s).

Since death is certain, the question of coping with its certainty is of major importance. One method of minimizing the negative value of the final event is to assume that it is not final and that life continues on some other plane. Whether this plane is reincarnation or the continued existence of the soul in a heaven or a hell (again using the Judeo-Christian nomenclature), there is scant recurring empirical evidence to prove the existence of such afterlife. This does not mean that there is no afterlife, but only that the acceptance of a particular belief in afterlife must be based on individual faith and the meaningfulness of such acceptance to individuals.§ Depending on the degree of faith in a particular dogma involving afterlife, the negative value of death

* Love and/or awe of the deity(s), and a structured set of beliefs, values, and behavior patterns are other factors. However, the set of beliefs about life and death that allow adherents to cope with fear of the unknown is generally considered to be a primary motivating factor in both primitive and modern religions.

† No attempt is made here to evaluate the "rightness" or validity of any religion or religious philosophy, but man's belief in such tenets may strongly influence his culture and behavior. Thus the examples used are to cite behavioral patterns.

‡ Both monotheism and polytheism are implied.

§ Other problems arise when one group tries to impose its beliefs on others. The problems of religious persecution and active evangelism are beyond the scope of interest here.

may be decreased and perhaps even reversed. In the latter case one may trade off this life against a better life in the next plane by achieving prescribed cultural and ethical goals at the expense of deprivation (although often undertaken willingly).

The chance behavior of natural threats to man has resulted in the development of a variety of conceptual responses to risk such as implied by fatalism and divine intervention. In the latter case, many religions attribute some or all acts of man or nature to divine intelligence with degrees of free will ranging from virtually none to total predestination. The Judeo-Christian "What has God wrought?" and the Islamic "By the will of Allah" are but literal expressions of such belief.

2.1.3 Controlling Man's Environment

Since the Renaissance and the Scientific Revolution, man has become increasingly aware of the possibility of controlling his environment. In the last hundred years the control of man's destiny has become a matter of major concern. Curing of disease to extend the quality and quantity of life, increasing the amount of food grown and the reliability of production, the search for peace, and the concern for the welfare of all people are examples of this phenomenon. Often the impetus for such activities arises from innate needs of people to help their fellow creatures for both ethical and religious reasons. As a result, the array of risk covers a wide spectrum of human experiences from premature death to loss of financial and aesthetic values. Next we must ask how the risks in the spectrum relate to each other and affect human behavior.*

2.2 SOME PSYCHOLOGICAL BASES FOR RISK

2.2.1 Pain and Suffering

Next to premature death, the avoidance of physical pain and suffering seems to be of primary importance in establishing behavior. Such pain and suffering includes hunger, thirst, injury, and disease. To some extent man is unable to evaluate the impact of pain until he actually experiences it. As a result, the possibility of pain is often ignored, and action to ameliorate pain while experiencing it dominates preventive measures. There are different degrees of pain, and one may choose to avoid a long-term annoyance by undertaking short-term intense pain (possibly underestimated in impact

* The epistemological problems of measurement of risk parameters and the aesthetic problems of value determination are discussed in subsequent chapters.

until experienced). If this were not so, one would not voluntarily submit to a painful operation such as a hemorrhoidectomy.

2.2.2 Motivation to Fill Needs

Man is motivated to fulfill his needs, whether the need is the avoidance of an undesirable condition such as premature death or pain, or the pursuit of some desired condition. The source of man's motivation to fill his perceived needs has been a major concern of psychological theory and research.

2.2.2.1 Psychoanalytic Theories of Personality

Freud and those who further developed and modified his theories were primarily studying abnormal human behavior and its causes. According to their theories, internal conflict and neuroses resulted from inability to fulfill physical needs—initially primitive pleasure-seeking drives of the id—and rational needs of the ego for social approval and self-esteem. Later modifications introduced conflict between the person and the environment, and added goal-seeking needs and aspirations. Further developments were the pointing out of the need to show superiority as proposed by Adler and the need to achieve organismic potentials (i.e., urges inherent to man the organism) called self-actualization.

2.2.2.2 Behavioral and Learning Theories of Personality

Behavioral theory is based on mechanistic stimulus-response learning behavior. Here, the environment controls behavior, and innate tendencies are relatively unimportant. Basic biological drives exist, but behavior is shaped by external stimuli in the form of rewards and punishment, fear being one of the most important drives in shaping learned responses.

Cognitive theories of learning are concerned with higher mental processes and the learning of information and generalizations derived from such information as well as particular responses. The theories emphasize the innate ability to organize processes within the individual.

2.2.2.3 Humanistic Theories of Personality and Needs

Humanistic or Third Force psychology is "now quite solidly established as a viable third alternative to objectivistic, behavioristic (mechanomorphic) psychology and to orthodox Freudianism."[3] This theory, originally developed by Maslow,[4] is based on the study of not only normal people but also superior individuals who have obtained a high state of human achievement called self-actualization. This work has been extended into the business world through the study of self-actualized people in organizations.[5]

Maslow defines motivation as the state of having an internal motive that incites the individual to some kind of action. By its very nature, motivation comes from within the individual and cannot be imposed on him. Man is a goal-seeker throughout his life. These goals are ubiquitous, serving as a measure of man's nature and forming his behavior. The action to achieve goals is a basic drive. Maslow's theory of personality and motivation converts these goals to need; man is motivated to reach a certain goal because he has an internally generated need to reach it.

Maslow categorizes and ranks basic sets of human needs into a conceptual hierarchy. This hierarchy is important to analysis of risk because it provides a basis for understanding how man values possible gains and unwanted consequences. This, coupled with the idea that healthy versus abnormal people are used to study behavior, makes the results more palatable for understanding normal, individual, societal risk behavior.

2.2.3 Hierarchy of Needs*

Maslow chose to categorize and rank sets of human needs into a conceptual hierarchy, beginning with the most primitive and urgent human needs and ranging upward to the apex of the hierarchy, self-actualization. Although there may be nuances and graduations within any given level of need, Maslow made the primary breakdown as follows in *Hierarchy of Needs.*

- Need for self-actualization
- Need for esteem
- Need for belongingness and love
- Safety needs
- Physiological needs

The physiological category refers to food, warmth, shelter, elimination, water, sleep, sexual fulfillment, and other bodily needs.

The safety needs include actual physical safety, as well as a feeling of being safe from physical and emotional injury; therefore, a feeling of emotional security as well as a feeling of freedom from illness.

Whereas physiological and safety needs are centered around the individual's own person, the need for belongingness and love represents the first social need. It is the need to feel a part of a group or the need to belong to and with someone else. It implies the needs to both give and receive love.

The need for esteem is based on the belief that a person has a funda-

* This material is derived from a National Industrial Conference Board article entitled "Behavioral Science",[6] and is used by permission.

mental requirement for self-respect and the esteem of others (except in extreme pathology). The need for esteem is divided into two subsets: first, there is the need for feeling a personal worth, adequacy, and competence; second, there is the need for respect, admiration, recognition, and status in the eyes of others.

Self-actualization is a more difficult concept to describe; it is the process whereby one realizes the real self and works toward the expression of the self by becoming what one is capable of becoming. Thus the need for self-actualization sets into motion the process of making actual a person's perception of his "self."

Maslow posits that these needs occur in a hierarchy of preeminence throughout a person's development and maturation. The graphic model of steps is used to underscore the ascending occurrence of each need. The hierarchy emphasizes the fundamental point that until one need is fulfilled, a person's behavior is not motivated by the next, higher level, need. For example, a person whose physiological needs are not taken care of, is not concerned with his safety. And until his physiological and safety needs are both fulfilled, he is not particularly interested in fulfilling his need for love.

Once a need is satisfied, it is no longer a motivating force (a hungry man is motivated toward food, but once he is well fed, he will no longer strive for food). The implications of the maxim, "a satisfied need is no longer a motivator of behavior," are probably among the most intriguing outgrowth of the need/motivation theory.

Maslow assumes that just as the needs appear in sequence as a person grows from infancy to adulthood, the pattern may be repeated as he encounters new experiences at various times in his life. From the standpoint of motivation, however, Maslow's theory assumes that individuals can be characterized as being primarily at an observable level at a given time or in a given set of circumstances.

The matter of time and circumstances is a crucial variable in understanding the hierarchical nature of need and motivation; that is, the needs are not static, but situational. Since the lower level needs are the most urgent ones, they must continually be satisfied in order for a person to be motivated toward higher level needs. But even when the lower level needs are satisfied, if they are threatened, they again become the stimulus for motivation. A man who is safe may risk his safety, even his life, if he becomes hungry enough. Or a person whose prime motivation was self esteem may drop down a level to seek belongingness if his sense of belongingness is threatened.

This concept is particularly pertinent to the study of risk, since it provides a basis for considering man's action in risk aversive situations where mortality, pain or suffering is not involved. A perceived threat to a perceived need is very real, and resulting human behavior to react to such threats involves attempts to adjust one's environment to meet his needs.

2.3 DIMENSIONS OF UNCERTAINTY

Uncertainty exists in the absence of information about past, present, or future events, values, or conditions. Although there are various degrees of uncertainty, the basis of the concept of uncertainty is the absence of information about parts of a system under consideration.* Thus the concept "degree of uncertainty" addresses that proportion of the information about a system not known, extending from a zero degree of uncertainty to a totally uncertain system (a unity degree of uncertainty). Most systems lie somewhere in between; that is, some information is known and some unknown.

2.3.1 Types of Uncertainty

There are two types of information about a system of concern. The types of information necessary to describe the system in a taxonomic sense represent the "degrees of freedom" of the system and consist of a set of variables that when determined exactly, totally describe the system. The second type of information involves, then, the measurement of the variables to determine their specific values. The uncertainty in describing the degrees of freedom of a system is defined here as the "descriptive uncertainty," and the uncertainty in specification of values is defined here as the "measurement uncertainty." The former refers to the specification of the variables in the system, rather than the value of a specific variable.† More explicitly, these terms can be defined as follows:

Descriptive Uncertainty. Absence of information relating to the identity of the variables that explicitly define a system. The inability to fully describe the "degrees of freedom" of a system.

Measurement Uncertainty. Absence of information relating to the

* See Glossary for a definition of "degree of uncertainty," "system," and other terms; "degree of uncertainty" is quite different from the idea of "certitude" and involves only consideration of information content.

† As an example, take the recent discovery that catalytic converters installed on autos to reduce pollution introduced a new variable—namely, sulfates—to the exhaust system. This new variable decreased the descriptive uncertainty of the pollution control system, but since the measurement of health effects produced by sulfates is not yet understood, the measurement uncertainty in the new variable is large.

specification of value assigned to each variable in a system. The
- inability to measure or assign values to variables in a system.*

2.3.2 Information from System Processes

To reduce the degree of uncertainty about a system, one attempts to gain information about various portions of the system. One way of getting information about the system is to learn the processes underlying system operation and to find out how information is generated for the system. For example, information may be gained through a random process where the concept of probability is useful to gather information about the system.

On the other hand, the process may involve the behavior of a rational opponent or some man-made or natural principle. The degree to which one can learn the governing process directly decreases the descriptive uncertainty about a system, since system variables and their relationships become better defined. Furthermore, the bounds of measurement uncertainty are described by the process that makes it possible to obtain information on specific values. For example, knowing that a process is random, one has a basis for a means for gaining information different from what would be required to deal with an intelligent opponent.

2.3.2.1 Processes Involving Rational Behavior of Opponents

In a process that entails the behavior of an intelligent opponent, the skill of the opponent in relation to the skill of the proponent is the matter of uncertainty. Thus a game of chess does not directly involve random processes, but the skills of the two individuals playing.† Moreover, most people are fallible and there is no guarantee that optimum strategies will be played. The whole field of game theory, as originally developed by von Neumann and Morgenstern[7] and extended by many others, such as Luce and Raiffa,[8] deals with situations in which the players are rational opponents.

Various processes can be attributed to opponents. For example, should the opponent act as a "rational, economic man," a whole process to

* The distinction between the descriptive and measurement uncertainty leads to what the author calls the "information paradox." One learns new things by uncovering new degrees of freedom. For each new degree of freedom, a measurement scale and specification of values on that scale is required. Thus as new degrees of freedom are added to one's knowledge, the measurement uncertainty increases geometrically. So, for a new degree of freedom F_i, involving a measurement scale of s_i divisions (it can be continuous rather than discrete), there are s_i possible conditions of information. Thus as the descriptive certainty increases $\Sigma_i F_i$, the measurement increases $\Sigma_i (F_i) s_i$. Therefore, the more one learns, the more uncertain one can become.

† Of course, one or more players could choose a random strategy and move according to a random pattern.

describe how man operates under economic conditions can be derived. Likewise, if one expects man to be a rational opponent in the manner that he maximizes his utility by some process, the basic concepts of game theory become operable. However there are no guarantees that an opponent will act "rationally." In fact, as a later chapter explains, many decisions are emotionally based or use value systems different from "economic man."

Playing against a rational opponent calls for skill, not luck, if the game does not involve any built-in degree of random behavior. In fact, if one is not skilled in understanding the total strategy of the game, he may choose to attempt to confound his component by introducing a random component to his strategy, such as flipping a coin to determine his next move.

Games involving rational opponents who are not artificially constrained to fixed rules of play exhibit extensive descriptive uncertainty as well as measurement uncertainty. It is, indeed, difficult to express the totality of variables a rational opponent may consider, although critical variables may be identified. Games such as chess are intriguing because all situations cannot be practically or meaningfully defined exhaustively. The stock market "game" is a challenge due to uncertainty, although there have been repeated attempts to "control" the market through manipulation of critical parameters.

Since all games are not totally defined and man does not always behave rationally, there is little certitude in the manner in which opponents will react in a game (unless the game is constrained to game theory limitations). At best, probabilities of the possible ways in which opponents behave may be all that can be obtained. This information may be derived from observance of previous behavior in similar situations with repeated trials or from a priori knowledge of how the opponent is expected to react. As such, these probabilities range from objective observation to subjective speculations; but by their existence they constitute some attempt at reducing uncertainty. As is shown subsequently, the existence of such probabilities of behavior and possible consequences results in risk behavior. Thus risk arises from uncertainty in man's behavior as well as from natural processes. The study of human behavior can contribute to reduction in both descriptive and measurement uncertainty.

2.3.2.2 Processes Involving Natural Phenomena

Processes involving natural phenomena are of two types: (1) processes that operate on the basis of empirical, natural laws, and (2) random* processes that imply statistical operations.

* Random is defined in Webster as: "(1) lacking or seeming to lack a regular plan, purpose, or pattern; (2a) marked by absence of bias, b. involving or resulting from randomization, c. having the same probability of occurrence of every other member of a set."[9]

Processes Involving Natural Laws. On the macroscopic levels, laws of nature such as Newton's laws of motion, Keppler's laws of planetary motion, Maxwell's laws of electricity and magnetism, are induced by empirical observations that are easily repeatable, from which hypotheses are deduced, and by the ability of the hypotheses to predict future natural behavior ascertained by further experiments. This process, referred to as the method of a scientific empiricism,* is the basis for natural science. Underlying processes may very well be statistical; the whole body of quantum theory† is based on such evidence. Nevertheless, at the level of pragmatism required to live with these laws and use them in every day life, the underlying statistical processes have minor impact. Yet it is an underlying precept of scientific positivism that it should be possible to discover all natural law. This implies that the descriptive uncertainty of the universe can be reduced to a value approaching zero. Whether such a goal is achievable is not considered here; but the process of understanding natural laws reduces uncertainty and does lead to an ability for man to control the environment.

The existence of measurement uncertainty is prescribed by scientific theory, especially at the quantum theory level. Here we have the Heisenberg uncertainty principle,[10] which at the subatomic level states that the energy required to measure position or momentum of an electron would itself perturb the position or momentum of that electron. As a result, according to Heisenberg, the position and momentum (energy) of an electron can be measured only to a minimum degree of uncertainty.‡

Miller and Starr have used this concept as the basis for an operational philosophy for decision making:

> The Heisenberg uncertainty principle coupled with the basic concepts of quantum mechanics reformulated truth as a probabilistic notion. By relating truth to risk and uncertainty, the fundamental connection between goodness and truth was shattered. Goodness remained a philosophical, theological, and personal matter. Individual truth came to be viewed as a property of cerebral-sensory systems; universal truth as approachable but ultimately unknowable. And so an operational philosophy of decisions developed, wherein the goodness of a decision would be measured by the extent to which its results satisfied the decision-maker's objectives. All natural systems have measurement uncertainty whether or not descriptive error exists. The ability to measure a variable to establish a particular value or relationship among variables

* Scientific empiricism: A philosophy that denies the existence of any ultimate difference in the sciences, strives for a unified science through a synthesis of scientific methodologies.[9]

† Quantum theory as originally developed by Max Planck, Ernest Schrödinger, Werner Heisenberg, and others.

‡ This also occurs in principle at macro levels, especially in behavioral system. The structure of a questionnaire may very well bias the measurement sought.

depends upon the precision [see Glossary] of the measurement system used (down to the limits imposed by the Uncertainty Principle).[11]

The scales[12] used for such measurement may be:

Nominal Scale (Identify–Taxonomy). A classification of items that can be distinguished from one another by one or more properties.

Ordinal Scale (Order–Rank). An ordering (ranking) of items by the degree they obtain some criterion.

Cardinal Scale (Interval). A continuous scale between two end points, neither of which is necessarily fixed.

Ratio Scale (Zero reference). A cardinal scale with one end point fixed by reference to an absolute physical end point (e.g., absolute zero on the Kelvin temperature scale), from which are developed other cardinal scales, all of which are related by simple ratios.

The actual measurement process at the limits of measurement precision involves errors that contribute to the inaccuracy of the measurement. Inaccuracy and its converse, namely accuracy, [see Glossary] often involve a statistical process, although bias and systematic errors may come into the picture as part of the measurement error. Since the statistical processes are random, this statistical approach to measurement, like many other processes, involves random behavior.

Processes Involving Random Behavior.* Random behavior as considered here implies a system without pattern, one in which all possible events in a set of events are equally possible. This follows directly from the verbal definition. Many physical systems exhibit random behavior; examples are supplied by the motion of atoms in a gas and by the electronic noise in insulators and semiconductors called white noise. Such systems usually occur at the micro level, but man has designed at the macro level systems that attempt to simulate random behavior (roulette wheels, lotteries, testing systems for statistical experiments, etc.).

In statistical testing systems, random processes serve as inputs into an experiment, where resulting observations on the system under test establish patterns from which hypotheses of system behavior are developed and tested. Such experiments reduce the descriptive uncertainty of the system but do not reduce measurement uncertainty. A process that obeys some statistical (probability) distribution provides an estimate of the most likely probabilities of occurrence for the next trial, but this trial can result in any

* Random behavior is one particular form of natural law of special interest. It may be considered the "absence of law."

possible value. Preknowledge of random outcome involves clairvoyance, and despite the active pursuit of such skills, there is considerable doubt regarding their existence or effectiveness.

In any case, probabilities of occurrence of consequences imply risk behavior.

2.3.3 Reduction of Uncertainty

In most systems of interest all three processes—human behavior and natural and random events—occur to some degree. The conduct of experiments and the gathering of data provide means of reducing uncertainty, but only to certain limits (Table 2.1).

Different approaches are required to reduce uncertainty in each process, but reduction of uncertainty is possible in all cases except for random measurement uncertainty for future events. In the random case, the study of historic data reduces descriptive uncertainty by ascertaining relationships among variables more accurately.

Reduction of uncertainty does not in itself reduce risk. Resulting information can be used to direct action to control risk, which is the objective sought. The value of information gained can be measured only by the degree of control of risk gained by separate, subsequent action.

A system that exhibits certainty of outcome does not involve risk, although it may involve unpleasantness. For example, sure knowledge of eventual death or taxation does not make these events any more palatable.

Table 2.1 *Descriptive and Measurement Uncertainty Limits for System Processes*

Process	Descriptive Uncertainty Reduction	Measurement Uncertainty Reduction
Behavioral	Limited by the ability to define all variables by which man behaves in given situations.	Limited by the degree of rational behavior of man as opposed to other types of behavior.
Natural deterministic	Theoretically unlimited, but limited by practicalities.	Limited only by precision of a measuring system down to the absolute limit of the uncertainty principle.
Natural random	Theoretically unlimited, but limited by practicalities.	Cannot be reduced by any method presently known and demonstrated for future events.

In fact the reverse may be true: the possibility of escape from such consequences may lead to some small measure of hope. The certainty of occurrence provides for maximum information for control action, but the degree of control possible varies from none to total. Certainty in the absence of control can lead to resignation and acceptance and the meeting of such losses with some semblance of dignity (if this is considered important personally or through group or cultural interplay). Of course suppression at the psychological level or substitution of other values for the outcome (e.g., assuming an "afterlife" in the face of impending death) are also possible.

These concepts are important because the increasing concern with risk and risk systems, currently in vogue, is aimed primarily at the control of risks, especially at the societal level. This is a relatively new concept in the history of man, concurrent with the scientific revolution.

2.4 DEFINITION OF RISK

2.4.1 Gains and Losses

The consequences of human or natural processes may result in either losses or gains. Furthermore, with opponents, one man's loss may be another man's gain. For example, a loss by a gambler at blackjack is the house's gain, and vice versa. With nature, the avoidance of unwanted consequences imposed by nature is sought and nature is thought to be indifferent to man's loss or gain.* Thus risk implies something unwanted or to be avoided. Risk is then associated with consequences that involve losses to the risk taker.†

Risk agents often willingly expose themselves to risks to obtain some possible gain, when in their individual deliberation, the possible gains outweigh the possible losses. If one substitutes "probable" for "possible" in the foregoing statement, quantitative balancing of probable gains and losses is possible within limitations of measurement uncertainty. This arises from the ability to express probabilities on cardinal scales between zero and unity.

On this basis we shall associate risk only with losses; that is, we say that man is risk aversive. Gains are synonymous with benefits and may be absolute, relative, or probable. In any case, we have risks that are sometimes taken to obtain desired possible gains. Action taken to reduce a risk can be considered a gain in the sense that possible loss is reduced.

* At least in any sense of consciousness. Man can and does affect his environment, often adversely; but conscious response of natural systems is a matter for metaphysicians and for those who debate the merits of scientific positivism and religion (and other philosophies).

† Risk takers are referred to as risk agents, and a more explicit definition of this term is given in later chapters.

2.4.2 Components of Risk

Risk, at the general level, involves two major components: (1) the existence of a possible unwanted consequence or loss, and (2) an uncertainty in the occurrence of that consequence which can be expressed in the form of a probability of occurrence. The consequence implies a negative value to a risk taker. The probabilities may be derived from the probable behavior of opponents and/or nature. Both the negative value of consequences and value assignment of probability of occurrence are measurable, but not necessarily independent. The magnitude of a probability can influence the magnitude of consequence value. For example, a particular natural disaster becomes the focus of increasing concern as its probability of occurrence goes from remote to impending.

2.4.3 General Risk Definitions

On the basis of the foregoing discussion, risk can be defined on a general level:

Risk is the potential for realization of unwanted, negative consequences of an event.

The causative event may be a single event, some combination of events, or a continuing process, and the consequences may affect individuals, groups of people, or society and its institutions, or they may affect physical and biological systems. The general definition is meant to be inclusive, not exclusive. The process of risk determination as discussed in the following chapter makes this definition more explicit. However, it is important that the control of risks be considered as well. The term "risk aversion" is used in this context:

Risk aversion is action taken to control risk.

The action taken to reduce risk may be motivated either through direct reduction of uncertainty (e.g., uncovering new risks) or by intuitive perception of risk takers.

Risk does not occur under conditions of certainty. Gains (benefits) and losses (costs) may be certain and can be traded off against each other, taking into account both present and future streams of benefits and costs through discount methods.* However the existence of uncertainty implies risk, adding a new dimension to losses and costs.† Thus risks and costs are

* Forecasts of the future involve uncertainty, but calculation of "opportunity costs" in the economic sense in the present provides considerable information for decisions.

† Uncertainty also implies "probable gains" in terms of benefit.

related, but they must be treated differently. Risk is a special subset of cost conditions.

2.4.4 Scope of Risk Analysis

The array of risks covers a wide variety of human experiences involving risks—personal or societal, man-made or natural, with consequences ranging from financial involvement through premature death. Risks may be strategic in the sense that they involve broad approaches for society or parts of society (foreign policy, military and defense risks, etc.). Conversely, they may be tactical, involving gambles to achieve some benefit, financial or otherwise (e.g., capital and securities markets, investments, commercial air travel, casino gambling).

The analytical approach to risk covers all the activities involving risk in theory, but the application of theory to actual situations has been considerably limited and only reasonably successful. The need to bridge the gap between the theoretical and practical aspects of risk involvement implies that the theoretical structures, though seemingly simple, are not adequate for practical problem solving.

Otway[13] has identified two aspects of risk assessment: risk estimation and risk evaluation.

> Risk estimation may be thought of as the identification of consequences of a decision and the subsequent estimation of the magnitude of associated risks.
>
> Risk evaluation is the complex process of anticipating the societal response to risks; . . . this could be termed the "acceptability of risk."[14]

However, R. W. Kates[15] has added a new dimension, risk identification, to the process. The three-part process contains a new parameter, risk identification (Figure 2.1). This approach seems to make considerable sense for three reasons:

1. risk identification involves reduction of descriptive uncertainty;
2. risk estimation involves reduction of measurement uncertainty; and
3. risk evaluation involves risk aversive action, which can result in risk reduction or risk acceptance.*

This classification system for risk assessment provides a convenient method for separating further discussion of the analytical aspects of risk, covered in subsequent chapters.

* Risk reduction often does not eliminate risk but reduces it to some "acceptable" level. In a given case risk may already be at an acceptable level, negating the need for further action, or risks may not be reduced below some given level, thereby forcing acceptance.

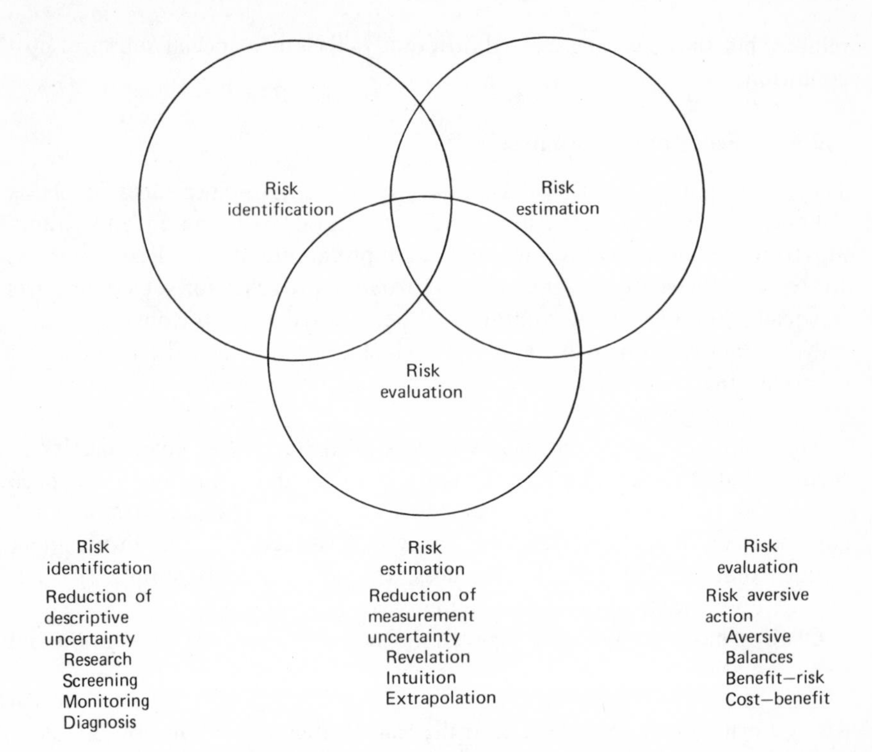

Figure 2.1. The elements of risk assessment.

REFERENCES

1. Allen Willett, *The Economic Theory of Risk and Insurance*. Philadelphia: University of Pennsylvania Press, 1951, p. 6. (first published 1901). Cited by Herbert S. Denenburg, Robert D. Eilers, Joseph J. Melone, and Robert A. Zelten, *Risk and Insurance*, 2nd ed. Englewood Cliffs, N.J.: Prentice-Hall, 1974, p. 4.
2. Frank Knight, *Risk, Uncertainty, and Profit*. Boston: Houghton Mifflin, 1921, p. 233. Cited by Denenburg et al. (note 1), p. 4.
3. Abraham H. Maslow, *Toward a Psychology of Being*, 2nd ed. Princeton, N.J.: Van Nostrand, 1968, p. iii.
4. Abraham H. Maslow, *Motivation and Personality*. New York: Harper & Row, 1954.
5. Abraham H. Maslow, *Eupsychian Management*. Homewood, Ill.: Irwin, 1965.
6. Harold M. F. Rush, "Behavioral Science: Studies in Personnel Policy," National Industrial Conference Board, Personnel Policy Study No. 216. New York: The Conference Board, 1969, p. 17.
7. John von Neumann and Oskar Morgenstern, *Theory of Games and Economic Behavior*. Princeton, N.J.: Princeton University Press, 1953.
8. R. Duncan Luce and Howard Raiffa, *Games and Decision: Introduction and Critical Survey*. New York: Wiley, 1959.

9. *Webster's Third New International Dictionary,* Unabridged. Springfield, Mass.: Merriam, 1971.
10. Werner Heisenberg, *Zeitschrift für Physik,* Vol. 33, 1925, p. 879.
11. David W. Miller and Martin K. Starr, *The Structure of Human Decisions.* Englewood Cliffs, N.J.: Prentice-Hall, 1967, pp. 23.
12. S. S. Stevens, "Measurement and Man," *Science,* Vol. 127, No. 3295, February 1938, pp. 383–389.
13. Harry J. Otway, "Risk Estimation and Evaluation," in *Proceedings of the IIASA Planning Conference on Energy Systems,* IIASA-PC-3. Laxenburg, Austria: International Institute for Applied Systems Analysis, 1973.
14. Harry J. Otway, "Risk Assessment and Social Choices," IIASA Research Memorandum RM-75-2. Laxenburg, Austria: International Institute for Applied Systems Analysis, February 1975, p. 5.
15. Robert W. Kates, "Risk Assessment of Environmental Hazard," SCOPE Report No. 8, International Council of Scientific Unions, Scientific Committee on Problems of the Environment, Paris, 1976.

3

The Process of Risk Determination

Risk determination is a process that covers both risk identification and risk estimation. It is aimed at defining a given level of risk, to permit risk evaluation, involving risk aversion and/or risk acceptance, to take place. This discussion is at the descriptive level and only covers the risk estimation process. A following chapter gives a more analytical view of the process.

3.1 THE FIVE STEPS OF RISK ESTIMATION

The process of risk estimation is basically a five-step process.

First (upper left-hand corner of Figure 3.1) there is a *causative event* or *events* that when properly defined, can have associated with it a probability of event occurrence. When such an event occurs, there are a number of possible outcomes that likewise can be defined, as well as the probability of the resultant outcome(s) also determined to varying degrees. These outcomes appear as the second step. In each case the event or outcome can be defined explictly, and a probability of occurrence can be determined to some level of precision as indicated in the figure.

As such, the causative event and the outcome do not involve risk because exposure to people, the environment, or institutions has not yet been taken into account. Thus the combination of causative events and outcomes by itself is limited to experiments, statistical design of experiments, and hypothesis testing. Experiments are carried out and outcomes observed to determine the behavior between the causative events and the outcomes. If such experiments involve exposure to people or the environment, then both conducting the experiments and testing the hypothesis entail risks.

For example, a laboratory experiment on measuring the acceleration of gravity calls for causative events, namely, a series of trials of dropping a weight from different heights. The outcomes are the measured distance and time of flight from each event. Such an experiment offers no risk to society in any direct sense, since the information is used to deduce a hypothesis that

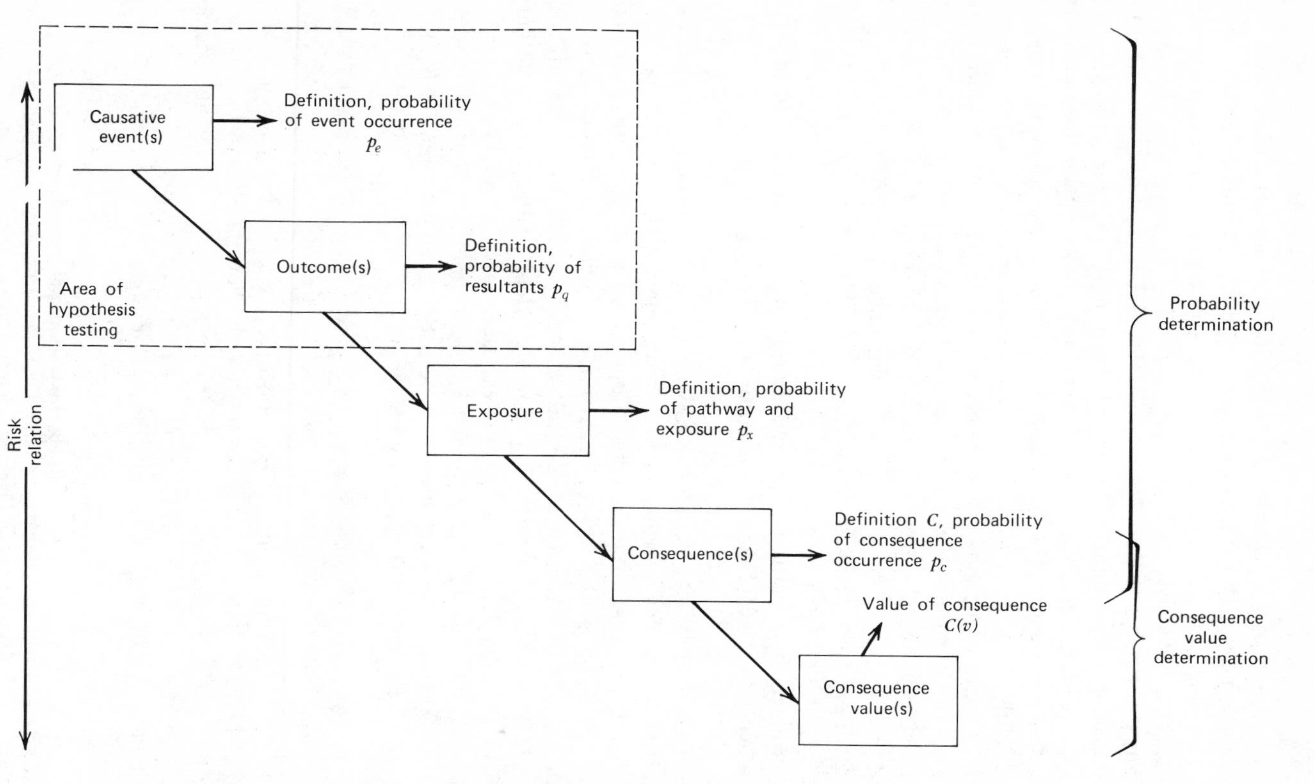

Figure 3.1. Process of risk estimation.

is then tested to predict time of response for other heights of drop. There may be some risk to the experimenter, however. He might drop a weight on his foot, and if he is a cautious experimenter he will attempt to minimize his exposure to such risk.

The third step in Figure 3.1 involves the exposure pathways by which outcomes can affect man, biota, or his environment. These pathways, which can be defined explicitly, have different probabilities of occurrence, leading to an array of exposures. For example, the outcome of an event* might be a tree falling in the forest. If there is no one in the forest, the exposure to the risk of being struck by the falling tree is zero. As long as there is some probability of someone being in the forest (pathway), the exposure (physical contact by the falling tree) is finite.† A description of all possible exposures and their respective probabilities of occurrence may be determined.

Each exposure results in an array of possible consequences that also can be explicitly defined and the probability of consequence occurrence determined. This is the fourth step. For example, when exposed to physical contact with the falling tree, the consequences range from no effect through various degrees of injury to death.

Risk determination does not end with definition of the consequences, but must also consider the value of the consequences to those people affected. It is this fifth step—the consequence value to those affected—that determines, along with the probability of occurrence, behavior in response to risk. For example, breaking a leg as a result of being struck by a tree in a forest would be viewed differently by people in different situations. A lumberjack might consider that he is paid to expose himself to such an event as an occupational hazard. An experienced woodsman and a tenderfoot might view the consequence value quite differently. Even death as a possible consequence is viewed differently by individuals.

3.1.1 Causative Events

Table 3.1, column I categorizes causative events by initiation and by temporal factors. Initiation of an event may be natural or man-made and may be purposeful to some end (e.g., to erect a pipeline), or it may be accidental in the sense that it derives from an unanticipated activity of man. The blowout of a tire on a rapidly traveling automobile is an example of an acci-

* The causative event might be a lightning strike, beavers at work, or decay.

† The classical argument of whether a sound occurs when a tree falls in a forest with nobody to hear it can be addressed by this approach. As long as there is a finite probability that a person could be in the forest to hear it (finite exposure pathway) the sound is real. All real situations have finite probability, albeit sometimes small.

Table 3.1 *Factors in the Structure of Risk Estimation*

I. Causative Event(s)	II. Outcome(s)	III. Exposure	IV. Consequence(s)	V. Consequence value(s)
A. Initiation 1. Man a. Purposeful b. Accidental c. Incidental 2. Nature a. Process • Random • Natural law b. Mode • Direct • Indirect B. Temporal Factors 1. Continuous a. Constant Level b. Varying Level • Periodic • Aperiodic 2. Discrete a. Singular b. Multiple • Sequential • Repetitive • Combinational	A. Aspect 1. Desirable (win–success) 2. Undesireable (loss–fault) 3. Indifferent (trivial) B. Temporal factors 1. Continuous a. Constant Level b. Varying Level • Periodic • Aperiodic 2. Discrete a. Singular b. Multiple • Sequential • Repetitive • Combinational	A. Controllability 1. Direct 2. Indirect 3. Uncontrollable B. Recipients 1. Individual a. Direct • Health • Financial • Social b. Indirect (institutional, etc.) 2. Population a. Direct b. Indirect 3. Environmental and social institutions C. Pathways 1. Degree of avoidability 2. Degree of voluntary choice 3. Nature of pathway	A. Summation of pathways 1. Total effects 2. Partial effects B. Exposure-consequence relationship 1. Temporal a. Immediate b. Delayed–latent 2. Recipient a. Identifiable b. Statistical	A. Individual risk agents B. Value groups C. Society as a whole 1. National 2. World D. Risk evaluators 1. Regulators 2. Ombudsman

dental causative event. The causative event may also be incidental to some other activity that is of particular interest. For example, the production of electric power by fossil fuels involves the incidental releases of certain pollutants to the atmosphere.

The causative events may be naturally initiated by some random process or natural law, or a combination of these. The mode of initiation may be direct in that the random process or natural law results in natural phenomena beyond man's control, or indirect in that it acts in conjunction with some activity of man that allows natural law to initiate causative events. For example, a hurricane is a direct natural event, whereas the flying of an airplane into a hurricane is only indirectly natural because man has come control of whether he flies into the hurricane. The indirect mode also refers to a situation in which man harnesses natural random processes (e.g., in games of chance).

The temporal factors indicate that the causative event may be continuous over time or it may be discrete. In the continuous case, it may be at a constant level or a varying level that is either periodic or aperiodic. The continuous release of pollutants from a factory is a case in point that may occur for either the constant or the varying level mode.

Most people are more familiar with discrete cases featuring singular events such as a failure of a piece of equipment, the spillage of a pollutant, or the collision of vehicles.

The causative events may also be multiple. They may be sequential (one event is followed by a subsequent event after some time), such as in the case of the two-hit model for cancer induction.* Multiple events may also be repetitive, such as the continued use of a roulette wheel or a daily sunrise. Multiple events may also be combinational in that several events must occur in combination for the resulting causative event to take place. As an example, for a person to receive an electrical shock from an appliance, the appliance must be turned on, it must contain a defective circuit, and it must have a return ground path (e.g., the person's wet feet) simultaneously. The resultant causative effect is the electrical shock, and the possible outcomes range from simple withdrawal to electrocution.

Causative events may be independent events or dependent on other events. As a rule, all causative events must be identified in a collectively exhaustive manner and interrelationships among events must be determined. Each event can be explicitly defined and a probability p_c assigned. Identification of new causative events or new interrelationships is called risk iden-

* In this model it is believed that a cell must be affected by two separate events before a cancer is caused. One event may be by contact with a carcinogen; another may be exposure to a quantity of ionizing radiation or any combination of two sequential events of this type.

tification. The definition and determination of interrelationships, and the probability of occurrence of events, is considered as risk estimation.*

3.1.2 Outcomes

As indicated in Table 3.1, column II, consideration of outcomes must cover both the aspect of the outcome and the temporal factors involved. The temporal factors are essentially the same as those in column I for causative events and are not treated in this discussion. However the aspect considerations are different. An aspect may be desirable, depending on who is receiving the outcome, in that it represents a win, a success, or some desirable attribute. On the other hand, it may be undesirable, representing a loss of some type, a fault in a piece of equipment, or something else unwanted. In many cases the output is indifferent to risk agents and can be considered trivial. For example, whenever a person crosses a street there is some risk of being hit by a car. Most of the time the undesirable outcome of being hit by a car does not occur. Although we may exercise caution in the act of crossing, once across we pay little attention to the experience and are indifferent to the successful outcome. Essentially, most outcomes are trivial, and concern is usually aimed at specific desirable or undesirable outcomes in most situations.

Outcomes are by definition independent conditions, and not only must the class of totality of outcomes be rendered collectively exhaustive by defining all possible outcomes explicitly, its members must be mutually exclusive as well. Assigned to each defined outcome is a probability of occurrence p_q; and the probabilities of all outcomes will sum to unity, assuming that causative events have actually occurred.† Discovery of new outcomes involves risk identification, and the determination of probabilities is risk estimation.

3.1.3 Exposure

3.1.3.1 General Aspects of Exposure

There are three main considerations for exposure (Table 3.1, column III). The first is the controllability of the exposure pathway, which ranges from directly and indirectly controllable by man to uncontrollable. For example, exposure to accidental risks from flying an airplane may be eliminated by

* The combination of events leading to final events can be represented by a decision tree structure showing the relationship among events. This tree structure is called an "event tree."

† When the outcomes are negatively valued and represent failure of a process, a decision tree structure showing interrelationships is called a "fault tree."

choosing not to fly. On the other hand, exposure to impact of a meteorite is generally uncontrollable, although it may be indirectly affected by choosing to live underground. In this manner, direct control implies control of the cause of the event and outcome, whereas indirect control involves the ability to avoid the exposure.

The second consideration is the recipients of the exposure. Individuals, populations, or environmental and social institutions may be affected by the exposure. In the latter case, the economic, welfare, and political institutions may become risk recipients. Recipients may directly receive the exposure in terms of their own health or financial or social situation, or they may receive it indirectly through impact on institutions that in turn affect the individuals involved. For example, although one may not wish to tamper with the voluntary aspect of smoking and its directly associated risks, the impact of increasing lung cancer on societal programs (e.g., costs of health insurance payments required to deal with this problem, especially for welfare and subsidized patients) indirectly imposes costs on individuals and the population as a whole.

The third consideration is the pathway itself—the degree of avoidability, the degree to which one voluntarily can or cannot accept the pathway, and, of course, the nature of the pathway.

3.1.3.2 Detailed Discussion of Exposure

First, a definition of each exposure pathway, leading to a defined exposure condition, must be developed. For example, an outcome of a crop-spraying activity may be an unexpected deposit of a pesticide such as DDT outside a protected area. The pathways, leading to exposure to man, can be by food intake due to concentration of the pesticide up the food chain; and through direct intake in drinking water, in dust on food, or by inhalation. In other words, there are multiple pathways leading to exposure. However each pathway has a magnitude of possible exposures and a probability of occurrence. The magnitude of the pathway is measured at unity probability and is modified by the measured or estimated probability of occurrence. The resultant exposure that can be defined explicitly is the summation of pathways leading to that measure. For example, exposure to DDT exists for both people and animals. The pathways just noted lead to exposure to both kinds of recipient.

This concept can best be understood by reference to Figure 3.2. The result of a causative event is a specific outcome W. The resulting exposure to different recipients (e.g., man and biota) is designated as X_k, where the index k indicates the recipient. The exposure pathways α for each recipient have a magnitude designated as $M_{\alpha k}(W)$ and a probability of occurrence

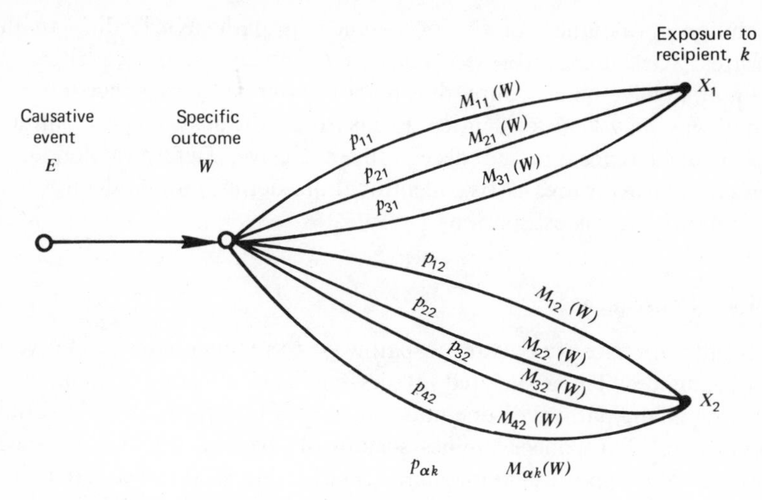

Figure 3.2. Elements of exposure pathways.

$p_{\alpha k}$. Note that the pathways to each exposure group are not identical. There are n pathways for each recipient group k.

The total of the magnitude of all pathways cannot exceed the magnitude of the outcome. Otherwise the exposure would contribute new risks, not limit them.

$$0 \leq \sum_{\alpha} \sum_{k} M_{\alpha k} \leq W \tag{3.1}$$

Conversely, the probability of each pathway may vary from zero to unity, and all probabilities may equal unity, such that to any exposure recipient,

$$0 \leq \sum_{\alpha=1}^{n_k} p_{\alpha k} \leq n_k \tag{3.2}$$

Then the total exposure to each recipient is expressed by

$$X_k = \sum_{\alpha=1}^{n_k} p_{\alpha k} M_{\alpha k}(W) \tag{3.3}$$

For qualitative discussions only a single exposure pathway need be considered:

$$p_k = \sum_{\alpha=1}^{n_k} p_{\alpha k} \tag{3.4}$$

In this manner each exposure condition can be defined explicitly and a probability p_k assigned to the occurrence of pathways leading to these exposures. These exposures need not be collectively exhaustive, since some cases may involve no exposure, thus no risk.* Furthermore, since more than one pathway may lead to identical exposure conditions, the probabilities of pathway occurrence are not necessarily exclusive. Determination of new pathways and exposures is risk identification; definition and determination of probabilities is risk estimation.

3.1.4 Consequences

The results of exposure by various pathways to recipients result in a variety of consequences. The associated factors are listed in Table 3.1, column IV; in looking at the pathways, one must consider all pathways that lead to like consequences. Thus one examines specifically partial effects from specific pathways. The exposure-consequence relationship is involved; first, in the temporal sense because the consequence may occur immediately after an outcome and subsequent exposure, or second, it may be a delayed, latent effect. Furthermore, the recipients may be identifiable members of the population or may be statistical subsets of large populations.

Each consequence may be defined explicitly in a mutually exclusive manner; and if "no consequence" is considered as a member of the set of consequences, they will be collectively exhaustive. The probability of each consequence p_c is a function of the probabilities of event, outcome, and exposure occurrence. This can be expressed in mathematical notation as follows:

$$p_c = f_c(p_e, p_q, p_k) \tag{3.5}$$

where the subscript c refers to the cth consequence, and f_c is the functional relationship leading to the cth consequence probability. The definition of the cth consequence is designated C_c.

The determination of new consequences is considered risk identification, whereas determination of C_c and p_c is risk estimation.

3.1.5 Consequence Values

A whole range of consequence values can occur for a specific consequence, depending on who is doing the evaluation. Consequence values range from objective measures of behavior to subjective action of individuals and groups, and this is illustrated in Table 3.1, column V. The evaluation of consequence value may be carried out by individual risk agents, by value

* If "no pathway and no exposure" are considered as members of the class of exposures, the set of exposures is collectively exhaustive.

groups, by society as a whole on a national or world level, or by risk evaluators acting for the groups previously mentioned. The risk evaluators can be those involved in government regulation or ombudsmen acting unofficially.

The consequence values are a function of the consequence definition and the probability of occurrence for each particular risk agent. Thus the relation for consequence value C to risk agent k is represented by, $C_{ck}(v)$ and is expressed as follows:

$$C_{ck}(v) = f_{ck}(p_{ck}, C_c) \tag{3.6}$$

Being subjective, consequence values are neither exhaustive nor exclusive, and they are dynamic and situation dependent. Identification of new consequence values can occur, but most of the process of setting values is considered risk estimation.

3.2 RISK ESTIMATION AS A TWO-PART PROCESS

The process of risk estimation involves two basic parts: a probability determination and a consequence value determination covering the areas indicated by the braces at the right-hand side of Figure 3.1. There is an overlap at step 4, Consequences.

Risk, then, is a function of the probability of the consequence and the value of the consequence to the risk taker.*

$$R = F[p_c, C(v)] \tag{3.7}$$

As can be seen, this definition is encompassed by the broad definition given in the previous chapter if the consequence value is measured in a negative sense.

If we look at the flow of causative event to consequence values and all the many aspects at each level, it is possible to put different kinds of risk in proper perspective. For example, a motor vehicle accident involves a discrete causative event that may be a mechanical failure, such as a tire blowout, or a human error, due perhaps to diminished reaction time, and the outcome may range from fender damage to complete destruction, exposing pedestrians or other nonriders. The consequences may range from a rise in the level of adrenaline of the participants to fatalities, and these results may have different values to different recipients. On the other hand, the risk may involve the dispersal of a continuous level of a carcinogen to the environment as a by-product of some industrial or natural process (the

* The parameters are not necessarily independent, since the probability value can influence the consequence value.

causative agent) in which there are pathways through the food chain to man that can result in the additional possibility of cancer, which is valued differently by different individuals and groups. For a classical gamble such as roulette, the causative event is the repetitive spinning of a ball on the roulette wheel, and the outcome is the final position of the ball. The exposure involves the amount and type of wager that an individual will pay, and the outcomes and exposures provide an array of consequences that again have different values to different gamblers under different conditions. In fact, the risk may be looked at from two points of view: first, that of the individual gambler, and second, that of the house, where the consequences are different because of the number of repetitions and the expected value of payoffs.

Basically, the structure of risk is the probability of an occurrence of a consequence in combination with the manner in which that consequence is valued. The implication of valuing consequences by various risk agents and risk evaluators implies that subjective value judgments are involved in the process of risk evaluation. However, it is true in the probability estimate as well.

3.3 OBJECTIVE AND SUBJECTIVE RISK PARAMETERS

The possibility of an occurrence of an event may be determined by direct measurement of repeated trials. When the number of trials is large, we can refer to the estimate of probability as objective because it represents an empirical estimation.* At the other extreme, if we have only estimates made from one or a few trials or totally by conjecture, the probability estimate is subjective. These definitions follow the classical ones of objective and subjective probability. In the subjective case, Bayesian approaches based on conditional probability and use of a priori information are useful. Between these two extremes there is another area, called synthesized probability. In this case, the probability of an event is not directly measured, but is modeled and estimated from similar objective probabilistic systems that are expected to act in a corresponding manner. For example, the so-called Rasmussen study[1] on reactor safety is an estimate of synthesized probabilities; the estimates were not actually measured, but were computed from tests on parts of the system and synthesized into a total model.

A consequence that is directly observable and measurable, and has a value that is expressed explicitly, is defined as an objective consequence

* In principle objective probability exists only in a mathematical sense. The past and present never repeat in the future, and all probabilities for empirical events are only estimates.

value. For example, the accounting value of the payoffs of a gambling establishment as far as the house is concerned, is an example of objective consequence values. At the other extreme—namely, subjective consequences—the value of a consequence to a particular risk agent is completely dependent on the risk agent's value system and situation. Between these two extremes, there is a level called "observable" consequence value. Here the measured, behavioral response of groups in society to objective or subjective consequences can be ascertained through study of their behavior.

A matrix of the nature of probability and the nature of consequence as just discussed appears in Table 3.2, with probability the left-hand column and the consequence along the top. The combination of probability and consequence defines risk. The objective probability–objective consequence case is the area of objective risk, where most classical studies of risk have taken place. They are the easiest to define and the easiest to work with. Synthesized probabilities or observable consequences result in areas called "modeled" risk. Here the model is not directly observable, but the correspondence of the model to reality determines its usefulness. The modeling may be developed through the probability estimate by the use of synthesized probability, through the valuation because of use of observable consequences, or through both synthesized probability and observable consequences, as shown in the table. All other risk areas appear as subjective, because of subjective estimates or subjective valuation or a combination of both.*

The classical area of science involving experiments and making empirical measurements deals primarily with objective risk. In the last few years, the idea of synthesized probabilities has developed extensively, and in the behavioral sciences the treatment of observable consequences as opposed to objective ones has assumed major importance. However, as one looks toward decision making in society the subjective risk area is perceived to be the reality. People base their decisions on subjective risk estimates, not on what is objective. In other words, the emotional aspects, rather than objective scientific knowledge are what drive people. As an example, the use of expected value techniques to estimate planned and accidental risks from reactors and the nuclear fuel cycle usually results in low estimates of risk for accidents. These estimates are not always accepted as true. They may or may not be credible, but such results seem anti-intuitive in terms of present behavior of those who object to nuclear energy on emotional terms.

The argument between subjective and objective reality extends back to

* Modeled risk or subjective risk involves estimation of probability or valuation of consequences or both; these are given in Table 3.2 in parentheses in each column where one of them predominates.

Table 3.2 *The Subjectivity and Objectivity of Risks from the Nature of Probability and Consequence Measures*

	Consequence		
Probability	*Objective Consequence:* event description that is directly observable and measurable	*Observable Consequence:* measured behavioral response of groups to objective or subjective consequences	*Subjective Consequence:* value of a consequence to a particular risk agent
Objective probability: measured by repeated trials	Objective risk	Modeled[a] risk (valuation)	Subjective risk (valuation)
Synthesized probability: modeled from similar objective probabilistic systems, but not measured	Modeled[a] risk (estimate)	Modeled[a] risk	Subjective risk
Subjective probability: estimated from few trials or through conjecture	Subjective risk (estimate)	Subjective risk (estimate)	Subjective risk

[a] Depending on correspondence of the model to reality, the modeled risk is closer to or further from objective risk, but it never reaches objective risk completely.

the ancient Greek philosophers' search for "truth." Socratic truth is objective, whereas Aristotelian truth is subjective. That Aristotelian truth is dominant in our society is easily evident from our legal system, where procedural matters (Aristotelian) dominate over substantive matters (Socratic). We release a person known to have committed a crime if his arrest did not proceed according to designated procedures.

The difference between objective and subjective behavior is well known to psychologists. The objective and subjective probabilities for reward and punishment as behavioral models have received considerable study. The objective probability for reward or punishment involves the actual value of reward or punishment. For example, the punishment for speeding on a certain portion of road may be loss of license and a high-value fine. But at a given time, if there are no enforcement officers on that highway, the objective probability of punishment is zero. However speed on these roads is often governed by the subjective probability of punishment in that the absence of knowledge of the placement of law enforcement officers involves some expectation that there might be one to observe your speeding. Thus behavior is generally based on the subjective assessment of reward and punishment.

If subjective risk estimates are the reality, they can be brought closer to objective risk measures only by educating people to the objective risks or by making the subjective estimates and valuations more explicit and visible, so people understand them better, a difficult communication processes at best.

3.4 VALUATION OF CONSEQUENCES

Assigning a value to a consequence, though a subjective act, is a distinct part of the process of risk estimation. However many of the same considerations and types of value judgment made also accompany the process of risk evaluation (i.e., determination of risk aversion and risk acceptance criteria). To assure that these processes are not confused, the process of "valuation" (i.e., the assignment of a value to a variable) always refers to the process of risk determination, whether the value assignment is to a consequence or a probability. "Evaluation" always refers to processes of risk aversion and acceptability.

The boundary between the processes of valuation and evaluation is, at best, fuzzy. Does the assignment of value imply resultant control or acceptance? Does the application of control or acceptance criteria affect the valuation of consequences? Undoubtedly, such interdependence exists and similar types of value judgment are involved. However for this analysis the two processes are considered to be relatively independent, since common factors are used for entirely different purposes.

Both processes imply the need for subjective value judgments; for consequence valuation, however, a set of factors has been identified that affects the manner in which individual, group, and societal value judgments are made. These factors are outlined in Table 3.3 and are dealt with in detail in subsequent chapters. The ensuing discussion briefly introduces and defines the terms.

Based on the definition of risk that involves the value of consequences and their probability of occurrence, the factors are classifiable by these

Table 3.3 *Factors in Risk Valuation*

- I. Factors involving types of consequence
  - A. Voluntary and involuntary risks
    1. Equity and inequity
    2. Degree of knowledge
    3. Avoidability and alternatives
    4. Imposition—exogenous and endogenous
  - B. Discounting in time
  - C. Spatial distribution and discounting of risks
    1. Geographic distribution of risk
    2. Identification of risk agents
    3. Spreading of risk
  - D. Controllability of risk
    1. Perceived degree of control
    2. Systemic control of risk
    3. Crisis management
- II. Factors involving the nature of consequences
  - A. Hierarchy of need fulfillment
  - B. Variation in cultural values
  - C. Common versus catastrophic risks
  - D. National defense
  - E. Natural versus man-originated events
  - F. Knowledge as a risk
- III. Other factors
  - A. Factors involving the magnitude of probability of occurrence of a consequence
    1. Low probability levels and thresholds
    2. Spatial distribution of risks and high probability levels
  - B. Situational factors
    1. Surprise and dissonant behavior
    2. Lifesaving systems
  - C. Propensity for risk taking
    1. Individual
    2. Group
    3. Conflict avoidance

parameters.* The major headings are factors involving (1) types of consequence, (2) nature of consequences, and (3) other factors.

Factors related to types of consequence first involve voluntary and involuntary risks. The definitions of these risks are complex and entail at least four factors: equity and inequity, degree of knowledge, avoidability of risk and alternatives, and the manner of imposition of the risk. However the terms are for the most part self-explanatory in that they are self-imposed or imposed externally to the risk agent. Discounting in time involves latency of consequences and the effects of time on risk valuation. Spatial distribution and discounting calls for consideration of whether the risk agent is identifiable, and whether the risk distributed to individuals or groups. The controllability of risks affects the valuation and evaluation as well, and it involves the perceived degree of control by individuals and the ability of social institutions to control risk.

Factors involving the nature of consequences cover the problem of valuing premature death at one extreme of the spectrum of consequences and aesthetic values at the other. Finally, the valuation of certain special situations is affected by their consequences. These involve military and preventive functions, lifesaving considerations, and cultural conditions when different cultural environments are at issue. Another factor reflects causative events and their source, whether man-made or natural (i.e., the latter involving acts of God).

Factors connected to the magnitude of the probability of occurrence involve the existence of natural low probability thresholds and consideration of near-certain risks when imposed on individuals or spread among many. There is also the problem posed because different individuals have different propensities for risktaking as basic behavioral conditions.

Such classification of risk factors furnishes a means of comparing different types of risk and risk behavior under different conditions. Studies of each of these factors can provide considerable insight into individual and societal behavior toward risk acceptance. In addition these factors also are involved in the process of risk evaluation.

REFERENCE

1. "Reactor Safety Study: An Assessment of Accident Risks in U.S. Commercial Nuclear Power Plants," (WASH-1400), NUREG-75/014. Washington, D.C. Nuclear Regulatory Commission, October 1975.

* Including the interrelationship between these parameters.

4

Risk Evaluation Considerations

Risk evaluation involves the related processes of risk acceptance and risk aversion. Risk acceptance implies that a risk taker is willing to accept some risk to obtain a gain or benefit, or if he cannot possibly avoid or control the risk, resignation and reluctant acceptance. The acceptance level is a reference level against which the new risk is determined and then compared. The level of the new risk is found through the risk determination process described previously. If the determined risk level is below the acceptance level, the new risk is deemed acceptable. If above, it is deemed unacceptable (if avoidable), and steps may be taken to control the risk. This control action is risk aversion, action taken to reduce the risk. It may consist of total avoidance of the risk, reduction of exposure to the risk, and control of causative events, outcomes, and consequences or processes that alter risk valuation, such as changing value systems, until an acceptable level is reached.*

Risk acceptance levels, to a great extent, involve subjective value judgments in the same manner as consequence valuation. Conversely, risk aversion entails control actions that are generally accompanied by changes in probabilities, consequences, and exposure pathways.

In a preliminary fashion, this chapter presents some major considerations in risk evaluation. Topics covered include the motivation for risk evaluation, the identity of the makers of evaluations, regulatory aspects of evaluation, and problems in the analysis of decisions based on societal behavior.

4.1 MOTIVATION FOR RISK EVALUATION

Every man, woman, and child on this planet is subject to a variety of risks throughout life. They have learned to live with risk, to accept it, to avoid it, and when possible, to control risk quite pragmatically.

* Figure 4.1 provides a hierarchal overview of the risk assessment terminology used here.

The increasing complexity of technology-based industrial society has resulted in growing concern for risks imposed by man-made technology. The *scale* of man's activities has increased sharply since the end of World War II. There has been a unique quantitative jump in industrial societies' ability to create and produce goods and services and, with it, create new demands for these goods and services. There has also been a *qualitative* change as technology has opened risks of new kinds: toxic substances not found in nature have been developed, the potential exists for creation of biological disasters through mutogenic manipulation of viruses, and planet-wide catastrophes have become possible in the event of nuclear war. Man has learned to compound substances unknown to nature that nature has no ability to counteract.[1] Thus, facing a bewildering array of risks that must be continually reconciled, man is motivated more than ever before to understand and avert risks. Some basic concepts can be stated to provide insight into man's behavior to measure and avoid risk.

4.1.1 The Nature of Man

Man is basically a risk aversive animal who usually seeks to avoid or minimize risk. However to achieve some perceived benefit, man will undertake

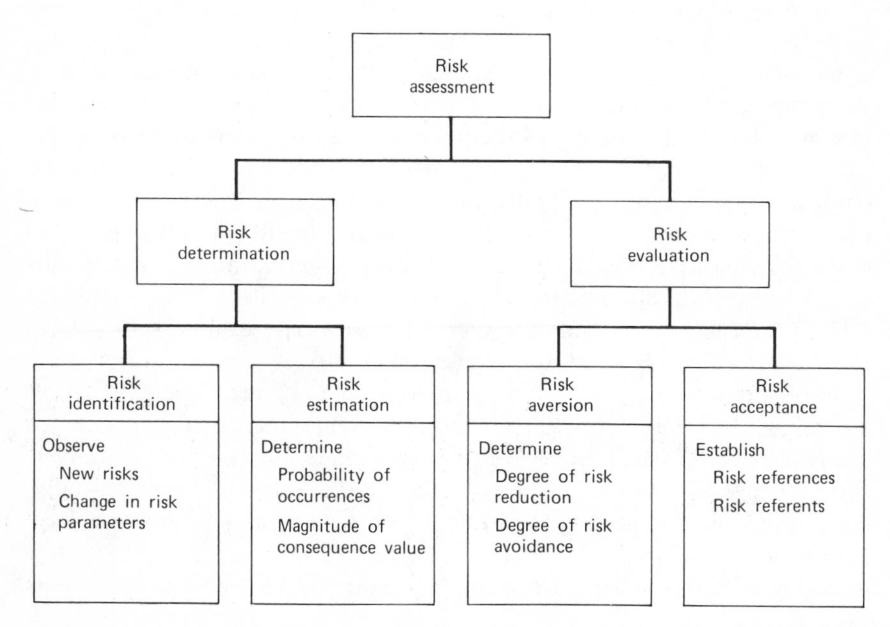

Figure 4.1. A hierarchy of risk assessment terminology [see glossary for definition of items].

to increase his risk. For example, men risk their lives with near certainty of premature death in war or peace to achieve the perceived benefits of freedom. This subjective, qualitative balancing of risks and benefits is an innate ability of man as a rational animal and serves as a model for analytic processes as undertaken here and elsewhere.

Although aversive to risk, man faces risks he cannot control, including the certainty of death. He evidently has built-in physiological and emotional blocks that permit him to ignore risks he can do nothing about and to go on living his life in a pragmatic manner. Man also rationalizes unpleasantness, including risks, by blanking them out: "out of sight, out of mind."

4.1.2 Perception of New Risk Situations

When risks, especially new ones, arise, man experiences anxiety and is motivated to worry and to invoke a risk analysis process, either qualitatively or quantitatively, either cursorily or in depth, resulting in a risk evaluation. This risk evaluation leads either to risk acceptance or risk aversion.

There are a variety of ways in which new risks are perceived and new risk situations occur.

4.1.2.1 Risk Identification

The process of risk identification* involves discovery and/or recognition of new risk parameters or new relationships among parameters, such as new events or new consequences, or a change in the event-consequence relationship. New causative events arise from research and development efforts in new areas. A good example is the problem of new commercial air traffic or supersonic transports, which some predict will cause changes in the stratosphere (outcome) that will expose man to higher levels of ultraviolet radiation, resulting in increases in skin cancer (consequences). In other cases the causative events may already have occurred and only new outcomes identified. Fluorocarbons, such as Freon have been used for refrigeration and spray cans for years. Yet only as recently as 1974 were these fluorocarbons found to have possible effects on the stratosphere, resulting in increased exposure to ultraviolet light and increased skin cancer induction. In both cases cited, there is some increased exposure to ultraviolet light and changes in the magnitude of consequences. New exposure pathways, and even new consequences, are discovered periodically. An example is mutagenetic experimentation with viruses by random fragmentation and recombination. It is possible that some random results may be highly contagious and toxic to man. The very conduct of the experiments

* The term "hazard identification" is often used interchangeably with risk identification.

involves risk, since there are possible accidental exposure pathways to man, and no idea exists of the toxicity of resultant releases, their contagion, or the ability of investigators to contain and control them. Another example is the recently discovered toxicity of vinyl chloride. Practices to minimize exposure have resulted. Thus new risks are perceived as a result of (1) new risks being generated, (2) change in the magnitude of an existing risk, (3) change in the perception of an existing risk, or any combination of these. In any case, these result in a decrease in descriptive uncertainty which can motivate risk aversion action, since an apparently new threat stimulates concern and possibly aversive action.

4.1.2.2 Changes in Values

The manner in which probabilities and consequences are valued for risks can undergo changes that alter the acceptance level of previously accepted risks. This involves reduction of measurement uncertainty and/or changing acceptable levels.

Changes in Risk Estimation Values. Better measures of probabilities of consequences can lead to changes in risk estimates in either direction, up or down. For example, improved understanding of the failure mechanisms and probability of failure of electronic components has led directly to risk aversive action in the design of extremely reliable systems, such as those used in the aerospace industry. Conversely, better estimates of the probability of developing lung cancer from smoking cigarettes has prompted many individuals to refrain from this habit.

In the same manner, consequence values may change. For example, an automobile driver may assign a rather low value to accident consequences under normal conditions. Immediately after witnessing a serious accident, however, his value may change radically and he will exert risk aversive action, tending to drive more cautiously than before, at least for a short time. Familiarity with a consequence leads to a reduction in value in many cases. Most people actively avoid high tension lines, but linemen tend to become less concerned by the danger as they ply their trade, often to their sorrow. Familiarity can generate contempt.

Consequence values change for a variety of reasons. Situations change, group behavior affects individual behavior, and societal behavior affects both groups and individuals.

Changes in Social Values. Social values are constantly undergoing modification. In the past, the changes have been rather slow, but this is no longer the case. One manifestation of this problem is what may be called "value vertigo."

Alvin Toffler, in his book *Future Shock,*[2] says that of the 800-odd lifetimes of 62 years that have passed since the beginning of man, most changes have occurred in the last one or two generations. At the same time, Jay Forrester and the Club of Rome[3] indicate that we are reaching a point on an exponential growth curve that has departed greatly from linear growth. There seems to be little question of the cumulative effects of technology on our society. The thesis of these students is that we are at a time in history of rapid change and unbridled growth.

There may be a more fundamental cultural problem. The value systems on which cultures are based have generally formed a stabilized reference system. However the value systems are affected by an exponential process that has a negative rather than a positive exponent. Examining the interval between major changes in historical value systems, we see that when plotted against calendar time, the time between changes approaches a negative exponential process with a decreasing mean. From the time of Abraham to the time of Christ is about 3000 years. About 1500 years later, Martin Luther appeared on the scene. Less than 400 years later, Darwin came along with his theory of evolution, which upset the religious value systems then in force. Less than 100 years later, Watson developed an understanding of DNA and the genetic code involving the basic understanding of reproduction, and less than 30 years after that, we are talking about the whole concept of genetic engineering. Thus these changes in value systems, the crux of the very nature of man, both metaphysically and psychologically, are occurring more rapidly on an approximate decreasing exponential base.

As any control engineer knows, when the reference system on which feedback control is based begins to vary with a time constant that is of the same order of magnitude as the time constant of delay of the system, instability results. Undamped oscillations can occur as the system attempts to hunt for a stable reference. This phenomenon may now be present in our society, since the reference values are changing at time constants less than that of a generation gap of 30 years. Therefore our society, and different parts of it, seems to be hunting for reference values on which to establish cultural control.

Loss of a reference system has a physiological analogue called vertigo. When one loses a physical reference, he becomes disoriented and physically ill. Since this analogy holds for the negative exponential growth process of value change intervals, I have named this process "value vertigo."

When one experiences physical or value vertigo, the seeking of a reference system is critical. To cure vertigo, one must reestablish the reference or find a new reference, physical or artificial. The seaman adapts to the dynamic motion of the ship as a dynamic reference to prevent seasickness, a form of vertigo. The pilot establishes his artificial horizon as an artificial reference to prevent the onset of vertigo. Those who are least experienced attempt to grab the first reference they can find to minimize their disturbance, but this is not as easily done for values.

The symptoms of value vertigo are evident. We have individuals and groups within society who are in a state of vertigo, or who have experienced vertigo and how attempt to latch onto the first acceptable reference system available to them. As a result, there is a wide variety of new value systems being formed, including the reestablishment of references to older ones. The latter case, is illustrated by the "back to Jesus" movement. In addition, people are resorting to old and new drugs, revolutionary politics, and a wide variety of cults. The rapid change of our value systems no longer allows generation after generation to be preconditioned, and nearly all of society begins to question our dynamically changing values. As individuals and groups of society go through a period of vertigo, they adopt a reference value system, not necessarily a desirable one from the "establishment" point of view, which at least for a time cures vertigo.

A negative exponential process is self-limiting, since it asymptotically approaches zero. However other processes may become dominant before the asymptote is neared. Is there a new process to govern our value systems that can help us overcome value vertigo? The first step toward discovering such a process is understanding the need for it.

What can be done to ameliorate value vertigo? There are several possibilities. First, we can slow down and begin to condition succeeding generations in a retrogressive manner. Already it may be too late to reverse the process. Second, we can do nothing and let nature take its course, hoping that mankind can adapt to change. Presumably, this implies synchronizing societal action with value changes and tracking the value shifts. Many in society are already doing this. We call it "faddism." Or third, we can substitute a new regulating process for the negative exponential function of value change. As a negative exponential function asymptotically approaches its limit, it is easy to conceive that newly recognized processes and functions may become dominant. The nature of this "all-prevailing" regulating process may well be the "survival of mankind" or the escalation up a hierarchy of "values" such as indicated by the Maslow[4] school of psychology or some other yet unrecognized process.

4.1.2.3 Increased Ability to Control Risks

As technology develops, new methods to control risk are being found: risks hitherto considered uncontrollable can now be controlled to various degrees, and more effective controls are found for risks that are already controllable. As an example of the first case, there is serious research in progress on means of controlling tornadoes, either through cloud seeding or explosive disruption of the funnel. The degree of success of such controls is not yet established, but they are being tested. In the latter case, new developments in building design (e.g., automatic sprinklers, compartmentalization) are resulting in changes in building codes and, it is hoped, better control of serious fires.

As more effective risk control techniques are developed, society tends to expect these controls to be applied to reduce risk.* There is considerable motivation to develop, then to implement, new and more effective controls. Those developing risk reduction techniques, such as a cancer prevention and cure, are motivated by a sense of duty to mankind, the desire to enhance professional reputation, possible profit, and any combination of these. On the other hand, society expects the best technology to be used, if it is not too expensive in terms of direct cost or inconvenience. In the auto industry, the high cost of air bags and the inconvenience of safety belt ignition interlocks are good examples of control techniques considered unacceptable by most people. Conversely, some people take large doses of vitamin C to prevent the common cold, although the efficacy of such control has not been established.

Such choices are made by individuals, groups, and societal institutions on a continuing basis. Different processes are involved in risk evaluation of this type.

4.2 WHO EVALUATES RISK?

The manner in which risks are evaluated varies and to a great extent depends on whether individual, group, or societal risk decisions are involved. The same is true of the identity of the evaluator.

4.2.1 Individuals

Any individual who is directly faced with a risk situation makes his own value judgments on risk determination and evaluation. All the factors listed in Table 3.3 come into play, especially the difference between voluntary and involuntary risks and the individual risk propensity of the risk taker. Individual behavior involving these factors is examined in detail in later chapters, but certain aspects deserve discussion at this point.

* It will be shown in later chapters that the perception of degree of control is dominant over the actual level of control.

4.2.1.1 Indirect Involuntary Risks

First, for an entirely voluntary decision, society has generally left the problem of risk evaluation to the risk taker, at least in a free democracy. Few such voluntary risks, however, do not have some indirect, involuntary effect on individuals other than the risk taker. For example, smoking has been considered a voluntary risk in the past. In fact, the Federal Trade Commission mandated warnings on each pack of cigarettes to notify every voluntary smoker of the risks, allowing him to base his decision to smoke on this knowledge. However exposure of nonsmokers to the effluents of smokers and the growing social welfare costs of increased rates of lung cancer must be considered involuntary risks. New regulations preventing smoking in public areas and segregation of smokers and nonsmokers on commercial airlines are examples of increasing concern for such imposed risks. Use of filters and low tar and nicotine tobacco, and research to find strains of tobacco that do not cause cancer, although motivated by desire for increased profits in some cases, indirectly tend to reduce residual welfare costs.

A number of situations indicate increasing concern for allowing voluntary freedom for such individual decisions in our "free" society as opposed to other political structures:

When progeny may be affected. Individual exposure to various agents (e.g., chemicals, radiation, marijuana) may have direct genetic effects that appear only in succeeding generations. The consequences are experienced in the future by others than the immediate risk taker.

When irreversible environmental effects occur. Irreversible effects involving other than the individual who causes the effect are of increasing moment. This is especially true when future generations are concerned. For example, the exhausting of our fossil fuel resources at the present rate of usage implies shortages for future generations.

When exposure of others to involuntary risk or cost is involved. This is exemplified by the case of smoking already discussed, as well as many others.

In one particular case, individual choice has been denied for millennia. There is nothing more personal than one's own soul. Yet evangelical religions seek to impose a particular dogma on individuals so that they can be "saved." During the Spanish Inquisition, salvation was forced on many. In our present society, imposition of religious views occurs through increased exposure to information and experience, but those left outside the fold are often the target of zealous evangelistic pressure.

4.2.1.2 Incompetent or Uninformed Risk Takers

A second condition, identified when society feels that an individual is not fit to make a balanced decision, is most evident with respect to minors and the mentally incompetent. However it has been expanded by the federal government* to cover occupational risks of some types, on the assumption that knowledge of risk is not always available to the risk taker and indirect involuntary risks may be involved. This is an indication of the increasing societal concern for those who may not be able to make reasonable voluntary decisions.

4.2.2 Groups and the Nature of Groups

It is well known by behavioral scientists [5, 6] that there are great differences between an individual's behavior and the influences on his actions of the groups to which he belongs. Thus it is particularly important to differentiate individual values from values of groups to which the individual belongs, and to understand how the value sets interact. Nevertheless, it is possible, within reasonable limits of accuracy, to identify individuals with similar values and to assign them to value groups[8] for convenience. A member of a value group belongs to a value group only to the extent that the values named are similar to the values of other members. There is no implied interaction among members, since they are not identified to one another. Thus a value group is an arrangement for measurement of values, as opposed to a group that influences member behavior and values.

4.2.2.1 Levels of Acceptable Risk for Value Groups

Different value groups have different levels of risk they consider acceptable, either because the benefits they receive offset the voluntary risk or because their status quo is such that they are indifferent to involuntay risks below a given magnitude. The determination of these levels of acceptability can be determined only by one of three techniques: (1) evaluation of historic data to ascertain levels of acceptability of existing risks, both voluntary and involuntary, (2) conducting experiments to measure the levels of acceptable risk for particular risks, and (3) introducing new risks and measuring the "squawk factor."†

Of these methods, the first one is perhaps easiest, since data are available and there is no bias inserted by overt action as required by the other two methods. For new risks for which historic data are not available, however,

* In the form of the Department of Labor's Occupational Safety and Health Administration.
† The "squawk factor" is the quantified vocalized response in opposition (number of telephone calls, telegrams, letters, etc.) received in response to an activity or proposed action.

combinations of all three are possible. In the first case, one attempts to compare the new risk with other, similar societal risks.

A subsequent section examines historic data in an attempt to draw some conclusions about acceptable levels of risk for the whole United States population as a value group.

4.2.2.2 Categorization of Group Behavior Toward Risk Evaluation

Kates[8] has suggested that group risk evaluation behavior can be generalized into three major categories of active group respondents to a risk situation. These categories seem particularly applicable and are reproduced here:

1. *"The Worry Bead School."* Hazard is different from the risk determined.*

Causes. Society's ability to cope limited: risk less than the threats perceived; may be applied to wrong hazards.

Risk Assessment Focus. Hazard identification; valuation and evaluation; societal response; medium-term effects.

Favored Method. Institutional mechanisms.

Favored Audience. Guardians (government regulators); risk takers.

2. *"Tip of the Iceberg School."* Hazard is always greater than the risk determined.

Causes. Consequences of technology are recent; when they happen, control is too late; increase in catastrophic risk.

Risk Assessment Focus. Hazard identification; consequence estimation; unsolved problems; long-term effects.

Favored Method. Scenarios.

Favored Audience. Guardians; risk takers.

3. *"Count the Bodies School."* Hazard is always less than the risk determined.

Causes. Hazards are determined and coped with, but expectations may change; we (as society) are ill informed and have a bias against technology and industry.

Risk Assessment Focus. Event estimation; valuation of consequences; coping mechanisms; short-term effects.

Favored Method. Quantification by reduction.

Favored Audience. Guardians; hazard makers.

Right to Freedom from Harmful Technology. Considered by as a fundamental "right" of society, especially by those who are oriented against technology.

* Note—some terms are changed to conform with definitions given in previous sections.

4.2.3 Society as an Institution

Society as a whole or major portions of society act as risk evaluation institutions informally and formally.

4.2.3.1 Informal Evaluation

Informal societal response occurs in a number of areas. One major area is the marketplace, where buyers avoid high costs, including risks. For example, public awareness of the special hazards of the rear engine Corvair automobile, given wide coverage by Ralph Nader, has led to the demise of the product.

Another area is withdrawal or avoidance of risk by a large-scale population. At a tourist resort, for example, increased exposure to disease such as encephalitis can lead to desertion of the resort.

Direct action in the form of protests, strikes, and even revolutionary activities is another area of response. Fortunately, these events often lead to more formal action in our society.

4.2.3.2 Formal Evaluation

The formal processes available for risk evaluation are the political processes, involving the legislative and executive branches of government, the judiciary system, and the administrative operations of the executive branch.

4.2.2.3 Social, Economic, and Political Aspects of Group Behavior

Dr. James A. Fay[9] of MIT has summarized some particular aspects of group action and behavior toward risk.

Social Aspects

Community Group. A sense of community concern against threat exists. The catastrophic threat outweighs the ordinary type of threat, since the former interferes with the ability of the community to deal effectively with the catastrophic situation. The community has a perception of inability to control or deal with the problem. Starr[10] calls this effect "group resilience," and indicates it is a highly nonlinear function of size of threat.

Fear of New Technology. There is a relationship between the visibility of risk presence of a technology and the objective level of risk (e.g., noise from aircraft flight paths reminds people of risk of crashes).

Economic Aspects

Alternative Technologies. Different groups have varying attitudes on how much to spend for risk reduction.

Political Aspects

Aggregation of Public Grievances. Dissimilar, but small groups join to gather strength for self-defense.

The court system and the electoral process have classically dealt with balancing of risks at individual and institutional levels after the damage from hazards has occurred. Amelioration of damages is a well-established, but complex, problem. However evaluation of anticipated risks is relatively new in government and is practiced primarily in the regulatory agencies of the executive branch.*

4.2.4 Valuing Agents and Risk Evaluators

The scope of risk evaluation is made more complex than the preceding definitions imply when there are many risk takers, each with his own set of subjective values and relationships to externalities. Furthermore, the assignment of risk often involves an evaluating agent making a judgment for a valuing agent, when all the individual judgments cannot be polled or reconciled. A government agency interpreting the needs and values of people in setting a regulation is an example of an evaluating agent as opposed to the valuing agents—namely, the people affected by the regulation. In making such judgments, a risk evaluator is always subject to question in terms of his knowledge and ability to make the judgments, the effect of his personal biases, and his fairness in making the decisions at hand. Often the risk evaluator attempts to determine criteria for public acceptability of risk by historically examining similar kinds of risk to see what levels have been accepted by society.

Thus a "valuing agent" can be defined as a person or group of persons who directly evaluate the consequence of risk to which the agent is subjected, whereas a "risk evaluator" is a person, group, or institution that seeks to interpret a valuing agent's risk for a particular purpose.

4.2.4.1 Regulatory Agencies as Risk Evaluators

When risks to various groups in society are deemed excessive in relation to the benefits derived from an activity, government regulatory agencies often intervene to protect the affected groups. Although intervention generally occurs in relation to such involuntary risks as exposure to hazardous pollutants and radiation, it also applies to voluntary risks that have involuntary aspects, such as cigarette smoking in public buildings, conditions that

* The Office of Technology Assessment established by Congress in 1975 is the initial effort by the legislative branch for to formalize direct consideration of future problems.

are subject to the occupational health and safety laws, and the sale and use of restricted items (e.g., fireworks). In some cases, such as proposed laws to require a driver to fasten his seat belt, intervention touches purely voluntary matters.

There seems to be little question about the appropriateness of regulatory agencies to control the imposed involuntary implications of some activities. However the regulation of purely voluntary risks is another matter. This is a constantly evolving situation and some perspective on the matter is useful.

4.2.4.2 Bureaucratic Evolution of the Risk Evaluation Role

The concept of the bureaucratic risk evaluator has evolved in the last hundred years in the United States, out of the development and growth of government regulatory agencies. Historically, the legislative mechanism in this country has been heavily influenced by an array of self-interest lobbies pursuing and seeking to preserve or improve their own gains. The development of regulatory legislation and regulating agencies resulted from the gross inequities that existed until public pressure forced Congressional action. The Sherman antitrust legislation is a good example of reaction to unfair financial control by a few through cartels and restraint of trade. Subsequently, agencies were formed to regulate particular industries in which monopoly or oligopoly were thought to be in the best interest of society because of increased efficiency. Over the years, many of these regulatory agencies became captive to the industry they were supposed to regulate.

In the last 20 or so years, the concerns of individual citizens who suffer from inequities have become increasingly important. This is due not only to the imposition of new kinds of risk involving new technologies, but to the increasing complexity of our society, as well as the rapid spread of information through instantaneous, total coverage by the press. These factors have led to the evolution and implementation of restrictive regulatory agencies, as compared to the permissive regulatory agencies* of the past, which were

* A permissive regulator is defined here as a regulatory agency or part of an agency that enables an industry to operate, issuing permits or licenses or establishing rate structures, in a manner to allow (permit) that industry to operate. Examples are the Nuclear Regulatory Commission, the Federal Power Commission, the Interstate Commerce Commission, and the Federal Communications Commission.

On the other hand, the restrictive regulator is empowered to prevent undue risks to certain groups in society or society as a whole. Parts of the Environmental Protection Agency (EPA), the Department of Health, Education, and Welfare (HEW), and the Occupational Safety and Health Administration (OSHA) are examples of such restrictive regulators.

It should be noted that some agencies perform both functions within their authority, but generally divisions within the agency separate these functions into different divisions of effort. Each type of regulator has its role, and it is important that these roles not be confused.

set up to permit an industry to operate under monopoly conditions. Restrictive regulatory agencies, aimed primarily at protecting the public from gross inequities, are found in such diverse fields as consumer protection, environmental protection, and standards of health and safety for the general population and in the work place.

The restrictive regulatory agency has become the risk evaluator for society in many instances. The role is not an easy one, since it implicitly recognizes that all inequities cannot always be resolved and that the general good of the total population must also be preserved. However it is desirable that the implementation of risk evaluation decisions be made in a visible, traceable manner and that input from all concerned be seriously considered and weighed in the decision.

Under what conditions is a regulatory agency empowered by Congress to act as a risk evaluator in particular areas? There is no explicit answer to this question, but it is evident that when groups in society feel they have suffered gross inequities, they attempt to seek relief through the courts, through public opinion, and sometimes with violent action. As individuals in the legislative body perceive the possibility of effects on their constituency and potential constituencies, the problems are focused and addressed, and legislative solutions are initiated. Usually the executive branch is empowered to carry out these mandates in a general manner, but sometimes Congress has specified the detailed mechanisms of implementation.* In any case, the role of the risk evaluator in the restrictive regulatory sense should be one of balancing inequities where possible and ensuring that the risks to be undertaken are spread out reasonably.

4.3 PROBLEMS IN RISK EVALUATION AND REGULATION

As indicated in the previous section, the role of a risk evaluator is to ameliorate risk inequities in society often in a regulatory setting. The amelioration of these risk inequities, by regulatory bodies implies the existence of some acceptable levels of risk. If these levels are achieved, the inequities are deemed small enough to be allowed. Otherwise, risk aversive action in the form of control is required.

Risks are controlled either by reducing the probability that a causative event will occur or by minimizing the exposure pathway, should an event occur. In the case of a causative event, an activity that may have some

* For example, Water Quality Act of 1972 expressly directs that water discharge control shall meet the requirements of "best practicable technology" by 1977 and "best available technology" by 1983.

beneficial aspects may be curtailed, banned, or eliminated to prevent or reduce its occurrence. For example, a way to reduce lung cancer is to make smoking in any form illegal. On the other hand, a new tobacco or tobacco substitute without any cancer-causing agents would also eliminate this causative event. Control of the exposure pathway through use of filters or alternate methods of smoking, such as cigars and pipes, allows the activity to take place, but at, hopefully, reduced risk. Furthermore, smoking involves both voluntary and involuntary risks; as previously mentioned, for example, both the smoker and the nonsmoker, who must suffer the smoke-filled rooms, are also obliged to help underwrite society's burden in treating premature morbidity and mortality of smokers.

A major philosophical question with practical implications is: should risks and risk causing activities be regulated, and if so, when? To answer this double-pronged question, we must balance the costs and benefits.

4.3.1 Societal Cost-Benefit Balances and Resultant Inequities

4.3.1.1 Cost-Benefit Overview

In the broadest sense, societal costs and benefits imply gains and losses to society as a whole or to specific groups within society, as opposed to more restrictive definitions featuring the minimization of environmental pollution or the reduction of threats. For this reason, and to avoid confusion, broad societal costs and benefits are referred to as societal gains and losses. These gains and losses are not always financial, and the hierarchial scale discussed in Section 4.3.3.2 provides one nomenclature for identification of different gains and losses. The scales for gains and losses can be identical in the sense that any parameter can sustain a gain or loss.

Risk is usually considered in terms of probable loss as one of these parameters. This arises because risk generally has a negative connotation. Theoretically there is no reason not to call the probability of beneficial occurrences "positive risks." To prevent confusion, however, the latter are referred to as "probable gains." Thus an event or practice with some probability of occurrence and a negative value of a consequence is termed a risk, whereas a probable occurrence of a consequence with a positive value is termed a probable gain.

When applying the same nominal identifying scales to both gains and losses, one would assume that the scales are identical in all respects. This is not true, however, when the differences between direct and indirect benefits, and direct and indirect costs in terms of gains and losses, are taken into account. Direct gains or benefits are explicitly identified as being received as a direct result of the activity involved. For example, to undertake opening a new manufacturing plant, the operator seeks to make new revenue from the venture, to his direct benefit. The buyer of the product would benefit indirectly, since a new source of supply would increase availability and might decrease price. The first case involves direct benefit; the second, indirect benefit.

Direct costs or losses, on the other hand, are explicit and can not be voluntarily avoided once the activity is undertaken. For example, once the new plant is undertaken, a commitment is made for direct investment of capital and operating funds. Indirect cost or loss may occur as a result of environmental pollution because of plant operation, and so on. The indirect cost is generally not borne by the recipient of the direct cost and benefits.

4.3.1.2 Balancing Gains and Losses

When gains and losses are balanced, both direct and indirect gains and direct and indirect losses must be included in the balance. The direct gains must be balanced against the direct losses and, in a similar manner, the indirect gains must be balanced against the indirect losses. When this is done, four categories of loss-gain balances result (Table 4.1).

No Contest Case—Unacceptable. The first two cases are referred to as "no contest" cases, since the decisions are decisively acceptable or unacceptable. In case 1 the direct losses exceed the direct gains and the indirect losses exceed the indirect gains. On this basis, the activity is completely unacceptable. Offensive biological warfare seems to provide a good example in our present peacetime posture. Development of such weapons only escalates the probability of use, and the direct losses of escalation far outweigh the direct gains of strategic advantage. The indirect losses far outweigh any indirect gains, since the loss may be extinction.

Table 4.1 *Categories of Loss-Gain Balances*[a]

Case	Direct Balance	Indirect Balance	Nature of Balance
1	$G_D < L_D$	$G_I < L_I$	Decisively unacceptable activity
2	$G_D > L_D$	$G_I > L_I$	Decisively acceptable activity
3	$G_D < L_D$	$G_I > L_I$	Unacceptable unless activity is subsidized
4	$G_D > L_D$	$G_I < L_I$	Unacceptable unless inequities ameliorated or allowed

[a] Symbols: G_D = direct gain (benefit), L_D = direct loss (cost), G_I = indirect gain (benefit), L_I = indirect loss (cost).

No Contest Case—Acceptable. In case 2 the reverse is true. Both the direct and indirect gains exceed the direct and indirect losses, respectively, and the decision is decisively acceptable. These cases are relatively straightforward in terms of decisions to be made and are of minimal concern. Compulsory education is a good example of this case. The direct benefactors are educators and students. The indirect benefactor is the total population. Direct losses are in terms of taxes, and indirect losses, such as having an aspiring population, are minimal. At present, the direct loss in taxes seems to be less than the gain of education of individuals. The total population gain of an educated society seems to outweigh the problems of aspirations of new groups. Of course no situation remains static, and many are questioning this decision as a result of the high proportion of taxes used for public education, given the degree and quality of education to be purchased.* This balance may not have existed at the time such decisions were made.

Subsidy Case. In case 3 the direct losses exceed the direct gains, but the indirect gains exceed the indirect losses. Here, those who receive the direct benefits would be unwilling to undertake the activity where the losses may exceed the gains. However if the indirect societal benefits are greater than the indirect societal losses, the activity may well be undertaken for the benefit of society as a whole. As a result, those directly seeking gains for the activity may have to be underwritten—that is, subsidized. When the indirect benefits are worth having in spite of the direct gain-loss imbalance, the government often subsidizes the activity. A case in point is the attempt to develop a cure for cancer. The cost of research is so large that the payoffs to the discoverer of any such cure would be insufficient to recuperate the funds invested. On the other hand, the payoffs to society as a whole would be very great indeed, and therefore the government chooses to subsidize cancer research. The researcher receives the direct benefits of such a subsidy and resultant research activity, but society is in line to receive indirect benefits, which are thought to outweigh the direct costs of subsidy.

The degree of difference between direct and indirect gain-loss balances determines if and to what extent subsidies might be justified.

The Inequity Case. In case 4 the direct gain exceeds the direct loss, but there are losses imposed on society as a whole that exceed any indirect gains that might be had. This is a case of gain-loss inequity, and the inequities must be either ameliorated or accepted before such an activity is warranted.

* The cost per pupil per hour of classroom time is increasing, and the quality of the product may be decreasing as symptomized by grade escalation and declining national scholastic test scores.

The inequity case illustrates the need for government regulation. It implies the necessity for a fair means of equitably spreading the costs, especially the indirect costs; or when these inequities cannot be resolved, a way must be found to determine acceptable levels for inequitable losses. The degree of difference between direct and indirect gain-loss balances determines if, and to what extent, regulation is justified. The accomplishment of this task, involving risks, is a major role of regulatory agencies. In this case, the development of acceptable levels of risk implies the existence of inequitable probable losses. Thus we must ask, how do regulatory agencies—or society, for that matter—develop acceptable risk levels?

4.3.1.3 Other Problems in Balancing Risks and Benefits

A number of other problems arise when risks and benefits are to be balanced analytically, as opposed to subjectively, for a risk taker faced with an immediate decision. One may ask why an analytic balance should be made in the first place. Why not allow each risk taker to make his own subjective judgment? The answer lies in the unequal assumption of risks and benefits by different groups in the population and the imposition of involuntary risks. If some action is to be taken to ameliorate imbalances—by government control and regulation, for example—methods are required to allow regulatory value judgments to be made in a visible, reconstructable manner.

Identification of Recipients. Those who undertake activities to receive benefits are not always those who receive the risks. As a result, involuntary risks are transferred to others who receive no direct benefit. Decisions allowing unequal sharing of risks and benefits for individuals and groups essentially become societal decisions and must be made by an arbiter. A good example is the "taking question" of the use of eminent domain.[11] Here the private ownership rights of an individual are usurped in the name of greater need of society, and an attempt is made to compensate the individual at risk. However compensation is the exception rather than the rule. For example, those living near approaches of airports that have begun to accommodate jet aircraft are not compensated for increased noise levels, which may be high enough to impair health.

In any case, identification of those who receive the benefits of an event and those who assume the risk must be made before any balancing. When an imbalance exists, the problem of redistribution of risk through public choice must be faced. Zeckhauser[12] has undertaken a study of this problem, especially the converse situation of how to spread risk and distribute it so that particular groups do not assume risks asymmetrically.

On the positive side, Zeckhauser indicates that risk spreading can improve planning when there are delays in settling uncertain situations; it can reduce anxiety through hedging for nonresolution of future uncertain situations, and it can achieve redistribution of future consequences through immediate contractual relationships (e.g., insurance) among parties at risk. On the negative side, some risks, such as those to intelligence and health, cannot be redistributed at any cost. In other cases, such as determining one's parents and birthrights, the consequences have already been adopted and past inequities cannot be spread equitably.

Zeckhauser has introduced time as a variable in assuming risks in that spreading of risks is most useful in reducing anxiety due to future uncertainties requiring present action. However time enters into risk-benefit balancing in other ways as well.

Timing of Risks and Benefits. It is becoming increasingly evident that many of the actions society takes to achieve short-term benefits impose risks that have long-term impact. In fact, the benefits sought for one generation impose risks and costs on subsequent generations.

Unfortunately, techniques for discounting future risks, just as the future worth of a dollar is discounted, are presently unavailable. Furthermore, the societal rate of discount and private rates differ. As one applies a discount rate to the value of a life, one must also bear in mind that society's or an individual's value of life will change over time in a manner that can offset the discount.

The reverse problem of balancing short-term risks against long-term benefits is one with which society is more familiar, and it depends on individual and societal propensity to recognize and accept deferred benefits. It is worth noting that those who are satisfied with their status quo recognize the concept of deferred benefits in a favorable light, whereas those who are dissatisfied have shorter views, especially when survival factors are involved.

4.3.2 Dynamics of Societal Balancing

Long-term impacts involve generations. However there are short- and medium-term effects, as well. Changes in values occur for a variety of reasons. Some of the major causes are listed below.

4.3.2.1 Situation Dynamics

Values change as situations change, and situations change dynamically.

4.3.2.2 Individual Versus Group Values

There are differences between an individual's own behavior and the influence of groups to which an individual belongs in valuing consequences.

Levels of Acceptable Risk for Value Groups. Different value groups arrive at different levels of risk that they consider acceptable. As indicated previously, many factors affect the way different value groups look at risk.

The "Squawk" Potential. One value group may attempt to influence other value groups when an issue can be promoted into one of national public concern. Because of our system of instantaneous communication and total press coverage, a potentially controversial issue is subject to magnification out of proportion to its importance. However it provides a vehicle for one group to sway others. This is a particularly critical condition in present society.

4.3.2.3 Squawk Potential and Credibility

Certain kinds of risk consequence have the potential to be valued emotionally rather than rationally. This becomes especially evident when one value group in society is against having specific voluntary or involuntary risk consequences imposed on them and seeks to influence others to prevent imposition by making it a major public issue, often involving unsubstantiated claims for impending disaster.

The "squawk" potential refers to a condition for an issue that is distasteful or unacceptable to a particular value group to become a major issue, blown up through dire predictions of consequences, based primarily on half-truths, but flamed by competing commercial news media. The objective is to stir up public opinion and persuade the emotions of diverse value groups to the point of view of the original value group, generating enough concern, perhaps hysteria, to affect elective governmental bodies, regulatory agencies, and the courts. On this basis, any issue has the potential for becoming a major "squawk."[13]

Value groups in society have always had the opportunity to effectively lobby out of proportion to their representation in society as a result of the inaction and lack of interest of the so-called silent majority. Today, however, instantaneous communications media have created an insatiable demand for news. Programming and media space have to be filled, and during times of lower news event activity, it often becomes necessary to manufacture news and/or magnify the significance of existing issues. As a result, featuring the unusual story or human interest items has become standard procedure, and dire predictions, exposés, and "bad news" seem to have more capacity to sell papers than good news. On this basis, the news and communications climate is indeed ripe for the minority value group to create a major issue out of almost any situation, with very little expenditure of resources. Though one may fault the media for selection of material for filling gaps and for overemphasizing some items at the expense of relatively straightforward reporting of information, the major blame lies elsewhere. The media could not "sell" squawks if the public were not interested in hearing about them; moreover, the squawks would not survive very long if there were creditible institutions to evaluate and respond to them in a manner reasonably acceptable to the public. The condition is even worse in times when the credibility of government is at question, as presently seems to be the case. It is often aggravated when the squawk potential is used, not by value groups with legitimate complaints but by those who seek to use this mechanism to their personal advantage. In the second category we have the political candidate trying to inflate an issue, the ambulance-chasing lawyer, the press creating shocking but misleading headlines, and finally those who are looking for publicity for its own sake, either to enhance personal worth on the marketplace or to appease the ego.

Squawks of this type give value groups suffering inequitable imposition of risks an opportunity that might not otherwise exist to have their case heard, but they place a heavy burden on society. The generation of exaggerated claims to initiate squawks to get the attention of the public and the press generally involves half-truths, unfortunately, however, the burden of proof is not on the value group making exaggerated predictions, but on the government or other agencies to disprove the claims, or better yet to evaluate the claims rationally. Such evaluation often requires considerable resources and sometimes years of effort. For example, in 1972 it appeared that the SST might contaminate the upper atmosphere. Today this seems to have been put aside as a negative premise. On the other hand, if the government pooh-poohs the claims or tries to disprove them without due consideration, the credibility of government itself suffers.

What can be done about this? One cannot or should not prevent a value group with a legitimate squawk from being heard. However half-truths must be nipped in the bud, and to the extent that a complaining value group has been able to rationally explain its case, resources might be made available to them for the continuation of such efforts, although there is no guarantee that the resources would be used properly. When squawks are made by one value group, an opposing value group will attempt to counter with its own half-truths unless a credible, rational case is made by the original squawkers. Unfortunately, two half-truths do not make a whole truth.

Government cannot resolve issues when its own credibility is suspect. However it seems likely that credibility can be restored by making all actions, deliberations, and value judgments visible and traceable. One may not agree with a value judgment or a decision, but if there is some assurance that all issues have been heard and dealt with fairly in making the decision, credibility will no longer be suspect, only judgment.

Tompkins disagrees with this approach and argues:

I do not think this presents either a valid argument or correct observation. The only credible point of view to those exercising their right to squawk is found among those spokesmen whose biases agree with theirs. Those who can be persuaded to the contrary are already persuaded. The rest are not interested enough to develop an opinion of their own. I see the argument in favor of a formal process for displaying value judgments as providing a mechanism by which the Government could hopefully explain and defend its position. Most decision makers I have met are more easily informed (or persuaded) by found logic and qualitatively stated value judgments to indicate how the speaker is "tilting" with respect to the balance among alternatives than they are by a presentation that can appear to be somewhat artificial. How the uncommitted sector of the public will react to the proposed methodology cannot be determined until it is tried.[14]

Conversely, Zeckhauser argues:

It should be understood that for many societal decisions the procedure by which the decision is made may be as important as the actual dollar numbers that are employed. Analysts frequently dismiss too quickly the value of having an equitable and widely accepted process. . . . Where process is important, analysts can take one of three tacks. They can labor earnestly to provide the inputs that are required by the processes. Alternatively, they can undertake investigation of the outcomes of the process. Has it been producing desirable results? Finally, they can examine the process itself, with the hope that by so doing they can improve its performance.

All three of these approaches have the virtue that they attempt to complement and inform presently accepted procedures for decision making rather than ignoring them. Not only will the analysis have greater impact than the more myopic approaches being cautioned against, the final outcome may be more attractive. Individuals may far prefer an outcome that they believe has been justly arrived at, to one where some unimpeachable but nevertheless distasteful calculations were the basis for decision.[15]

Though agreeing with Tompkins that one cannot anticipate how the public will react to such a methodology until it is tried, the author feels that the Zeckhauser approach at least offers hope that rational action is possible. If not, at worst, little will be lost in the attempt.

Since societal risks are one of the subjects for which squawks are often generated, information and judgment must often be qualitative and subjective; but the process of making decisions can be made visible and traceable.

4.3.3 Measurement Problems in Consequence Valuation and Evaluation

4.3.3.1 Measurement Scales

Difficulties arise when value is assigned to consequences because values themselves are often intangible. Unfortunately, most operational techniques for dealing with decision theory[16, 17] require cardinal values [see Glossary] to be assigned to consequences in the form of utility functions.

Whereas the use of cardinal utility leads to elegant mathematical solutions of expected utility and expected value, decision theory solutions are limited to those few problems in which cardinal utility can be assigned. Unfortunately, the real problems of interest in risk deal with intangibles, and these problems are not so easily handled by decision theory techniques because of great uncertainties in converting intangible value scales to cardinal ones. The author contends that in conversion from intangible values to cardinal scales, there is a limited, intrinsic level of precision that is meaningful to the valuer. As a result, there is always uncertainty in such valuation, and that uncertainty must be made visible and examined in any valuing process.

One cardinal scale that is often used for valuing consequences is dollars. However it has been well demonstrated by many authors, such as Friedman and Savage,[18] that though dollars may be linear on a measurement scale, the utility of money may be quite nonlinear. As a result, dollar scales are limited in use to economic problems as a general rule, and even the utility of money is often inadequately applied in such problems.

4.3.3.2 Hierarchy of Scales

Associated with any action there is usually a wide variety of consequences many of them intangible. Identification of different kinds of risk consequence is necessary, along with some estimation of the kinds of scale that may be used. To address this problem, the author has developed a hierarchy of risk consequences based on the conceptual hierarchy of needs developed by Maslow.[4] Here the highest priority need is survival, which has been broken down into premature death, avoidable illness, and other survival factors (Table 4.2). Each major category of need is dominant over those below it as long as the level at that need remains unfulfilled. Once it is fulfilled, the lower level needs become dominant in turn, although higher level needs can preempt lower level needs at any given time. The hierarchy continues through exhaustible resources (survival and security factors), physical security, belonging, egocentric needs, and self-actualization. In this case, self-actualization refers to the quality of life and the desire for "the good life," as well as doing things just for the sake of doing them. At each level for the risk consequence, as the table indicates, different variables and measurement scales are possible. Although scales for these are discussed in detail later, it should be noted here that for the needs at the middle of the scale, dollars or the utility of money are particularly useful; but at either end of the scale, dollars or the utility of money become inappropriate. At the high end of the scale, death and injury are involved. Since essentially life

Table 4.2 *Hierarchy of Risk Consequences*

A. Premature death (avoidance)
   1. Catastrophic—chronic
   2. Avoidable—unavoidable
   3. Committed—uncommitted
   4. Indentifiable individual vs. statistical
   5. Knowledgeable vs. unknowledgeable risk

B. Illness and disability (avoidable)
   1. Major—minor
   2. Temporary—permanent
   3. Immediate—latent
   4. Degree of suffering, dehabilitation, discomfort, etc.

C. Survival factors
   1. Hunger avoidance
   2. Shelter availability
   3. Environmental control (heat, light, etc.)
   4. Procreation—individual, population

D. Exhaustible resources
   1. Energy
   2. Space (land)
   3. Environmental quality
   4. Minerals, including water

E. Security: protection of rights, property, wealth

F. Belonging—love
   1. Family
   2. Clique
   3. Group
   4. Society

G. Egocentric
   1. Power
   2. Wealth
   3. Fame
   4. Opportunity for advancement
   5. Conspicuous consumption

H. Self-actualization
   1. Leisure
   2. Aesthetics
   3. Ethics
   4. Opportunity for choice
   5. Freedom of action
   6. Knowledge
   7. Quality of life

and health cannot be "bought," dollars by themselves are an ineffective and incomplete measure of the value of a life.* At the other end of the scale are factors that affect the quality of life, and again these factors are difficult to put in monetary terms. How much is a scenic vista, or the potential of being able to look at one, worth to an individual?

In spite of this, it is sometimes necessary to use cardinal scales to represent these intangibles. Yet the author maintains that this can be done only with a finite level of precision; the uncertainty therefore is great, but it can be specified in explicit terms.

There is no question that it is difficult to measure values of risk consequences associated with the more intangible values (value of life, quality of life, etc.). As shown previously, several other factors must be considered in

* Actuarial measures are for the benefit of insurance companies, not individual risk takers directly.

making such measurements and understanding the level of uncertainty that exists for these measurements. In any case, it is important to specify the uncertainty and to learn how to use it as a parameter.

REFERENCES

1. E. F. Schumacher, *Small Is Beautiful.* New York: Harper Colophon Books, Harper & Row, 1973, p. 17.
2. Alvin Toffler, *Future Shock.* New York: Random House, 1970, p. 14.
3. Jay Forrester, *World Dynamics.* Cambridge, Mass.: Wright Aiken, 1971.
4. Abraham H. Maslow, *Motivation and Personality.* New York: Harper & Row, 1954.
5. George C. Homans, *The Human Group.* New York: Harcourt, Brace Jovanovich, 1950.
6. Herbert A. Simon, *Models of Man.* New York: Bailey, 1957.
7. W. D. Rowe, "Decision Making with Uncertain Utility Functions," Doctoral thesis, American University, Washington, D.C., 1973, pp. 29–30.
8. Robert W. Kates, "Risk Assessment of Environmental Hazard," SCOPE Report No. 8, International Council of Scientific Unions, Scientific Committee on Problems of the Environment, Paris, 1976.
9. James A. Fay, at the Engineering Foundation Conference on Risk Benefit Methodology and Application, Asilomar, Calif., September 21, 1975.
10. Chauncey Starr, at the conference cited in note 9.
11. Fred Basselman, David Callies, and John Banta, *The Taking Issue.* President's Council on Environmental Quality. Washington, D.C.: Government Printing Office, 1973, Stock No. 4111-00017.
12. Richard Zeckhauser, "Risk Spreading and Distribution," in *The Political Economy of Income Distributions* edited by Hochman and Peterson. New York: Columbia University Press, 1973, pp. 206–228.
13. J. P. Davis has pointed out that a "squawk" could concern exaggerated benefits as well as "impending disaster." Private communication, July 19, 1975.
14. Paul C. Tompkins, personal communication, June 6, 1975.
15. Richard Zeckhauser, "Processes for Valuing Lives," discussion paper, Public Health Program, Kennedy School of Government, Harvard University, January 1975, pp. 7–8.
16. Robert Schlaifer, *Analysis of Decisions Under Uncertainty.* New York: McGraw-Hill, 1969.
17. John von Neumann and Oskar Morgenstern, *Theory of Games and Economic Behavior,* March 1953.
18. Milton Friedman and L. J. Savage, "The Utility Analysis of Choices Involving Risk," *Journal of Political Economy,* Vol. 56, 1948, pp. 279–304.

5

Risk Evaluation Methods

Valuation of consequences and the process of risk evaluation are similar in that both involve subjective values. Furthermore, individual decisions on risk, especially voluntary ones, do not often differentiate formally or consciously between consequence valuation and risk evaluation. Consequence values and benefits are balanced in a relative manner, that is, relative to the values and situation of the individual risk agent. There is no question that many personal decisions involving risk necessitate great deliberation, anxiety, and soul searching. Such important decisions are addressed in a later chapter on individual propensity for risk.

The problem of risk evaluation, involving a risk evaluator making decisions for others, is of primary interest here. Decisions made by risk evaluators may be formal or informal. In the formal case we have legislative proceedings resulting in laws, courts basing decisions on the law and legal precedents, and regulatory agencies acting to make risk decisions on a group or societal basis. The informal case involves the dependence of individual risk agents, singly or as a group, to rely on others they trust to make decisions for them.

5.1 INFORMAL RISK EVALUATION PROCESS

The informal risk evaluation process comes into play when an individual feels he cannot, by himself, make a voluntary risk decision, or faces an involuntary risk decision with which he cannot cope. A greater source of perceived expertise in coping with such decisions is sought, as well as the idea of sharing or pooling risk situations with others in like circumstances.

5.1.1 Expert Advice

The help of experts is often solicited by those who must make decisions. Some experts such as doctors, lawyers, and clergy, are particularly competent to aid people because of their experience and training. Still others, however, have only some perceived degree of expertise and serve as "father figure" or leader. In either case, an individual seeks the expert, advice, then decides to accept it. The degree to which the risk agent considers the risk evaluator a "true" expert and the extent to which the advice is acceptable to the risk agent determines how fully the advice is accepted. For example, if a doctor advised a minor operation to save one's life, a patient would tend to accept that opinion. But if a major, risky operation is said to be necessary, the patient may seek advice from other doctors. Where there is hope, there is a perception of probability of escape from unwanted consequences.

5.1.2 Risk Decision Sharing

Instead of seeking expert advice, a risk agent may seek to share his decision with his peers. This occurs when others face the same type of risk situation or a similar one. Both passive and active group behaviors result.

Passive behavior involves an individual sharing his risk problem with others* to gain sympathetic understanding, catharsis of the psyche, and the knowledge that he (the risk agent) is not singled out, but is one of a group. This behavior can make involuntary risks more palatable, especially when there is no control over the risk.

Active behavior involves the formation or joining of a value group to take concerted action to alter the risk, implying some possibility of control over the risk. The control of risk may or may not actually exist. But if such control is perceived subjectively, it can help allay the risk impact.

5.2 FORMAL RISK EVALUATION PROCESSES

5.2.1 Steps in Evaluating Risk

When a new project is undertaken in which risks of a particular type may be unevenly imposed on society or parts of society to achieve some particular benefits, a process to establish an acceptable level of risk is necessary. This process is very simple in concept, but difficult and complex in practice. It involves four basic steps.

1. *Establish a Risk Referent Level.* Establish a level of risk for each category of risk consequences involved that is deemed acceptable on some basis. This "reference level of acceptable risk" is determined independently of the activity to be evaluated.

* This would include sharing with a "father figure."

2. *Determine the Level of Risk Associated with the New Program and Alternatives.* Identify and estimate the level of risk involved in the new program and possible alternative programs.

3. *Compare the Risk with the Referent Within the Limits of Error of the Estimate and Referent.*

Risk Estimate Less or Equal to the Referent. Risk is acceptable, and the best cost-benefit balance among alternatives can be used to decide which alternative should be selected.

Risk Estimate Greater than the Risk Referent. Either abandon the program or take risk aversive action.

4. *Risk Aversive Action.* If risk is acceptable, further reduction of risk to a level "as low as practicable* can be sought. This implies a higher level of safety at minor cost increase. If risk is unacceptable, then alternate programs involving lower levels of risk or risk exposure can be selected. These alternatives include increased safety, decreased exposure or other means to reduce risk. Two particular cases are important:

* "As low as practicable (ALAP) is a concept attributed to health physics practice in the radiation protection field. A definition from the draft ALAP Guidance of the Environmental Protection Agency of 9/23/75 is: "When the term as low as practicable (ALAP) was first introduced into the radiation protection field, it was assumed to have a relatively straightforward interpretation: needless exposures to radiation should be avoided, no matter how low. When translated into good health physics practice, this rule of thumb was intended to insure exposures that were as low as practicable."

"Over time, two factors appear to have influenced the evolution of a more sophisticated and analytical definition of ALAP. First, was the realization that exposures to radiation even at very low doses probably created risk of adverse health effects the cost of which should be justified by compensating benefits. Secondly, it became apparent that systems of control to reduce planned exposures to low levels of ionizing radiation could cost significant amounts of money, and these costs of controls should be offset by comparable benefits of reduced risk of adverse health effects."

"Neither the health physicist nor health physics procedures could be expected to encompass such broad cost and benefit considerations in controlling radiation exposures. Decisions on establishing ALAP controls rose to the level of management or regulatory bodies, and economic and social criteria were given greater prominence in ALAP considerations. New general definitions of ALAP emerged incorporating two basic criteria that would have to be passed before an exposure could be accepted as ALAP. The first criterion was that any controlled exposure to radiation ought to be justified by benefits that outweigh the risks of adverse health effects associated with the exposures. For example, if the x-raying of welds in shipyard involves certain unavoidable exposures to radiation, the adverse health risks of those exposures should at least be balanced by benefits from the use of the radiation to detect flaws. The measure of benefits in this case would be the value of the advantage of using x-rays over an alternate quality control procedure. The second criterion that developed was that radiation exposures should be reduced to the point where the marginal costs of additional controls would not be offset by the benefits of the marginal risks of health effects that would be avoided."

Risk is Controlled. Find an alternative at or below the referent level or abandon the project.

Risk is Uncontrollable. Find the lowest practicable risk exposure alternative.

5.2.2 Steps in Determining a Risk Referent

All the foregoing steps are fairly well established, at least in concept, in the body of cost-benefit methodology, except for the setting of a risk referent. Little effort has been expended in this area, which involves several steps and alternatives.

5.2.2.1 Establish a Risk Reference

Find some reference against which to measure risk acceptability for different types of societal risk. There are several possibilities.

Absolute. Level of risk presently occurring at which no control is possible (e.g., probability of the sun exploding).

Relative. Find consequences whose probabilities are low enough to be ignored, (e.g., probability of meteorite damage to an individual).

Relative. Use the present behavior of society as a reference (method of revealed preferences).

Relative. Use present behavior and extrapolated behavior patterns as a reference.

5.2.2.2 Establish a Risk Referent

The risk reference applies to some overall value of risk to society. It does not take into consideration how great an increase in that risk is society willing to accept to achieve the benefits of a given activity. Therefore the reference level must, be modified to allow the development of a referent for a particular type of activity. This modification of the reference level for a specific activity thus, is, differentiated from the reference level by calling it a referent. For example, if the total risk of a given type from all activities that produce these kinds of risk is x, this is the reference level. For a risk with some small benefit, doubling the risk reference would be unacceptable, and a smaller proportional amount would be needed. This smaller value becomes the specific risk referent for the new program and its benefit. It may be set by a number of methods, such as the following:

- Value judgment based on intuitive estimation.
- Value judgment based on gross balancing of cost and benefits and other factors.

- Value judgment based on detailed, quantitative balancing of costs and benefits.

5.2.2.3 *Determine Whether Referent Should Be Single or Multiple Valued*

Table 4.2 gave a hierarchical classification of different kinds of risk consequence, leading to the question of whether a risk referent should be a single overall value or index or a set of different references for different consequences. For example, premature death, morbidity, property damage, and aesthetic values each can be referenced separately and the risk estimate for each type consequence compared to its particular referent. In this case, all balances would have to be judged acceptable to warrant the activity or specific risk aversion action. On the other hand, weights could be assigned to each class and an overall index derived. The problem lies in determining whose judgment should be used to set the weights.

In the author's opinion it is necessary to make balances at the multiple consequence level, even if a single index is used additionally. Single indexes often tend to mask the underlying problems that multiple balances may identify. In other words, the single index acts as a smoothing function at a decision level so high that the use of this index is often unwarranted. For example, the Stanford Research Institute undertook a study for a MITRE Corporation Environmental Workshop comparing nuclear energy with coal energy.[1] This study used statistical expected value for consequences as the overall index for comparing coal and nuclear fuel as two alternatives. The single index leads to a seemingly anti-intuitive result; namely, that the nuclear option is superior in all respects to the coal option. This occurs because dollars are the single value used and every consequence is weighted into a dollar value. Their use of expected value is also limiting. It masks the distribution of risk and omits information. The authors concluded that financial cost exceeded social costs by a wide margin. Considering the intensity of the nuclear debate at the present time, such conclusions (which may be reasonable in terms of consequences alone) do not relate to the problem, which involves some very subjective consequence values expressed by widely separated value groups as described by the three schools in Chapter 4.

In any case, the choice of single or multiple referent values must be made with full understanding of the need to assure that sufficient information for high-level decision making is available.

5.2.2.4 *Ascribe an Acceptable Level of Precision for the Balancing Process*

The processes of risk determination and establishing a risk referent involve value judgments and subjective information in part. As a result, both systems have considerable measurement uncertainty. Any balance made can have no more precision [see Glossary] than that of its least precise, balance, and greater precision should not be expected. However, some level of precision that would be acceptable can be prescribed, and overlapping error bands in a balance may be considered unacceptable or acceptable, depending on the precision prescription. For example, if the balance is within a factor of 2 or an order of magnitude, it might well be judged acceptable.

Problems of precision and measurement uncertainty in decision making are covered in subsequent chapters.

5.3 APPROACHES TO ESTABLISHING RISK REFERENCES

A variety of approaches can be used to establish risk references and acceptable levels of risk at the societal level. Some important ones considered here derive the reference from the risk system itself, from unavoidable natural risks acting as thresholds, and from societal behavior as a whole.

5.3.1 Exposure-Consequence Relationships

A number of exposure-consequence relationships connect consequences and their values to various levels of exposure to risk. Both individuals and populations can experience these relationships. The exposure magnitude, assuming the exposure occurs, gives rise to a level of consequence magnitude at some level of probability. Some of the general forms of this relationship are illustrated in Figure 5.1, where the abscissa represents increasing exposure and the ordinate represents the level of consequence (probability, magnitude, or both) that can result from that exposure. For reference, a case in point might be the relationship between an exposure to a toxic effluent for a certain time and the resultant health effects of increasing concentrations of the effluent.

Threshold Relationship. Curve 1 of figure 5.1 illustrates a threshold condition: even though exposure may take place, there is no risk unless the particular threshold is exceeded. This is illustrated by the point at which curve 1 leaves the abscissa. Curve 1 is linear above the threshold, but it may take another shape.

Breakpoint Relationship. Curve 2 illustrates a nonthreshold case, but there is a distinct breakpoint in the relationship. These curves are linear below and above the breakpoint, but they may again be of some other form.

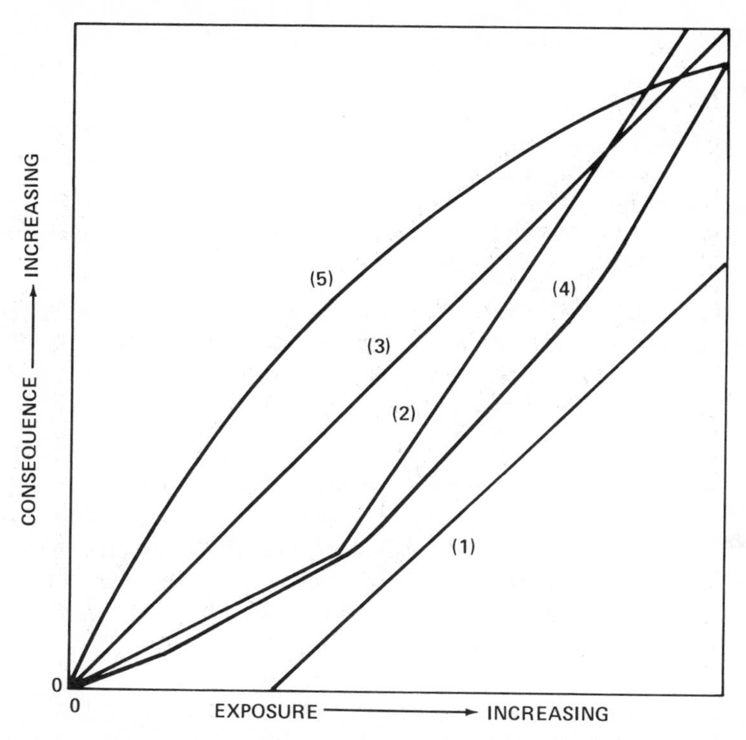

Figure 5.1. Relationships between exposure and consequences.

Linear-Nonthreshold Relationship. Curve 3 is the classical linear exposure-consequence relationship when no threshold exists. Zero risk occurs only at zero exposure. This is the condition that is assumed to exist for exposure from radiation for regulatory purposes. It is obtained by extrapolation from data at high levels of exposure down to zero to estimate risks at low levels of exposure.

Nonthreshold-Lowered Sensitivity Relationship. Curve 4 is a nonthreshold relationship showing lower sensitivity to risk at lower exposure levels.

Nonthreshold-High Sensitivity Relationship. Curve 5 is the reverse of curve 4, where there is increasing sensitivity at lower levels of risk exposure. Curve 5 is illustrative of the condition that occurs when more sensitive members of a population are affected by risk exposure conditions.

5.3.2 Systemic Criteria for Risk Reference

The process of risk determination leads to methods of developing risk references and risk acceptance by the nature of the pertinent exposure-consequence relationship. Since such references are derived from the risk initiating system and its process, they may be called systemic criteria. These involve the exposure-consequence relationships.

5.3.2.1 Internal Criteria

Curves 1 and 2 in Figure 5.1 represent conditions whereby criteria may be determined by considering only the risk parameters that are internal to the system; that is, by considering only risk by itself.

Risk Thresholds. Curve 1 allows one to set a threshold based on no consequences to exposed populations as long as the exposure is kept below the threshold. Therefore the threshold, possibly with a safety factor built in, provides an acceptable level of risk (i.e., no risk).

Breakpoints. The lower level of consequence increase below the breakpoint may or may not be acceptable (curve 2). In any case, however, levels set below the breakpoint are more effective in minimizing risk than levels set above it. The acceptability of risk below the breakpoint must be treated similarly to the remaining curves. The breakpoint does provide a rationale for setting acceptable risk levels based only on risk in many cases.

5.3.2.2 External Criteria

When the exposure-consequence relationship is continuous through the origin, acceptable levels of risk cannot be set using systemic risk criteria by themselves. It is necessary to use outside references to establish risk acceptance levels. A number of paradigms* be used.

Cost-Effectiveness Paradigm. The cost effectiveness of risk reduction is a paradigm that has many aspects. It is often called cost-benefit analysis in a narrow sense, since the benefit considered is that of risk reduction. Various actions to reduce risk may be ordered on the basis of the ratio of the magnitude of risk reduced and the magnitude of the cost of risk reduction. When smoothed, the resultant curve is concave upward (Figure 5.2). The curve reveals that the problem of assigning risk has simply been transferred to a new parameter, the cost effectiveness of risk reduction. However both internal and external criteria still must be used to determine the acceptable level of cost effectiveness.

1. Internal Criteria. Those associated with the shape of the cost-effectiveness curve.

Breakpoints. Discontinuities can provide a rationale for selection of cost-effectiveness acceptance levels.

* A paradigm is a structured set of concepts, definitions, classifications, axioms, and assumptions used in providing a conceptual framework for studying a given problem.

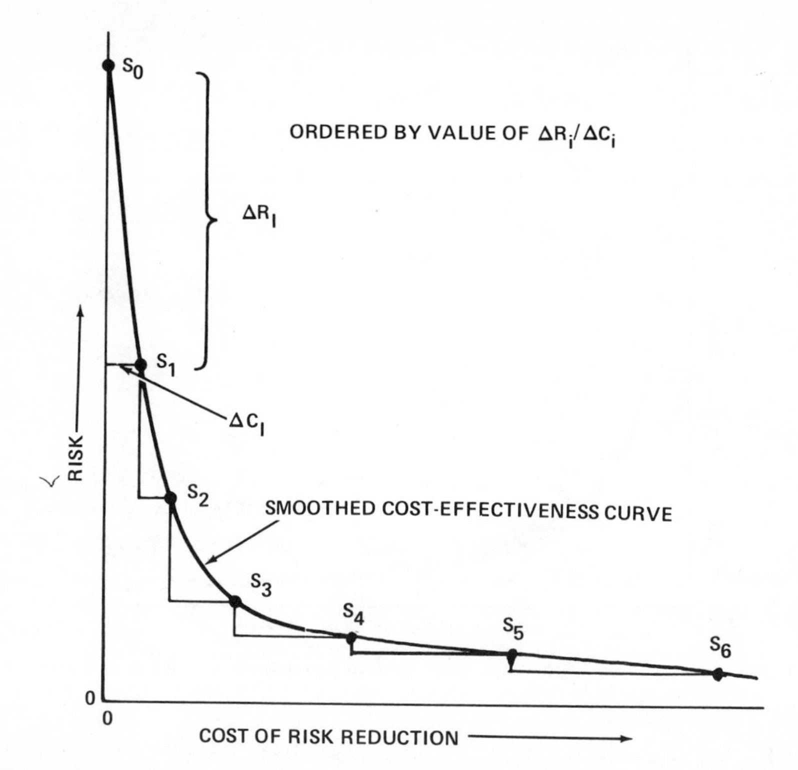

Figure 5.2. Cost effectiveness of risk reduction ordered relationship for discrete actions S_i = Risk reduction action. ΔR_i = Change in risk for S_i. ΔC_i = change in cost for S_i.

Unit Slope. The scales for each axis can sometimes be normalized so that the scales are identical. If this is possible, then when the slope of the curve is equal to unity, the marginal cost of increased reduction is equal to the marginal benefit of the risk reduction.

2. *External Criteria.* Require some external referent. A number of these are compared in Figure 5.3.

No Risk Reduction. A point on the curve where no funds are spent for risk reduction.

Zero Risk. A point on the curve that is dependent on the definition of zero risk but represents, generally, a very high cost solution. Figure 5.3 shows both a "zero risk" definition and actual zero risk.

As Low as Practicable. There are a number of definitions for this concept. The first implies a relative level of acceptance based on societal risk as a whole. In this case, when the incremental cost per risk averted is equivalent to similar costs for similar risks to society, the system will be

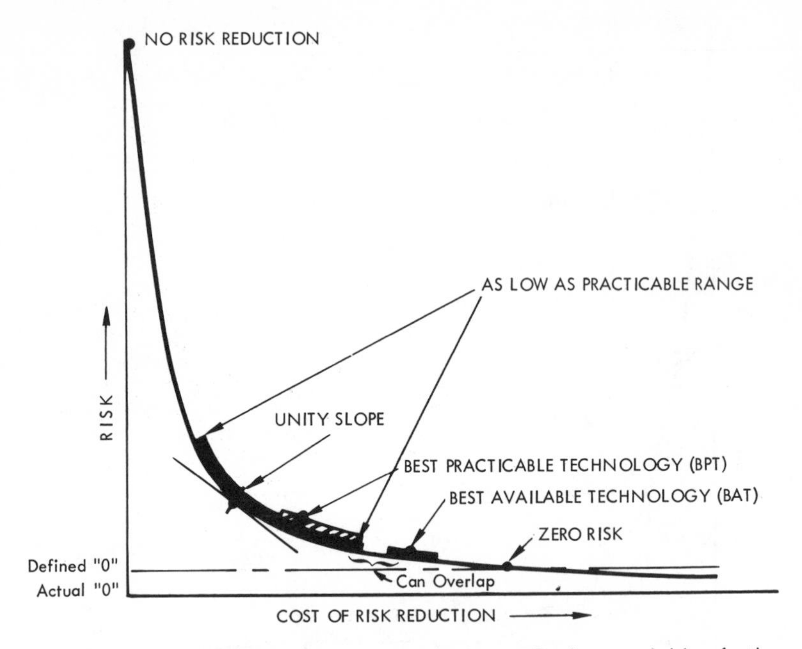

Figure 5.3. Some criteria for acceptance levels of cost effectiveness of risk reduction.

ALAP. An alternate definition implies a relative risk for the particular activity in question; that is, when the incremental cost per risk averted is such that a very large expenditure must be made for a relatively small decrease in risk as compared to previous risk reduction steps, the activity causing the risk is ALAP.

Best Practicable Technology. Another technology definition as is used in the Water Quality Act of 1972. In this case, best practicable technology involves finding the average practice of the best industry processes for effluent or risk control.

Best Available Technology. Best available technology is further out on the curve and hinges only on the requirement that a particular process has at least been demonstrated.

In all cases for cost effectiveness of risk reduction, a referent is required, either internal or external, to set acceptable levels of cost effectiveness of risk reduction. As a result, this paradigm faces the same types of problem encountered by risk acceptance levels except that risk is not considered directly. Economic factors are added, since the cost effectiveness of risk reduction is used as the primary parameter.

5.3.3 Natural Risk Levels as Thresholds for a Risk Reference

Another paradigm utilizes an absolute risk reference that is derived from an examination of the unavoidable natural risks to which society is subject. Examples are the probability that the sun may explode in one's lifetime and the risk of being hit by a meteor falling to earth. These risks are unavoidable and uncontrollable, at least with present technology; therefore they have been accepted by society and are generally ignored. Thresholds set from such risks are very low, but certainly acceptable.

Another variety of natural risk is directly controllable, such as that from lightning; here lightning rod technology is extremely effective in avoiding lighting strikes. For another set of natural risks, exposure to the risks can be avoided—for example, by choosing not to live where floods or hurricanes are subject to occur. Certain aspects of risk from natural background radiation fall into this category. Thresholds may be derived from these levels, but they are perhaps less useful than those of the absolute risks for which man is faced. Basically, the paradigm assumes that man experiences certain levels of risk that are acceptable, since they are acts of God. People learn how to live with risks they cannot avoid or control.

A reference value is not itself an acceptable level of risk, but only a reference, particularly in this case, since natural risks are considered to be acts of God and are valued differently from man-originated risks.

5.3.4 Societal Behavior as a Risk Reference

The behavior of society as a whole provides another paradigm for use as a risk reference. The basic premise is "society is as society does." This implies that societal behavior is acceptable, regardless of whether it is "right" or "good." At any given time, it is possible to use statistical measures to determine what society is doing or has done on an aggregated basis. This is standard statistical practice, and sets of this type of data are more fully discussed and described later.

Since society is dynamic and has a variety of explicit and implicit goals, however, a static measure at a given point in time is inadequate to describe acceptable "behavior" by itself. A combination of actual trends from historical data and identified goals (in the broadest sense) to determine where society is heading is needed to supply a more meaningful definition of societal behavior. As a general rule, society is attempting to minimize loss at the high end of scale hierarchy in Table 4.2 and to maximize gains at the low end. Thus threats of premature death are to be minimized, whereas gains in the quality of life (self-actualization) are to be maximized.

There are no absolute references here, and different societies have different norms of behavior, resulting in different value systems. Furthermore, subgroups in society may be at odds with overall societal values in part or *in toto*. However the political process provides a means for change, and change resulting from this process is reflected in changes in societal behavior.

Societal behavioral systems are the result of the collective behavior of all the individuals in the population acting individually, in a variety of groupings, and reacting continuously to new pressures. Analysis is indeed difficult, and whereas some effort has been made in analysis of individual behavior, the behavior and influence of groups, and the aggregate behavior of society, there is an absence of significant progress in these areas and in their interrelationships, at least in comparison with progress in the harder sciences. Nevertheless, analyses of societal behavior and of decisions resulting from such efforts are possible within limitations and can yield useful results without precise inputs.

5.4 ANALYSIS OF SOCIETAL BEHAVIORAL DECISIONS

Social decisions are most often made intuitively rather than on the basis of an analytical, objective study. Because of the body of personal experience that has been built into individuals from birth, many decisions are made automatically as a result of previous trial-and-error situations, other learning experiences, and personal capabilities. Most decisions are made without benefit of formal analysis. For example, a decision by a young person to take up smoking usually involves some aspect of seeking the guise of maturity at a time of actual immaturity, without regard to the risks of smoking. The short-term importance of this goal makes analysis of the risks, including habit-forming tendencies, inappropriate to the risk agent under these conditions.

Another reason for the difficulty in analyzing such decisions lies in the problem of measuring decision parameters. Most of the parameters are related to intuitive concepts, such as values, and cannot be measured objectively. The subjective scales needed to provide such measurements are, by necessity, limited in precision. The scales can be precise only insofar as quantification is meaningful, and the imprecision of language to express small differences in a meaningful way is generally not the fault of the language, but the inability of individuals to assign real meaning to the differences. For example, it is sufficient to say: "I like ham sandwiches better than cheese sandwiches." Conversely, to say "I like ham sandwiches

2.95432 times more than I like cheese sandwiches," is not only an overstatement of the condition, but may well be false.*

Systems analysis techniques, such as operations research, decision theory, and probability theory, are often ineffective in addressing real problems with imprecise scales. These techniques are aimed at the manipulation of numbers, often in elegant fashion, and require infinitely precise scales for such manipulation to be meaningful. Real problems with imprecise scales are often constrained to do-able problems, but as a result of the applied constraints, they no longer represent real situations. Thus "system analysis" may be characterized as "a set of solutions looking for problems it can solve." Most behavioral problems are not in the "solvable" class. Pragmatism rather than elegance is needed in these cases.

5.4.1 Value Judgments as Gross Measures

If behavioral scales are imprecise, one convenient method of providing measures is through subjective value judgments. Since meaning is in the eye of the beholder, all individual value judgments are "true" ones. Only when value judgments of one person are to be compared to others and used as a consensus does the "correctness" of a value judgment become a cause for concern. The determination of "collective" value judgments is thus of prime significance. Such value judgments impinge on society in every aspect of life. The agreement on the result of empirical observation of a physical measurement of a parameter is a value judgment because the interpretation of the observation is personal, although the "collective variance" may be small. On the other extreme are value judgments based solely on emotional experiences. Somewhere between these two extremes are value judgments that involve acceptance by society of certain types of risk. These can be used by regulatory agencies, among other users, for guidance in making equitable decisions. At best, these value judgments provide only gross measures of aggregate behavior, are relatively imprecise, and have use only when described in a manner agreeable to all parties to the value judgment.

The precision of such judgments is limited and the description of judgment conditions is as important as the measure itself when cost-benefit (gain-loss) evaluations are involved. The scale must be sufficiently precise and described explicitly enough to permit all involved to understand the scale, but it must be gross enough to reasonably resemble the aggregation

* If the actual precision is one significant figure (i.e., 3 times better), then 10,000 possibilities exist at the prescribed precision of six significant figures, of which only one is assumed to be correct, and it is not even of academic interest.

errors involved in scale generation. For example, gross statements such as "benefits far outweigh costs" and "benefits marginally outweigh costs" may represent the upper limit of precision in gaining acceptance of such value judgments meaningfully.

5.4.2 Analysis with Gross Value Judgments

As long as value judgment scales are meaningful to those affected and are universally interpreted, they are useful in analyzing decisions affecting society. However the analysis, though sometimes resulting in decisive answers, is no more precise than the inputs. The objective is to get decisive answers that are generally acceptable, regardless of the precision involved.

Results of such analyses are valid only in the sense that a sizable portion of society accepts them knowingly. The ability to display the analysis in an open, visible, traceable, repeatable manner is a requirement of a valid analysis. Both the types of value judgment made and the actual values assigned must be exposed, discussed, and argued.

Under these conditions, decisions resulting from such analyses are never "right," only "accepted."

REFERENCE

1. Stephen M. Barranger, Bernie R. Judd, and O. Warren North, "The Economic and Social Costs of Coal and Nuclear Electric Generation: A Framework for the Coal and Nuclear Fuel Cycles," STI Project MSU-4133, as discussed in *'Proceedings of Quantitative Environmental Comparison of Coal and Nuclear Generation and Their Associated Fuel Cycles Workshop,"* MTR-7010, Vol. I, McLean, Va.: August 1975, pp. 15–94.

6

Examples of Risk Acceptance Decisions

Two examples of risk acceptance decisions are described now to furnish perspective on the process. The first example involves a decision, made as an individual or based on expert advice, about whether to get an influenza shot. The second involves the action of a regulator in a similar type of decision—whether smallpox vaccinations should be mandatory.

Both examples are primarily illustrative, and the degrees of uncertainty are not specifically carried through the examples.

6.1 AN INDIVIDUAL OR EXPERT ADVICE DECISION

Influenza is a contagious disease to which the entire United States (and world) population is exposed at one time or another. In the United States, vaccination against influenza is available to most of the population through private physicians, employers, local health departments, and a variety of other sources; in many cases there is no direct financial cost to the recipient.* Exposure and resultant infection are involuntary. However individuals who sustain risks of vaccine side effects to attain the benefit of protection against possible debilitating disease are acting voluntarily, and the choice is usually an individual one, sometimes preceded by medical advice.

The mass media provide a great deal of information on influenza epidemics and vaccination recommendations.[1] Thus the public is generally well informed, at least qualitatively. Although quantification of individual decision is dependent on the quality of data available, the exercise can provide some insight on the manner in which decisions are made qualitatively, and

* This analysis was completed prior to the development of a potential "swine flu" epidemic and the rapid development, manufacture, and distribution of a vaccine. This action does not change the illustrative value of this example. Moreover, the example has anticipated epidemics of new strains of influenza.

whether quantitative analysis improves such decisions. In carrying out such an analysis here, we can also show (1) the types of decision involved, (2) the kinds of information required and the limitations in obtaining it, (3) the structure of the decision, and (4) how the quantitative information can be used.

On the other hand, this analysis does not purport to be a complete, accurate, and final analysis. On the contrary, though data have been made available to the author, the absence of concrete information in many areas frequently has made it necessary to rely on the value judgment of experts in the contagious disease area.

All the data supplied here and the value judgments involved have been supplied by Charles A. Hoke, Jr., MD, U.S. Department of Health, Education, and Welfare, Center for Disease Control, in Atlanta.* All value judgments involving epidemiology are those of Dr. Hoke and do not reflect any organizational assessment on policy or risk factors related to influenza or influenza vaccine. All such personal judgments are identified.

6.1.1 Background—Risk Identification Process

Influenza is a contagious viral disease whose viruses are mutagenetic, and new strains with renewed virulence periodically appear. When a new strain appears (technically when a type A_n virus is altered to become a new type A_{n+1} virus), prior immunization either through previous infection and resultant immunity or prophylaxis is ineffective. The result is a severe epidemic of influenza. In years when type A_n viral contagion is unusually widespread, a mild epidemic results.

Data on influenza go back nearly 50 years, but pre-1943 data must be discarded because major changes in control of mortality from side effects and complications of influenza have resulted from new drugs and antibiotics. The scope of an epidemic is measured by observing increases in death rates from all causes when influenza is prevalent. As an example, data received from Dr. Hoke on the 1968–1969 influenza epidemic are reproduced in Table 6.1.† In this epidemic, those over 55 had a higher excess death rate

* Personal communication from Dr. Hoke, Medical Epidemiologist, Viral Diseases Division, Bureau of Epidemiology, Center for Disease Control, to Dr. Lambrose Lois, acting in the author's behalf, January 9, 1975.

† These data are directly observed. However the degree of immunization from previous infection or vaccination is not known at any given time. If there were no immmunization, the death rates would presumably be higher. The problem is that a decision on vaccination affects the observed results.

Table 6.1 *Age-Specific Death Rates for December 1968–January 1969*

Age	Excess Deaths due to All Causes	Estimated Population for July 1, 1969	Excess Deaths from Influenza[a] (100,000 pop.)
1–14	256	55,810,000	0.46
15–24	−13	34,214,000	−0.04
25–34	350	24,390,000	1.4
35–44	1,555	23,176,000	6.7
45–54	4,084	23,154,000	17.6
55–64	8,770	18,212,000	48.1
65–74	12,360	11,954,000	103.3
75–84	5,011	6,227,000	80.4
85+	1,381	1,289,000	93.1
Totals	33,754	198,426,000	17.01

[a] Estimated for each population interval separately, including a separate estimate for the total population.
Source: Dr. Charles A. Hoke, Jr.,

from influenza than the average age group, with the 65–74-year-old range reaching a peak. Dr. Hoke describes the effects of the disease as follows:

From the point of view of the individual sufferer, influenza's effects can range from a mild sore throat to a few days of febrile upper respiratory tract infection followed by several weeks of gradual recuperation to death. The outcome generally depends on the health of the individual before he is infected. Well people tend to develop lesser disease while the elderly or the chronically ill tend to be at much higher risk of dying from complications of influenza. However, there are many exceptions to this generalization and medical science really has no good understanding of why these variations exist. Vaccine recommendations are, incidentally, based on the increased risk to these groups. One important exception to the principle that the previous state of health determines likelihood of complications occurs in some small children affected by influenza-B. Occasionally, a child will develop Reye's syndrome, a combination of liver degeneration and encephalopathy following an epidemic of influenza-B. We do not, unfortunately, have any tenable hypothesis as to a causal link. Reye's syndrome, though rare in itself, carries about 1 in 2 to 1 in 3 chance of death.[2]

6.1.2 Risk Estimation Process for Involuntary Risk

Flu contagion is an involuntary event. This part of the risk estimation process deals only with the involuntary risk aspect.

6.1.2.1 Exposed Populations

From these comments, three different populations of interest can be identified: (1) the total population, (2) the elderly, and the chronically ill (high risk) population, and (3) a less sensitive (low risk) population, where group 3 = group 1 − group 2. The present use of influenza vaccine is predicted as stated, on protection of the group 2 population. Populations 1 and 3 are also examined here.

The frequency of epidemics is hard to estimate, but severe epidemics occur about once every 10 years, with mild epidemics (based on excess death rate in the population) every two or three years. It may well be that if a criterion of number of cases of severe and/or mild influenza reported by doctors were used to define an epidemic instead of death rate, mild epidemics might be found to occur more often. On the other hand, through efforts of the World Health Organization, the Center for Disease Control, and other groups, early recognition of the mutagenetic occurrence of a new type of virus (i.e., a type A_n virus changes to an A_{n+1} type) provides several months of lead time in forecasting a major epidemic. Whether this is enough time in which new vaccines can be developed and distributed is still an open question.*

6.1.2.2 Exposure Potential

Exposure to contagious disease depends to a great extent on an individual's geographic location, since epidemics can occur in specific areas at any time. The change of exposure varies with the type of epidemic (or the absence of one, as well).

> Epidemics are not easily predictable. In 28 of the last 40 years, in the U.S., influenza epidemics extensive enough to cause perceptible excess mortality have been observed. During an influenza epidemic, as many as 20% of individuals in an affected area might be infected. Commonly, outbreaks affect some areas and leave others untouched. The effects of influenza epidemics are felt at all levels of function in a community. Businesses and schools may be closed, and hospitals are short-staffed when help is most needed.[2]

Estimates on the exposure to influenza are purely guesswork, and prediction of areas in which outbreaks will occur is not generally possible.

6.1.2.3 Severity of Disease

Dr. Hoke has developed a utility scale to express the severity of a case of influenza. The unit of utility is the FLU, and 1.0 FLU is equivalent to a severe, full-blown case of influenza. Dr. Hoke's FLU scale is shown below.

* The recent "swine flu" effort supports the contention that such control is indeed possible, if not universally acceptable.

Influenza Intensity	FLU Utility*
Mortality	100 FLU
Morbidity	
Full-blown case	1.0 FLU
Mild case	0.5 FLU
Vaccine reaction	
Chills and fever	0.9 FLU
Mild reaction	0.2 FLU
Sore arm	0.1 FLU

The influenza intensity scale represents a consequence description; and the FLU utility is, one assignment of value to the consequence. Other values could be assigned. For example, death might be 1000 FLU instead of 100, and anything less than 0.1 FLU might be "zero."

6.1.2.4 *Probability of Disease Occurrence*

Estimates of the probability of occurrence for different consequences for a severe epidemic year appear in Figure 6.1*a* along with data for a mild year (Figure 6.1*b*). The risk of severe and mild cases is shown in conjunction with the percentage of the total population experiencing these symptoms. Mortality is about 4×10^{-4} fat/yr/ind in the total population.†

6.1.3 Vaccination—Risk Aversion Process

Up to the present, proof that vaccine will prevent epidemic influenza is lacking; it is probably an oversimplification to compare the risk due to influenza directly with the risk due to vaccine. Furthermore, no one has ever demonstrated that influenza vaccine will reduce mortality during an epidemic, though it seems logical to expect that it would.[2]

6.1.3.1 *Risk of Influenza Vaccine*

Although individual studies documenting reactions to vaccine have been reported over the last 15–20 years, the vaccine composition has constantly changed and many studies were not comparable in design, making conclusions tenous at best. However, up to several years ago, influenza vaccine was commonly associated with the occurrence of an acute influenza like syndrome of shaking chills and fever in many individuals. The overall frequency of reactivity was not established, though figures in the 10% range were frequently quoted. During the past several years, the Bureau of Bio-

* Dr. Hoke's value judgment.

† The precise units fatalities per year per individual (fat/yr/ind) bear an element of redundancy. Henceforth "fat/yr" is used, and exposure to an individual is understood.

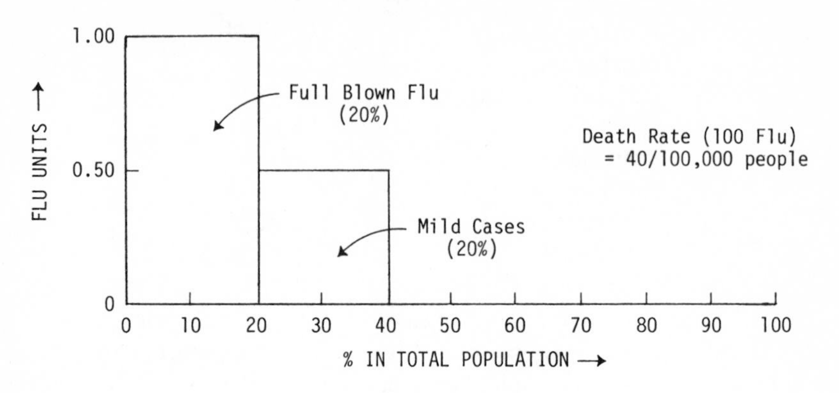

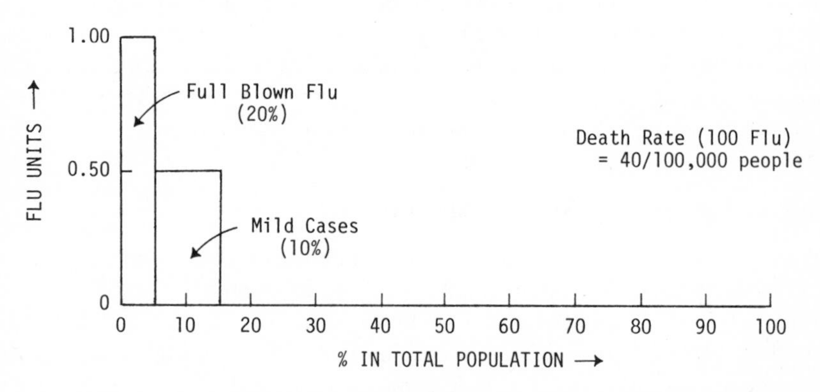

Figure 6.1. Risk of illness from influenza. *Source:* Charles A. Hoke, Jr., M.D.

logics has encouraged the six manufacturers of influenza vaccine to reduce the amount of nonessential products of vaccine production. The current vaccines, though quite pure, appear to be associated with significant side effects judging from our data. I am currently developing as yet unpublished data on the immediate reactivity of influenza vaccine. In addition, the literature has several reports of serious side effects, such encephalopathy, anaphylaxis, and idiopathic thrombocytopenic purpura. The causal relationship to vaccination is impossible to establish. A statistical relationship has not been demonstrated either. My recent studies on a few groups indicate that from 5% to 30% of individuals receiving influenza vaccine will develop a syndrome of shaking chills and fever. In 15% of vaccinees, this syndrome would equal about ".90 FLU." In another 15% of vaccinees, the syndrome would be about ".2 FLU" in severity in that people are ill for only 24 hours at most. About

20% of people develop a sore arm after the shot, which I would estimate is about "0.1 FLU."[2]

The overall plot is shown in Figure 6.2. In conclusion, if one assumes these figures to be reasonable risk estimates, it seems as though the risk of illness from vaccination is certainly similar to the risk of illness from influenza from an individual's point of view.[2]

6.1.3.2 Degree of Protection

The flu vaccine gives about an 80% chance of protection from influenza.[2]

6.1.4 Risk Estimation Process—Voluntary and Involuntary Risk

6.1.4.1 Decision Tree for Influenza Vaccination Decision

Figure 6.3 is a decision tree laying out the situational conditions for influenza exposure to the total population. Three different populations are examined in this illustration; the total population, a high risk population, and a low risk population. Only the total population is represented in the illustration, but results for the the other populations are presented. Figure 6.3 gives the starting points for all three, which have identical branches but different probabilities. As an example, the first branch takes into account the difference in death rate for high risk and low risk populations. Whereas the total population has a death rate of 40 per 100,000 people, the high risk group is closer to 100/100,000 and the low risk group is 10/100,000. The a_i values that represent the pre-decisional death rates are higher than these values. The a_i's have been adjusted so that the final death rates (the observable data) are resultant values in the decisional analysis. This assures that

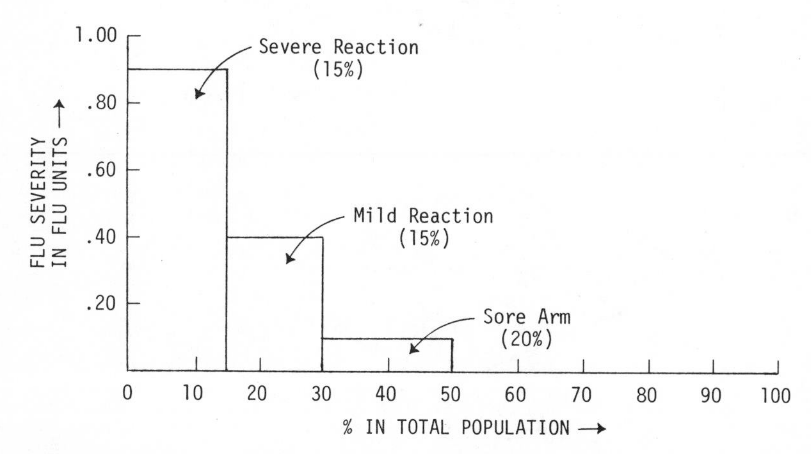

Figure 6.2. Risk of reaction from flu vaccine. *Source:* Charles A. Hoke, Jr., M.D.

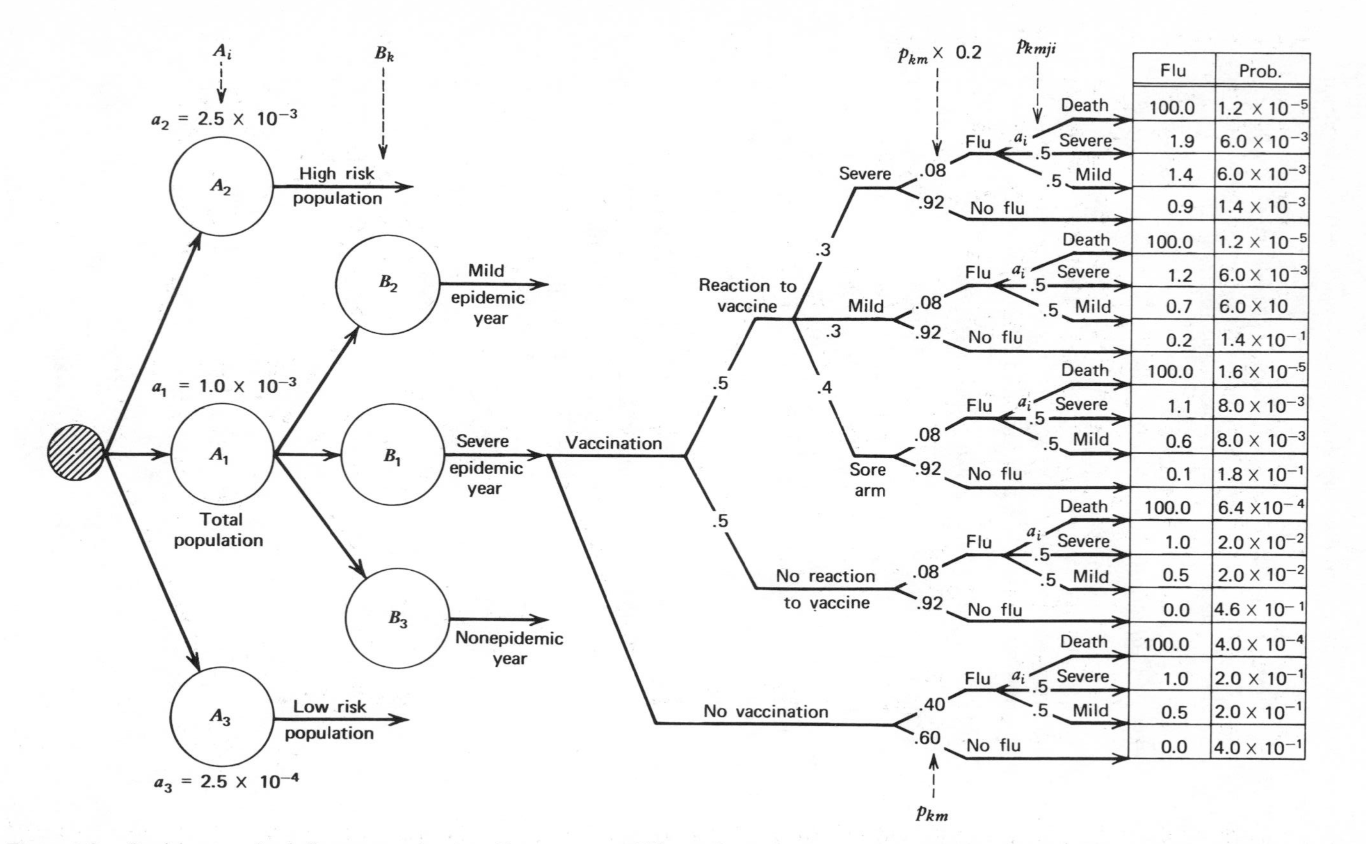

Flu	Prob.
100.0	1.2×10^{-5}
1.9	6.0×10^{-3}
1.4	6.0×10^{-3}
0.9	1.4×10^{-3}
100.0	1.2×10^{-5}
1.2	6.0×10^{-3}
0.7	6.0×10
0.2	1.4×10^{-1}
100.0	1.6×10^{-5}
1.1	8.0×10^{-3}
0.6	8.0×10^{-3}
0.1	1.8×10^{-1}
100.0	6.4×10^{-4}
1.0	2.0×10^{-2}
0.5	2.0×10^{-2}
0.0	4.6×10^{-1}
100.0	4.0×10^{-4}
1.0	2.0×10^{-1}
0.5	2.0×10^{-1}
0.0	4.0×10^{-1}

Figure 6.3. Decision tree for influenza vaccination. For purposes of illustration only the case for a severe epidemic year and the total population is shown. For a mild or nonepidemic year identical areas exist, but $p_{212} = .33$, $p_{2131} = .67$, $p_{3121} = .10$, $p_{3131} = .90$, $p_{21} = .15$, $p_{22} = .85$, $p_{31} = .05$, $p_{32} = .95$. For different populations the a_i's change, but the tree remains the same.

the analysis reflects what is actually observed, and the a_i's are thus pseudo-death rates in that they reflect what the death rate should be at this preliminary stage.

The second branch takes the severity of epidemics into account. If one is concerned with yearly conditions only, prediction of the threat for a given year may be used to determine which branch is pertinent. Otherwise, historical data provide a frequency of occurrence as shown by the probabilities of occurrence, $p_1 - p_3$.

The next branch represents the decision of whether to be vaccinated. For the case of vaccination, prophylaxis is only 80% effective; thus there is a chance that both influenza and a vaccination reaction will occur.

The event branches of the decision tree for a severe, mild, or nonepidemic year are followed by two subseqent decision branches which represent a decision for vaccination or for no vaccination. For both trees, a critical probability, that of getting a case of influenza without vaccination versus not getting it, is denoted by $p_{km..}$. The subscript k denotes the type of epidemic year and the subscript m has two values: $m = 1$ implies getting a case of influenza and $m = 2$ implies not getting it. This probability is a function of the type of epidemic year such that:

Type of Year	Getting Influenza	Not Getting Influenza
Epidemic	$p_{11} = .40$	$p_{12} = .60$
Mild	$p_{21} = .15$	$p_{22} = .85$
Nonepidemic	$p_{31} = .05$	$p_{32} = .95$

Furthermore, if one does catch influenza, the probability of a severe case over a mild case is a function of the type of epidemic year. The probability of p_{kmji} expresses this, where $j = 1$ represents death, $j = 2$ a severe case, and $j = 3$ a mild case, and $m = 1$ in all cases, by definition. Since

$$\sum_{j=1}^{3} P_{k1ji} = 1 \tag{6.1}$$

and one would only expect a death after a severe case of influenza

$$p_{k11i} = a_1 \tag{6.2}$$

$$p_{k12i} = 1 - p_{k13i} - a_i \cong 1 - p_{k13i} \tag{6.3}$$

	p_{k12i} Severe Case	p_{k13i} Mild Case
Epidemic year, $k = 1$	.50	.50
Mild year, $k = 2$	.33	.67
Nonepidemic year, $k = 3$	.10	.90

Vaccination works independent of population or type of epidemic year (assuming effective immunization) such that there is a 50% chance of receiving a reaction: 30% have a severe reaction, 30% a mild one, and 40% a sore arm; this is the distribution shown in Figure 6.2. If vaccination is given, the chance of getting the influenza is reduced to 20% over that without vaccination. The new probabilities are

$$\text{Getting influenza} \quad = .2 \times p_{km} \tag{6.4}$$

$$\text{Not getting influenza} \quad = 1 - .2 \times p_{km} \tag{6.5}$$

These probabilities and branches define all possible outcomes, and the compound probability of a consequence and the description must be calculated. These are given in Table 6.2, where the consequences have been aggregated along with the probability of occurrence for all three epidemic conditions, all three populations at risk,* and the cases for vaccination and no vaccination. The FLU utility scale is also shown with an estimate of average days lost for each condition.†

6.1.5 Risk Evaluation Process

Several methods can be used to evaluate the risk of vaccination against no vaccination; the first is that for expected utility.

6.1.5.1 Expected FLU

Dr. Hoke's FLU scale provides a utility scale for computing expected FLU (i.e., the product of the probability of occurrence and the FLU utility of the consequence). These have been calculated and appear in Table 6.3, which is called a decision matrix.

In a given type of epidemic year, the expected FLU utility (describing a measure of undesirability) for vaccination exceeds that for no vaccination, except in years of severe epidemic. In this case, the balance of expected utility indicates that it is desirable to be vaccinated regardless of the population group to which one belongs. This, of course, assumes that proper A_{n+1} vaccine is available.

On this basis, one would opt for a vaccination in an epidemic year, but not for a mild epidemic or a nonepidemic year. It should be noted that this decision is made without any knowledge of previous immunity of this risk agent through earlier vaccination or disease. Furthermore, the immunity in the total population is not generally known (except perhaps from the difference in observed cases in severe and mild epidemic years, when new

* The population at risk varies only in the death rate.

† Day loss is a measure of interference with work, leisure, and so on.

Table 6.2 *Aggregation of Consequences, Their Utility, and Probability of Occurrence*

Consequence Aggregation			Probability of Occurrence					
			Epidemic Year		Mild Epidemic Year		Nonepidemic Year	
Consequence Description	Flu Value	Avg. Days Lost	Vaccination	No Vaccination	Vaccination	No Vaccination	Vaccination	No Vaccination
Deaths								
High risk population	100	—	2.0×10^{-4}	1.0×10^{-3}	7.5×10^{-5}	3.8×10^{-4}	2.5×10^{-5}	1.3×10^{-4}
Total population	100	—	8.0×10^{-5}	4.0×10^{-4}	3.0×10^{-5}	1.5×10^{-4}	1.0×10^{-5}	5.0×10^{-5}
Low risk population	100	—	2.0×10^{-5}	1.0×10^{-4}	7.5×10^{-6}	3.8×10^{-5}	2.5×10^{-6}	1.3×10^{-5}
Illness and reaction								
Severe flu + severe reaction	1.9	9	6.0×10^{-3}		1.5×10^{-3}		1.5×10^{-4}	
Severe flu + mild reaction	1.2	5	6.0×10^{-3}		1.5×10^{-3}		1.5×10^{-4}	
Mild flu + severe reaction	1.4	7	6.0×10^{-3}		3.0×10^{-3}		1.4×10^{-3}	
Mild flu + mild reaction	0.7	4	6.0×10^{-3}		3.0×10^{-3}		1.4×10^{-3}	
Severe flu + sore arm	1.1	4	8.0×10^{-3}		2.0×10^{-3}		2.0×10^{-4}	
Mild flu + sore arm	0.6	3	8.0×10^{-3}		4.0×10^{-3}		1.8×10^{-3}	
Illness—no reaction								
Severe flu	1.0	5	.02	.20	4.9×10^{-3}	4.9×10^{-2}	5.0×10^{-4}	5.0×10^{-3}
Mild flu	0.5	3	.02	.20	.01	.01	4.5×10^{-3}	4.5×10^{-2}
Vaccine reaction—no flu								
Severe reaction	0.9	4	.14		.15		.15	
Mild reaction	0.2	1	.14		.15		.15	
Sore arm	0.1	0	.18		.19		.20	
No effects	0	0						

Table 6.3 *Expected Flu Values and (Expected Days Lost) - shaded area indicates a decision for a vaccination based upon minimum expected flu values.*

KIND OF YEAR	i = 1 TOTAL POPULATION		i = 2 HIGH RISK POPULATION		i = 3 LOW RISK POPULATION	
	VAC.	NO VAC.	VAC.	NO VAC.	VAC.	NO VAC.
k = 1 Epidemic year (prob = 0.1)	.254 (1.07)	.340 (1.60)	.266	.400	.248	.310
k = 2 Mild epidemic year (prob = .3)	.213 (0.88)	.164 (0.55)	.218	.187	.211	.153
k = 3 Non-epidemic year (prob = .6)	.194 (0.79)	.033 (0.16)	.196	.041	.193	.029
Any year	.206 (0.84)	.103 (0.42)	.210	.121	.204	.094

strains imply no short-term previous immunization) at any given time. To the extent that previous immunity might reduce vaccination reactions, it also might reduce the risk of contracting the disease, making vaccination unnecessary.

The FLU scale is arbitrary, especially in the assignment of 100 FLU for death. If, for example, death were assigned a FLU value of 1000, vaccination would also be indicated for mild epidemic years for the total population as well as for the high risk group. Vaccination would still be unwarranted for the low risk population in mild epidemic years and for all populations in nonepidemic years.

6.1.5.2 Death

If death alone were the only consideration, the 80% effectiveness of vaccination in preventing death would warrant 100% vaccination.

6.1.5.3 An Employer's View

An employer may undertake to provide free or subsidized vaccination for employees, to minimize days lost on the job. Table 6.2 contains a rough estimate of the average days lost in illness for different influenza and vaccination reaction conditions. The figures are based on a seven-day week, but the results will not be different for a five-day week because both vaccination and no vaccination values would be reduced by the same factor. The employer is only interested (on a statistical profit-loss basis) in days lost, not in death directly. Thus the condition holds only for the total population. The figures in parentheses in Table 6.3 represent the expected days lost. On this basis, vaccination is warranted only in an epidemic year.

6.1.5.4 Actual Response

What actually happens in society is described by Dr. Hoke.

> Approximately 14,000,000 people voluntarily receive the vaccine each year. You may be aware that vaccine is recommended for certain groups who are at high risk from influenza. If the same calculations were performed for them, the risk of illness related to influenza would be far higher. It so happens that older people also have been said to have fewer reactions to vaccine than those younger.[2]

6.1.6 Limitations of the Analysis

This analysis is not meant to be conclusive, and it has made visible a number of limitations that must be considered: (1) no account is taken of sensitivity differences among different populations to severity of influenza and vaccine reaction, (2) history of immunity from previous infection or vaccination is not considered for either individuals or populations, (3) effectiveness of different vaccines for different virus strains has not been assessed (4) value judgments (Dr. Hoke's) have been used in place of actual data in many steps, (5) subjective judgments, such as the utility scale, have not been tested on a widespread basis, and (6) the exercise itself has not been evaluated.

The analysis itself has made visible (1) the array of decisions involved and how they interact with each other, (2) the kinds of information required and limitations involved, and (3) the basic structure of the decision. These were the stated objectives. Furthermore, the quantitative information does not contradict intuitive conclusions one might expect on a qualitative basis. As Dr. Hoke has stated:

> Significant gaps in our knowledge exist, however. The acceptance of influenza vaccine in spite of them indicates a substantial willingness to accept a risk. Influenza vaccination theoretically benefits society and the individuals in it by lessening the impact of this potentially devastating epidemic disease.[2]

The quantitative analysis thus reinforces the intuitive action of a substantial part of the population and provides insight on how such decisions are affected by bits of information.

6.2 A SOCIETAL REGULATORY DECISION

An actual decision made by regulatory authorities is outlined here for illustrative purposes. In this case, the whole decision tree is not shown, only the basic information involved. The objective is to provide a real case of the evaluation process being carried out at a different level.

6.2.1 Risk Determination

The highly contagious disease of smallpox (variola) is no longer endemic in the United States, and no challenges to the immunity of the U.S. population have occurred since 1949.[3] Until 1971, approximately 15 million smallpox vaccinations were performed each year in this country. Of these, 6 million were primary vaccinations.[3] Our defense against smallpox until 1972 was based on four principles: (1) routine vaccination of the population, (2) vaccination of travelers, (3) inspection of vaccination certificates of travelers returning to or entering the country, and (4) investigation of suspect smallpox cases, with rapid isolation and control of smallpox importations.[3]

The risk of infection and spread of smallpox in the United States is such that one can expect one importation of smallpox every 12 years.[4] Two of every three importations will cause spread of the disease, for which an average of 23 cases will occur in each outbreak, and a death-to-case ratio of one-third will result in eight deaths per outbreak.[5] Based on a population of 2×10^8 people, and a frequency of occurrence of 0.44 fatality per year, the individual risk is 2.2×10^{-9} fat/yr.

Although vaccination is effective in preventing smallpox when properly administered, there are risks associated with vaccination. There are approximately seven deaths per year associated with vaccination, and in 1968 close to 500 cases involving morbidity were reported.[6] For adult primary vaccinations, this results in a death rate of three per million (i.e., 3×10^{-6}), with a significantly higher rate for infants under one year of age.

6.2.2 Risk Evaluation

The question addressed by the Department of Health, Education, and Welfare, the regulating agency, consisted of two alternatives: (1) continue the present policy of compulsory measures as they relate to routine smallpox vaccination, and (2) immunize personnel involved in health services and all travelers only, dropping compulsory measures. In the latter case, only about 1 to 4 million primary adult vaccinations would be given each year, whereas 6 to 15 million, including children, would be vaccinated in the first case.

If one assumes that the population has the opportunity to travel and to be vaccinated (either as a traveler or routinely), the individual risk rates are now lower than 9×10^{-8} fat/yr for routine vaccination and 2.1×10^{-8} for travelers only. A change in policy from routine vaccination to vaccination of travelers and health personnel (i.e., dropping the first of the four principles of smallpox defense, but retaining the other three) results in a reduction in individual risk of about 7×10^{-8} fat/yr. This risk avoidance alternative avoids present risks by more than an order of magnitude from the risk of smallpox importation of 2.2×10^{-9} fat/yr. Thus the selection of the second alternative, to drop routine vaccination, was made by the Public Health Service in 1972.[7] This was a regulatory decision involving involuntary risk to the population.

6.3 LIMITATIONS OF THE EXAMPLES

Both cases given as examples involve a limited level of risk evaluation, the choice between two alternative actions—vaccination or not. The balances can be made without recourse to cost-benefit analysis, since the alternative in both cases reflects changing levels of mortality and morbidity. Both sides of the balance were made in the same terms and did not consider cost in the classical sense.

Such decisions, though certainly real, do not reflect the broader types of decision that society often faces when there are many alternatives, many value groups, and all forms of risks, costs, and benefits. It is this broader range of decisions that concerns us now, although the examples serve to introduce basic concepts using empirical situations.

REFERENCES

1. Duncan Spencer, *Washington Star,* January 31, 1975, p. A-1.
2. Private communication from Dr. Hoke, December 13, 1974.
3. J. Michael Lane, J. Donald Miller, and John M. Neff, "Smallpox and Smallpox Vaccination Policy," *Annual Review of Medicine,* Vol. 22, 1971, pp. 251–272.
4. "Vaccination Against Smallpox in the United States, A Reevaluation of the Risks and Benefits." Atlanta: U.S. Department of Health, Education, and Welfare, Public Health Service, Center for Disease Control, revised February 1972.
5. Lane et al.. *op. cit.,* p. 66.
6. Lane et al., *op. cit.,* p. 266.
7. "Vaccination Against Smallpox in the United States," p. 17.

7

A Risk Estimation Format

To examine the concept of the estimation process in detail, a format for considering risk estimation in a definitive form has been developed. First, it is assumed that risk is meaningful only when associated with the occurrence of a causative event or set of causative events with a nonzero probability of realization (i.e., a probabilistic event). The description of the event and its probability distribution form a classical probability-event space based on ordinary probability theory such that a number of different outcomes can result with differing probabilities of occurrence. This event-outcome space does not involve risk unless there is an exposure pathway to people, biota, or the environment.

An outcome-exposure-consequence space relates the possible outcomes to possible recipients in the form of consequence descriptions and probabilities of occurrence. Each consequence has assigned to it a value that is meaningful to a particular valuer (valuing agent) who may potentially experience the consequence. There will be (*a*) different values for a given consequence for differing valuing agents, and (*b*) different values for a given consequence, which are dependent on external circumstances involving time and situational conditions for the same valuing agent. Furthermore, a given value of a consequence is not independent of (1) the probability of an event, (2) the description of an event, (3) the probability of a consequence of an event, or (4) the magnitude of the consequence.

For this reason, we consider an *event-outcome space domain,* an *exposure-consequence domain,* a *consequence-value domain,* and then a relationship among these domains to express risk.

7.1 EVENT SPACE DOMAIN

7.1.1 Events-Outcome

Risk always involves the occurrence or potential occurrence of some event. An event is defined by its complete description and the probability of its occurrence. The sum of the probability of the occurrence of an event and the probability of its nonoccurrence, which may imply the occurrence of alternate events, must equal unity by definition of the concept of probability. Thus a given event, denoted by the symbol E_i, where i is an index indicating different events, is determined by its description D_i and its probability of occurrence p_i.*

$$E_i = [D_i, p_i] \tag{7.1}$$

The event space involved consists of the universe U, which is the sum total of n events occurring or not occurring, as shown in the following equation:

$$U = \sum_{i=1}^{n} (E_i + \bar{E}_i) = \sum_{i=1}^{n} \{[D_i, p_i] + [\bar{D}_i, (1 - p_i)]\} \tag{7.2}$$

where $\bar{E}_i$ represents the negative (complement) of the event or its description. As the number of possible events increases, classical probability theory provides a structure for consideration of more complex event spaces and their analysis.

For a given risk situation the causative events must be collectively exhaustive; that is, all possible events that can affect the risk situation must be included. This exhaustive set of events is the universe U for the particular risk situation. The universe may not represent the actual risk situation for two reasons.

1. *Lack of Knowledge.* In any real situation there is no absolute way of knowing if all possible causative events affecting a risk situation have been included. The identification of new causative events that should be in the universe is part of the process of risk identification.
2. *Pragmatic Limits.* For the purpose of analyzing a risk situation, large numbers of causative events may be unwieldy. Those events which have little effect on the risk situation may be omitted to simplify the analysis. However the resulting universe is different from the universe for the real situation. The limited universe is a model for which the error in its utilization in place of the exhaustive case will be small. A tradeoff between ease and cost of analysis against the meaningfulness of the analysis is involved.

It would be desirable to have the events in a universe independent of each other and mutually exclusive. Such conditions imply complete knowledge of underlying processes, the totality of events, and their relationships. Though simplifying analysis, such situations are rare empirically. For example, two

* As is discussed later, this "probability of occurrence" is not necessarily a random function; it may be a definite functional relationship.

cars traveling in opposite directions on a highway meet as they hit a large pothole, blowing out a tire on each car. The events of meeting and each vehicle blowing a tire on a pothole at the site of the meeting result in a variety of outcomes, including different conditions of collision, accident, and misses. However an underlying causative process might be that both cars had tires from the same defective batch of tires. Only the tires' defectiveness caused a blowout at the pothole. Thus it is important that both independent and dependent events be measurable and analyzable. However the description of the events must be mutually exclusive, to ensure that no event overlaps any other event in meaning.

Equation 7.2 is the condition for collectively exhaustive events. From this

$$[D_i, p_i] + [\bar{D}_i, (1 - p_i)] = 1 \tag{7.3}$$

and the universe contains n such events descriptions. However the descriptions are mutually exclusive, such that

$$\begin{aligned} D_i \cap D_j &= \phi \qquad i \neq j \\ D_i \cap D_j &= 1 \qquad i = j \end{aligned} \tag{7.4}$$

where ϕ is the empty set, $\cap$ represents the intersection, D_i AND D_j.*

As described in Chapter 3, these causative events may be naturally occurring or man-made, discrete or continuous. The events may be sequentially or combinationally related in the sense that some events are conditional on others. These conditions are illustrated in Figure 7.1. Five events, designated by their description as D_1 through D_5 are shown as simple† causative events, result in three compound events, labeled $\mathbf{D}_6$, $\mathbf{D}_7$, $\mathbf{D}_8$.‡ These events are dependent and conditional on the simple events; $\mathbf{D}_6$ depends on the joint occurrence of D_1 AND D_2, either simultaneously or sequentially. The probability of occurrence of $\mathbf{D}_6$ is the intersection of p_1 AND p_2. Similarly, $\mathbf{D}_8$ is the result of the union of D_4 and D_5 and will occur if D_4 OR D_5 OR both occur. The resultant set of events that may be of interest (E_A, E_B, E_C, E_D) may be either simple or compound, as shown. The events D_1 through D_5 appear as random events. They need not be. They may follow some natural law or be originated by man in a controlled manner. In this case the probabilities p_1 through p_2 must be replaced by other functional relationships, designated as f_i.

As an example, consider the conduct of an operation to genetically

* Conventional set theory notation is used here where "intersection" is a logical AND, and "union" is a logical, inclusive OR.

† "Simple" is used in the context that to the best of our knowledge, these events are independent, or can be considered to be independent without extensive error. "Compound" implies dependence.

‡ A compound event is represented by $\mathbf{D}_i$.

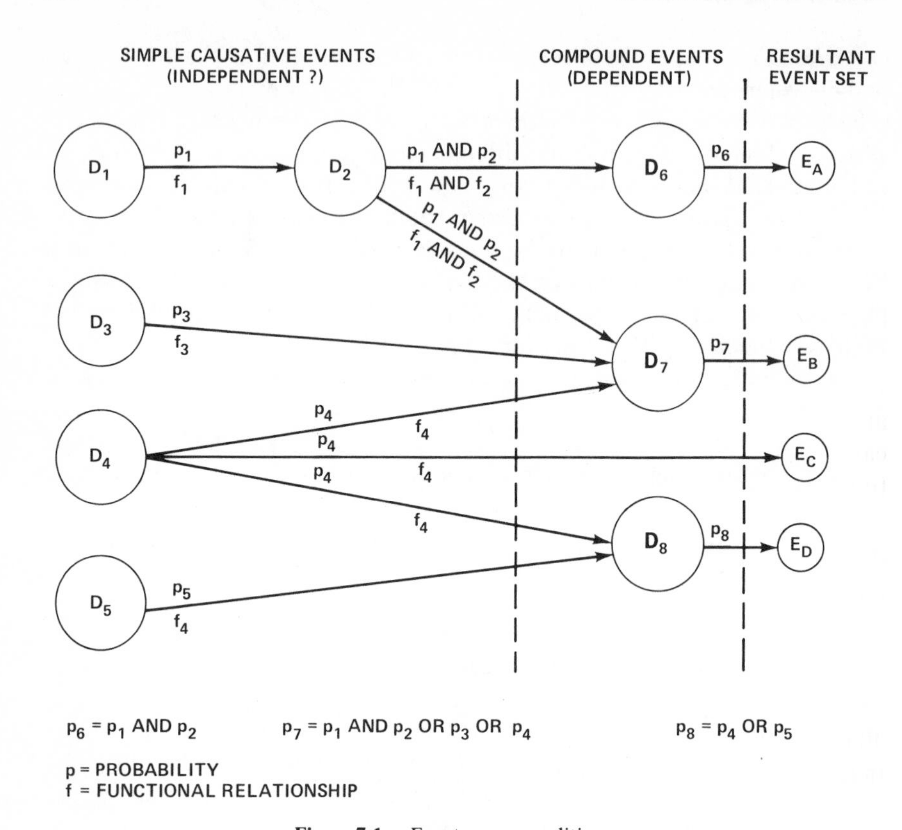

Figure 7.1. Event space conditions.

synthesize new viruses by random breaking and joining of DNA molecules. This research activity requires a carefully controlled environment and containment; but new contagions and possibly harmful viruses have a small, but finite probability of contaminating the populace. An example,* if possible events that can fit Figure 7.1 occur, is as follows:

- D_1 New virus strain is developed.
- D_2 Glove box containment rupture.
- D_3 Container rupture.
- D_4 Improper sterilization.
- D_5 Careless operator.

* A strictly hypothetical situation for illustration.

Dependent events are:

$\mathbf{D}_6 = E_A$ Worker contaminated.
$\mathbf{D}_7 = E_B$ Virus escapes inner containment.
$\mathbf{D}_8 = E_C$ Inner containment contaminated.
E_D Virus killed.

Each event has a number of possible outcomes associated with it. Each risk situation may entail one or more such events.

7.1.2 Outcomes

If and when each event occurs, it will result in a particular outcome. There is a set of possible outcomes from which a particular outcome occurs, each with its own probability of occurrence. An outcome is denoted in a similar manner to an event by its description Q_j and its probability of occurrence p_j, such that an outcome W_j is designated

$$W_j = [Q_j, p_j] \tag{7.5}$$

For a particular event E_i, denoted by the resultant events in Figure 7.1, for example, there will be j possible outcomes. The description of the outcomes must be collectively exhaustive and mutually exclusive, and the sum of the probabilities of the j outcomes must be unity:

$$\sum_{j=1}^{m_i} p_{ij} = 1 \tag{7.6}$$

where m_i is the total number of possible outcomes for event E_i, and p_{ij} is the probability that if event E_i occurs, the outcome j will result.

The double-index notation takes into account the possibility that the same outcome may occur as a result of separate events. This is illustrated in Figure 7.2, where the causative events are assumed to be certain. The probability of occurrence of a certain outcome is

$$W_j = \left[Q_j, \sum_{i=1}^{n} p_{ij}\right] \tag{7.7}$$

where n is the total number of events. When the events are not certain and obey random processes, the probability of occurrence of events must also be considered:

$$W_j = \left[Q_j, \sum_{i=1}^{n} p_i \times p_{ij}\right] \tag{7.8}$$

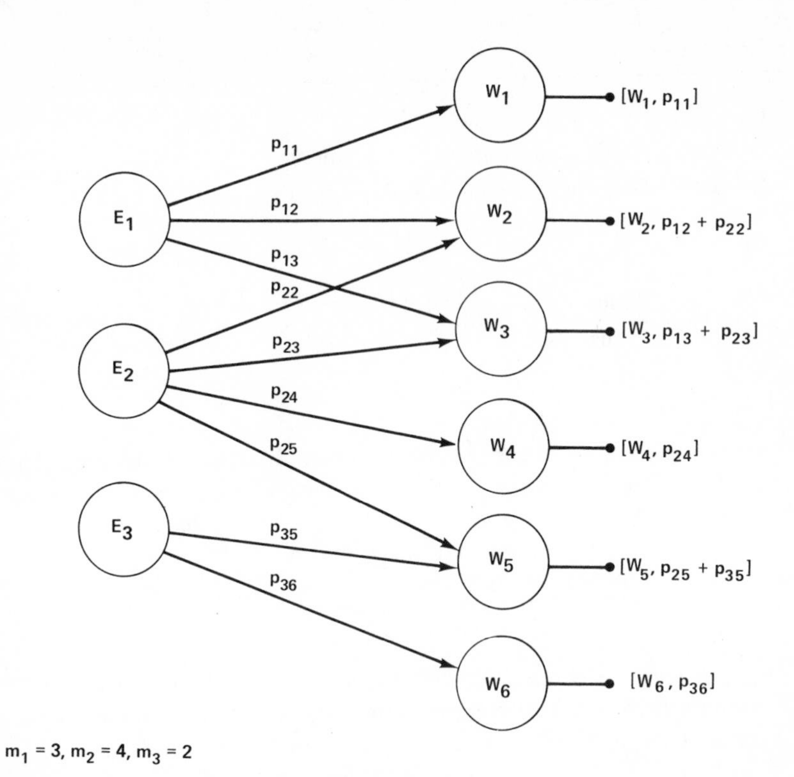

Figure 7.2. An event-outcome space (events are assumed to be certain): $m_1 = 3$, $m_2 = 4$, $m_3 = 2$.

7.1.3 Continuously Occurring Events

There is a class of events whose description involves the specification of an event continuing over all time of interest instead of over a short time interval. The planned continuous release of pollutants to the environment from a polluting source and the accidental spill of pollutants are examples of two different types of description.

The description of the event may specify a time-dependent relationship for the magnitude of the event's descriptive parameters. That is, the description allows a planned pollution release to change over time by some varying function, which is the usual real world case. The specification of a continuously occurring event may be denoted by

$$E_i = [D_i(t), p_i] \tag{7.9}$$

where $D_i(t)$ implies a time-dependent event description, and the probability of occurrence may be unity within the total time period in question, but involves uncertainty with respect to the occurrence at any given time. Thus the magnitude of the occurrence is expressed by a time-dependent description.* The probability may be less than unity if the time of the event is not certain (or represented by a functional relationship, f_i.)

The specification of time-dependent functions is often difficult, and simplifying assumptions are often made whereby the integral of the magnitude over the period is assumed as a single event in time. A variety of other possibilities exist for simplifying descriptions. In general, one often sacrifices precision in expressing parameters for simplification of use and manipulation. However loss of precision and resultant inaccuracy must not be ignored.

7.1.4 Event-Outcome Space

When events are uncertain and may be either independent or dependent as in Figure 7.1, and have a set of outcomes as in Figure 7.2, an event-outcome space for a given experiment or empirical situation can be specified. If different types of outcome are involved, it is possible for more than one outcome to occur from the occurrence of a set of events.

To illustrate this condition, let us extend the example begun earlier. Three different sets of outcomes, each set mutually exclusive and collectively exhaustive, are shown:

Set 1 Event E_A—Glove box exterior contaminated.
- Q_1 —with unknown virus
- Q_2 —with known, harmful virus
- Q_3 —with known, nontoxic virus

Set 2 Events E_C OR E_D—Virus killed OR inner containment contaminated.
- Q_4 —Virus destroyed
- Q_5 —Other viruses contaminated
- Q_6 —Virus unharmed

Set 3 Event E_B—Virus Escapes Inner Containment.
- Q_7 —Outer containment breached—Known, nontoxic virus
- Q_8 —Outer containment breached—Known, toxic virus
- Q_9 —Outer containment breached—Unknown virus
- Q_{10}—Outer containment breached—virus destroyed
- Q_{11}—Outer containment not breached

* That is, the complete, long-term probability may be known, but the instantaneous probability can be expressed only by some probabilistic function.

It is possible for more than one outcome to occur depending on the events. For example, Q_3, Q_4, Q_{11} are highly probable.

The universe of the experiment is the totality of outcomes

$$U = \sum_{j=1}^{m} W_j = \sum_{j=1}^{m} \left[Q_j, \sum_{i=1}^{n} p_i p_{ij} \right] \tag{7.10}$$

When no exposure to man, biota, the environment, or human institutions occurs, there is no risk. The event-outcome space is an experimental space that may or may not involve empirical situations.

As such, formal probabilistic and statistical methods may be used for analysis, scientific and statistical inference, and testing of hypotheses without necessarily involving risk.

Sets 1 and 3 may involve exposure to humans, and so on. Set 2 might involve only the institutional structure of the experiment.

7.1.5 Uncertainty in the Event-Outcome Space Domain

In specifying an event-outcome space, there is descriptive uncertainty in a number of instances and measurement uncertainty in others.

7.1.5.1 Descriptive Uncertainty

The description of both events and outcomes involves uncertainty in determination of whether they are mutually exclusive or collectively exhaustive, and the degree of independence of events.

Inclusion Uncertainty. Degree to which a set of events and outcomes is collectively exhaustive. Implies unknown events or outcomes.

Specification Uncertainty. Degree to which descriptions of events and outcomes are mutually exclusive, respectively, and the degree to which dependent relationships among events are known.

Modeling Error. Degree to which an experimental universe differs from the universe it is modeling because of omission of events, outcomes, and relationships for simplification of analysis.

7.1.5.2 Measurement Uncertainty

Measurement uncertainty lies primarily in the inability to measure probabilities (or functions) accurately. Both a priori and a posteriori probabilities have measurement errors. Repeated measurement results in statistical convergence of variability but says nothing about the exact value of the next future trial.

7.1.5.3 *Uncertainty Specification*

If q_i is the descriptive uncertainty encompassing the three above-named types and b_i and b_{ij} are the measurement uncertainties for the probability of occurrence for events and outcomes, respectively, the uncertainty of universe U is

$$U \pm u = \sum_{j=1}^{m} \left[Q_j \pm q_j, \sum_{i=1}^{n} (p_i \pm b_i)(p_{ij} \pm b_{ij}) \right] \pm \mathrm{u} \tag{7.11}$$

where

$$u = f(q_j, p_i, p_{ij}) \tag{7.12}$$

is the modeling error, and

$$\sum_{i=1}^{n} (p_i \pm b_i) = 1 \tag{7.13}$$

and

$$\sum_{j=1}^{m} (p_{ij} \pm b_{ij}) = 1 \tag{7.14}$$

to preserve the nature of probability limits.* The error term b_i contains both descriptive uncertainty (specification D_i's) and measurement uncertainty (estimation of p_i) for n causative events.†

7.2 THE EXPOSURE–CONSEQUENCE DOMAIN

7.2.1 Exposure Pathways

The exposure pathway domain is illustrated in Figure 7.3 for two outcomes and three recipients of exposure. The exposure from an outcome may have multiple pathways to the same recipient for different modes of exposure.

* Alternatively, one can deal with expected values, $\hat{E}$, such that

$$\hat{E}\left[\sum_i p_i\right] = 1 \quad \text{or} \quad \sum_i \hat{E}(b_i) = 0$$

$$\hat{E}\left[\sum_j p_{ij}\right] = 1 \quad \text{or} \quad \sum_j \hat{E}(b_{ij}) = 0$$

† The maximum uncertainty is expressed by a uniform distribution of the range of uncertainty for both probability assignments and event descriptions. The substitution of specific probability distributions in place of uniform distributions is a direct means for reducing uncertainty by adding more information.

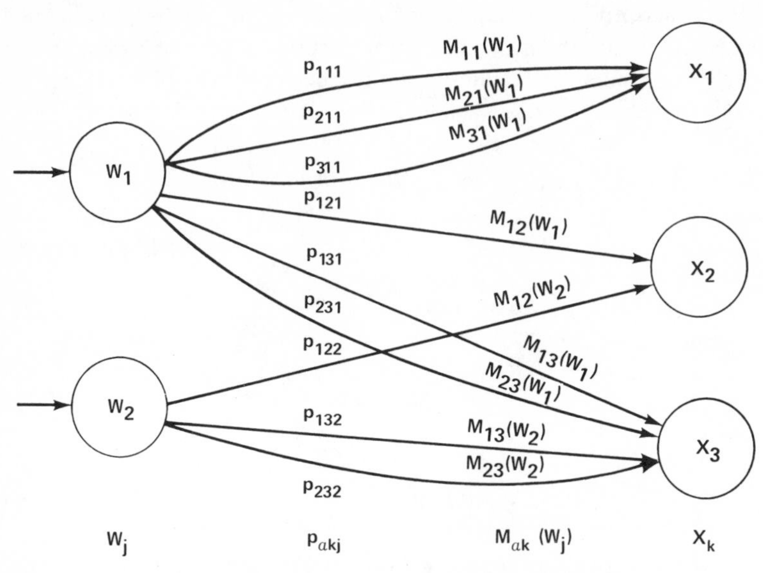

Figure 7.3. Outcome-exposure pathways.

These modes are designated by the subscript α.* The recipient is denoted X_k and the outcome W_j.

For each pathway there is a pathway magnitude designated as $M_{\alpha k}(W)$, which can vary from zero to unity such that

$$0 \leq M_{\alpha k}(W_j) \leq 1 \tag{7.15}$$

The total magnitudes of pathways for a given outcome to a specific recipient must also be less than or equal to unity, since the exposure to a recipient cannot exceed the outcome. For n_{jk} pathways between a specific outcome j and a specific recipient k, the total exposure cannot exceed unity.

$$\sum_{\alpha=1}^{n_{jk}} M_{\alpha k}(W_j) \leq 1 \tag{7.16}$$

The totality can be less than unity because pathways may diminish the impact of an outcome.

The exposure may or may not occur if an outcome takes place when a random process is involved. Thus a probability of occurrence for each

* A Greek subscript is used to indicate that the pathways are subdivisions of the total pathway and are not just indices.

pathway exists as $p_{\alpha kj}$. For certain exposure this probability is set at unity. For negligible exposure, $p_{\alpha kj}$ is near or at zero. For a process other than random, some $f_{\alpha kj}$ may replace the probability.

$$0 \leq p_{\alpha kj} \leq 1 \tag{7.17}$$

$$0 \leq f_{\alpha kj} \leq 1 \tag{7.18}$$

For each recipient X_k of α pathways from outcome j, we write

$$0 \leq \sum_{\alpha=1}^{n_{jk}} p_{\alpha kj} \leq n_{jk} \tag{7.19}$$

This means that each probability can range over the total probability range and some pathways may be certain. In fact, all pathways may be certain. Then

$$\sum_{\alpha=1}^{n_{jk}} p_{\alpha kj} = 1 \tag{7.20}$$

Each recipient receives an exposure from a given outcome such that

$$X_{kj} = \sum_{\alpha=1}^{n_{jk}} p_{\alpha kj} M_{\alpha k}(W_j) \tag{7.21}$$

The total exposure to a recipient is the sum of exposures from different outcomes. This assumes that exposures from the different outcomes are of like nature, that is, the outcomes are similar. (If they are not similar, an exposure pathway structure is required for each type of outcome, such as the three sets in the earlier examples.)

$$X_k = \sum_{j=1}^{n} \sum_{\alpha=1}^{n_{jk}} p_{\alpha kj} M_{\alpha k}(W_j) = \sum_{j=1}^{n} X_{kj} \tag{7.22}$$

and the probability of occurrence $p(X_k)$ from equation (7.7)

$$p(X_k) = \sum_{i=1}^{n} p_i p_{ij} X_k \tag{7.23}$$

As an illustration, recipient X_k might be a worker using the glove box in the earlier example. He will not be exposed unless the contaminating virus enters his system after an accident. This can occur by skin absorption. Such factors involve the magnitude of the pathway. The probability of the pathway occurring is the chance of, say, a sneeze from a diseased subject spreading it. likewise X_2 and X_3 might be members of the general population.

7.2.2 Consequences

When an outcome results in exposure to a risk recipient (agent), a whole spectrum of possible consequences may occur. The same consequence may result from one or more outcomes. The consequences have a unique description, designated C_c, where the index represents the cth consequence, and a probability of occurrence p_c. The set of consequences descriptions must be mutually exclusive and collectively exhaustive. The relationship among outcomes, exposure pathways, and consequences is schematized in Figure 7.4, an extension of Figure 7.3, where recipients now experience consequences as a result of exposures. There is only one set of consequences, but this can be experienced in whole or in part by one or more recipients. Thus each recipient has his own set of consequences for the total set. This subset is designated by C_{ck}, which represents the cth consequence to the kth recipient.

As an example, the worker or a member of the population exposed to a virus might have the following array on possible consequences.

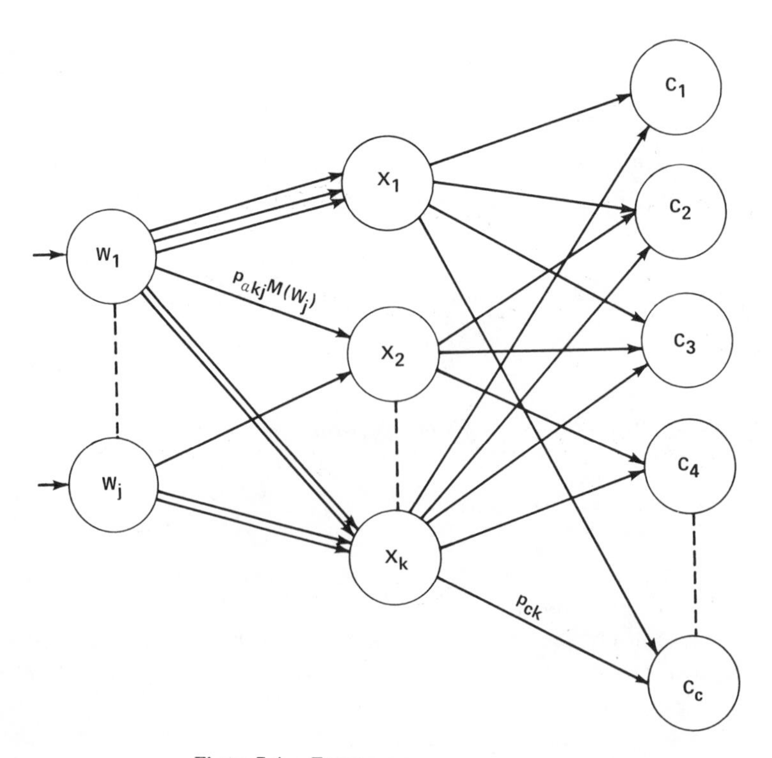

Figure 7.4. Exposure-consequence space.

C_1 Rapid death and contagion.
C_2 Rapid death and no contagion.
C_3 Significant life shortening and contagion.
C_4 Significant life shortening and no contagion.
C_5 Severe illness.
C_6 Mild illness.
C_7 No effect.

These items are illustrative, not a complete set. Someone vaccinated against the virus might only experience C_6 or C_7. Others might be subject to all.

The probability of occurrence of a particular consequence involves a condition of both the probability of exposure and the probability that a particular consequence will occur. For each recipient, the probability of one of the set of consequences occurring must be unity

$$\sum_{c=1}^{n_k} p_{ck} = 1 \tag{7.24}$$

where n_k is the number of consequences in the kth recipient's consequence subset. The total probability of a given consequence to the kth recipient is

$$p_c(k) = p_{ck}p(X_k) = p_{ck}\sum_j p(X_{kj}) \tag{7.25}$$

where all exposure pathways to a recipient as reflected in (7.22) are indicated.

The total probability of a consequence p_c occurring to all recipients is the sum of probabilities to each recipient.

$$p_c = \sum_k p_{ck}p(X_k) \tag{7.26}$$

When k is larger than 1, consequences to groups and society can be ascertained and more than one consequence can occur simultaneously.

7.2.3 The Event-Outcome , Exposure-Consequence Space

The overall action of causative events acting to affect consequences is represented by a sequence of equations as follows. The causative events defined as D_i occur with probability p_i, result in outcome j, and amount to the set of possible outcomes.

$$p(W_j) = \sum_{i=1}^{n} p_i p_{ij} \tag{7.27}$$

Both the definitions of events D_i and outcomes Q_j are implied. The three equations

$$p(X_k) = \sum_{i=1}^{n} p_i p_{ij} X_k \tag{7.23}$$

$$p(X_k) = \sum_{i=1}^{n} p_i p_{ij} \sum_{j=1}^{n} \sum_{\alpha=1}^{n_{jk}} p_{\alpha kj} M_{\alpha k}(W_j) \tag{7.28}$$

$$p_c = \sum_{k} p_{ck} p(X_k) \tag{7.26}$$

result in an overall probability of occurrence of a consequence.

$$p_c = \sum_{k} \sum_{i} \sum_{j} \sum_{\alpha} p_{ck} p_i p_{ij} p_{\alpha kj} M_{\alpha k}(W_j) \tag{7.29}$$

Thus p_c together with the consequence definition C_c defines the event-consequence space domain.

7.2.4 Uncertainty in the Exposure-Consequence Domain

As in the event-outcome space, uncertainty exists in the exposure-consequence domain.

7.2.4.1 *Descriptive Uncertainty*

Inclusion Uncertainty. The identification of relevant pathways and consequences as well as identification of all recipients involves uncertainty of the degree of coverage.

Specification Uncertainty. Definition of pathways and consequences may not always be mutually exclusive or collectively exhaustive.

Modeling Error. Simplification by constraining pathways and consequences to given bounds for ease of analysis is often necessary.

7.2.4.2 *Measurement Uncertainty*

Probability Estimation. Measurement errors exist for all probabilities shown in (7.27).

Pathway Magnitude Measurement. The magnitude of exposure pathways is often measurable empirically. Like all such measurements, these measurements involve problems of instrumentation, data collection and processing, and interpretation.

Each term in (7.27) has an error term associated with it in the manner illustrated in (7.11) and (7.12). Similarly, there are error terms for specification of D_i, Q_j, X_k, and C_c. Thus an overall error term u is

$$u = f(D_i, Q_j, X_k, C_c, p_i, p_{ij}, p_{\alpha kj}, p_{ck}, M_{\alpha k}, (W_j)) \qquad (7.30)$$

but is not enumerated further.

7.3 CONSEQUENCE–VALUE DOMAIN

7.3.1 Valuing Agents

Each consequence is valued by those who may be affected by it. The consequence itself is not meaningful, since it is merely a description. Of concern is the measure of value of the consequence occurrence to the risk taker. Different groups in society as well as individuals may assign differing values to the same consequence.

The value of a given consequence to a particular valuing agent is represented by the symbol ν_{ck}, which is the value of the cth consequence to the kth valuing agent. The value ν_{ck} is not necessarily a constant and may vary over time and situation, even for an individual, and it may even be a function of the probability magnitude for the consequence occurrence and the magnitude of the consequence itself. Assignment of value to a consequence is highly subjective, and is, along with uncertainty of probability assignment, a source of major uncertainty in evaluating the value of risk.

7.3.2 Risk Evaluators

The scope of risk is made more complex because there are many risk takers, each with his own set of subjective values and relationships to externalities. Furthermore, the assignment of risk often involves an evaluating agent making a judgment for a valuing agent. A government agency interpreting the needs and values of people in setting a regulation is an example of an evaluating agent as opposed to the valuing agents—namely, the people affected by the regulation. In this case, γ's functions are used to differentiate the interpretation of a valuing agent's assessment of risk by the risk evaluator as opposed to that of the risk agent directly. In making such judgments, the risk evaluator is always subject to questions about his knowledge and ability to make the judgments, the effect of his personal biases, and his fairness in acting as a judge. Often the risk evaluator attempts to determine criteria for public acceptability of risk by examining historically similar kinds of risk to see what levels have been acceptable to society.

Thus we define a "valuing agent" as a person, or group of persons, who directly evaluates the consequence of risk to which he is subjected, and a "risk evaluator" as a person, group, or institution that seeks to make an interpretation of a valuing agent's risk.

Figure 7.5 includes factor γ_{ck} for each ν_{ck} relationship of consequence to a valuing agent for this purpose. The γ factor represents only one of a variety of methods that a risk evaluator might use to assign varying relevance to differing groups of valuing agents. For example, the γ_{ck}'s might represent the fraction of the total population involved, indicating that the risk evaluator might favor the value of large groups over small.

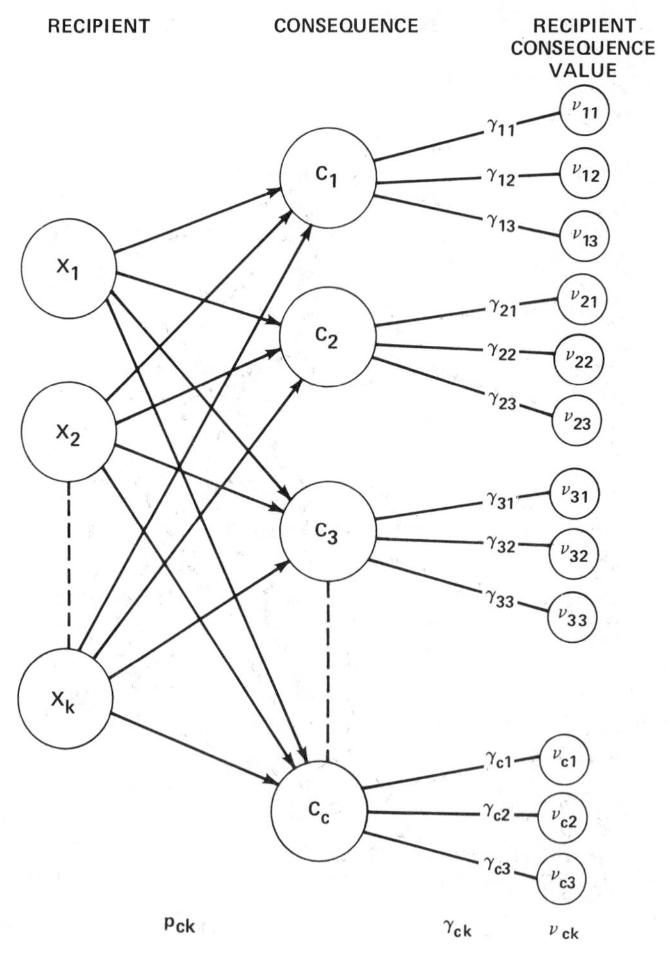

Figure 7.5. Value-consequence domain.

Alternately, the assignment might be the political judgment of the evaluator regarding which group should be favored.

The object of the separate notation is to differentiate between the value assigned to a consequence by a valuing agent and the importance of that agent to the evaluator involved in making a decision on risk assessment. For example, an ecologist acting as a valuing agent might place high negative value on a consequence that results in destruction of an aesthetic natural resource, whereas a decision maker may make a value judgment to the effect that the ecologist's value may have little weight, since many others may have assigned a low negative value to the same consequence. In this way, the γ's serve as interpretive modifiers of the valuing agent's judgment.

As an alternate notation, it may be decided that the symbol ν_{ck} will always refer to the kth valuing agent, and the symbol $\nu_{ck}(\gamma)$ will represent a risk evaluator's interpretation of the valuing agent's judgment. A subscript applied to γ may be used to identify different risk evaluators.

7.4 RISK IN TERMS OF THE RELATIONSHIP BETWEEN PROBABILITY AND CONSEQUENCE VALUE

Risk of a particular undertaking R is evaluated by examining the values of consequences and the probabilities of consequence occurrence for that undertaking. Thus the risk is a function of the value of a consequence ν and its probability of occurrence ρ.*

$$R = f(\rho, \nu) \tag{7.31}$$

One function often used to express the risk of an undertaking is the Bayesian concept of expected value. The products of the probability and consequence values are summed to provide an expected value of the risk of the undertaking, which may be compared with the expected value of alternative undertakings, including no action. For an undertaking with n consequences, expected value of risk (EVR) is computed by

$$\text{EVR} = \sum_{n} \rho_n \nu_n \tag{7.32}$$

The variances or standard deviations of the measure, based on historical or a priori information, are measures of the dispersion around the central value. For the same expected value, a risk averter would prefer lower values of the standard deviation over higher values.

* The Greek letter rho (ρ) indicates the compound probability of occurrence of a consequence having a given value.

This concept assumes that ρ_n and ν_n are independent of each other. It is shown subsequently that this is seldom if ever the case. *As a result, the concept of expected value of risk has limited application.* The selection of a useful risk function is a key problem in risk analysis and is investigated in more detail later.

The consideration of event, consequence, and value domains provides a means to examine closely the risk relationship, to understand the factors in establishing risk.

Thus

$$R = f[p_c, \nu_{ck}] \qquad (7.33)$$

provides a general expression for risk as a function of a consequence value by a valuing agent and its probability of occurrence, irrespective of the events involved. If a risk evaluator is involved, (7.33) would take one of two alternate forms, depending on the notation used:

$$R = f[p_c, \gamma_{ck}\nu_{ck}] \qquad (7.34)$$

where a risk evaluator's bias or value concept is added on, or

$$F = f[p_c, \nu_{ck}(\gamma)] \qquad (7.35)$$

where the indication is that the risk is interpreted by the risk evaluator, not the risk taker.

The probability of the consequence p_c has five component factors representing the chain of events, outcomes, exposures, consequences, and value assignments. This chain of events, outcomes, exposures, consequences, and value assignments must be preserved in consideration of risk determination, particularly when the assignment of value to a consequence by a valuing agent is functionally dependent on the assignment of probabilities and description of consequences.

The assignment of value is subjective and probably has a wider range of uncertainty than the other parameters. Therefore the problem of assigning value must be considered in depth.

Part B

FACTORS IN RISK VALUATION AND EVALUATION

Many factors affect the subjective valuation of risk consequences and evaluation of risk action. This part addresses many of these factors qualitatively, to provide insight into how and why they affect risk perception. Part D quantifies these factors.

Four chapters make up this part. Chapter 8 covers factors associated with the type of consequence; whether it is a voluntary or involuntary risk situation, an immediate or delayed consequence, whether a statistical or identifiable risk agent is involved, and the degree to which the risk is controllable.

Chapter 9 deals with the factors related to the nature of the consequence itself and how it is valued. Classes of consequences, variation in cultural values, magnitude of consequences, origination of the risk event, and the degree of knowledge of risks are examined.

Chapter 10 treats a variety of other factors such as the magnitude of the probability of occurrence as it affects consequence valuation, situational factors, and the psychological behavior of groups and individuals in terms of their propensity to take risks.

The individual propensity to take risk is a major variable in any individual risk situation, and some further insights into this dimension are developed in Chapter 11.

8

Factors Involving Types of Consequence

The factors involving types of consequence are identifiable conditions that directly influence the valuing of risks. This influence is observable from the manner in which societal and group behavior responds to these different types of risk factor. This chapter discusses qualitatively the four most significant factors: (1) involuntary risk, (2) discounting in time, (3) spatial distribution and discounting, and (4) controllability of risks. A subsequent section attempts to quantify some of these factors.

8.1 VOLUNTARY AND INVOLUNTARY RISKS

Considerable confusion exists with respect to the meaning of voluntary and involuntary risks. In general, a voluntary risk involves some motivation for gain by the risk taker, whereas an involuntary risk is imposed on a risk taker without regard to his own assessment of benefits or options. In the latter case, the problem of avoidability of risk is often confused with the definition of voluntary and involuntary risks. This becomes particularly apparent when one considers a person who has an option to move away from the source of the risk (e.g., a house near the approach runways of a busy airport). Although the airport may have been constructed long after the person settled in the runway area, he—the person at risk—does have the option to move at any time at some cost. Furthermore, although the risk from a flood or a tidal condition may be natural in origin, people do live in disaster-prone areas and are well aware of the risks. Are these risks voluntary? A rather vigorous analysis of the conceptual aspects of voluntary and involuntary risk is necessary to address this question and to provide some insight to the problem.

The concept of voluntary and involuntary risk is indeed complex. At least three major factors, each in itself complex, affect the meaning of voluntary and involuntary risk. These are (*a*) equity of risk and benefit distribution, (*b*) the avoidability of risk and availability of alternatives, and (*c*) the manner in which risk is imposed on the risk taker. Each of these three factors is defined in detail, then a combination of them is used to address voluntary and involuntary concepts.

8.1.1 Equitable and inequitable Risks

One aspect of voluntary and involuntary risks is based on identification of who takes the risks and who gets the benefits. In this manner, one may attempt to define an equitable* risk as the case in which the risk taker stands to benefit directly from the event involved and is, therefore, able to balance the risks against the benefits to determine whether he should assume the risk. On the other hand, the case of one who is subject to risk but receives none of the concomitant benefits (or at best, diffuse indirect benefits), can be defined as an inequitable risk, since there is no risk-benefit balance. A risk agent subjected to an inequitable risk can only balance the imposed risk against his cost of avoiding the risk. It is assumed that all risks, except for natural planetary and astronomical events, are avoidable or at least can be minimized at some cost. Thus the ability to balance risks and benefits and thereby take action to avoid the risk, if so desired, is a voluntary process, whereas cases that balance risk against avoidance (fleeing, i.e., "quitting the game") without supplementary benefit is an involuntary process. The degree of avoidability of a risk does indeed affect the valuation of the risk, but is addressed as a separate factor.

The foregoing definition is inadequate, however, unless the degree of knowledge the risk taker has about the risks he is subject to is considered simultaneously. The degree of knowledge about risks falls into four classifications: (1) risk completely known to the risk agent, (2) risk is covertly withheld from the risk agent (i.e., the imposer of risk seeks to hide this risk from the recipient risk agent), (3) risk information is overtly available, but the risk agent makes no attempt to use or acquire this information, and (4) the risks are uncertain and undefined to all parties; no information is available.

When the factors of benefit attainment and degree of knowledge are combined, eight classifications of risk result (Table 8.1). Essentially, the inequitable risks are imposed on the risk agent when he receives no benefit from the condition that imposes the risk. However the covert information case—when the risk agent also achieves some benefit—is also inequitable, since he is not fully apprised of the means to balance risk and benefits personally.

* The word "equitable" is used in the sense of "fair to all concerned—without prejudice, favor, or rigor entailing undue hardship." *Webster's Third New International Dictionary* (G. & C Merriman-Webster, Springfield, Mass., 1971).

Table 8.1 *Definition of Equitable and Inequitable Risks*

For a Given Risk	Risk Agent Receives Direct Benefit	No Direct Benefit Received by Risk Agent
Risk known (overt)	Informed equitable risk	Informed inequitable risk
Risks unknown—hidden (covert)	Deceptive inequitable risk	Exploited inequitable risk
Risks unknown—available (overt)	Unwary equitable risk	Unwary inequitable risk
Risk unknown—unavailable to all (uncertainty)	Unknown equitable risk	Unknown inequitable risk

8.1.1.1 Informed Equitable Risk

The risk agent knowingly accepts the risk to obtain direct benefits. For example, a passenger on an airplane accepts some small risk of an accident for the direct convenience of rapid travel over long distances.

8.1.1.2 Informed Inequitable Risk

The risk agent is knowledgeable about the risk imposed and gets no direct benefit. For example, someone living close to an airport approach is subject to an increased probability (over other areas) of an airplane crashing into his house. Danger to nearby residents from low level radiation emissions from a nuclear energy facility is another example.

8.1.1.3 Deceptive Inequitable Risk

Although the risk agent receives some benefit from the activity undertaken, the primary recipient of the benefits covertly seeks to keep adverse information from the risk agent. The covert action is not necessarily malicious, but may be undertaken to minimize disruptions. A case in point is the transportation of nuclear materials by passenger aircraft (at least to the present time), when the passengers may be unaware of the extra risks imposed by shipments on the aircraft. The shipper and receiver of the shipments get the direct benefits of the shipment. A passenger's choice in balancing risk of flying versus convenience in negated by the deception.*

* This statement does not imply a conscious effort to deceive, nor does it convey approval or disapproval of the system of shipment of nuclear materials, which may have wider tradeoffs of risks, costs, and benefits to society; it illustrates, however, the nature of deceptive inequitable risk to an individual.

8.1.1.4 Exploited Inequitable Risk

Not only does the risk agent receive no benefit from taking a risk, but the level and/or nature of the risk is purposefully withheld from him; for example, exposing a specific population to a disease (or withholding a cure for an already exposed population) for the purposes of gathering medical and epidemiological information, without informing the exposed population of the risks, is inequitable exploitation of the risk agents.

8.1.1.5 Unwary Equitable Risk

The risk agent receives benefits but is unaware of the risks involved (or degree of risk) through his own indifference, negligence, or unwillingness to make the effort to obtain the information on risk when it is readily available. Gambling without knowing the house percentages for different games is one example.

8.1.1.6 Unwary Inequitable Risk

The risk agent is unaware that he is assuming a risk from which he receives no benefit, although he could find out about the risks if he so chose. There is no attempt to withhold information from the risk agent, and his concern for identifying his risks controls whether a risk remains an unwary one or an informed one. There are many people who prefer to remain ignorant of the risks because increased anxiety, resulting from new risk information, may be more undesirable to them than the risk consequences themselves.

8.1.1.7 Uncertain Equitable Risk

Obvious benefits notwithstanding, the risks may be uncertain. Taking a drug that has undetermined side effects is an example.

8.1.1.8 Uncertain Inequitable Risk

The risk agent receives no direct benefits and the risks are uncertain. The use of pesticides that may have cancer-causing potential, such as aldrin and dieldrin, is an example. The pesticide user benefits directly, the consumer only indirectly at best, and the latter assumes any risks that might exist, although those risks are unknown.

In summary, an equitable risk involves freedom of choice to accept or reject a condition that imposes risk through a balancing of the risks and benefits to the risk agent. The choice of alternatives can, of course, be so limited that only a go–no go decision is possible. For example, there is only one possible way to travel from the East Coast of the United States to the West Coast in less than eight hours—fly by jet airplane. Thus one can go or not go. The negative benefits (costs) of not going are perceived to exceed the risks in flying in most cases. However the risk is assumed in a "go" decision, to achieve specific benefits.

Conversely, if the risk agent receives no direct benefit from a condition imposing a risk, or if the information to make risk-benefit balance has been withheld, the risk is inequitable and the risk agent can decide to minimize the imposed inequitable risk if he chooses to do so at some cost. However he is acting to avoid a risk that is imposed on him without benefit to him. Thus inequitable risks involve no choice to accept or reject the condition that causes a risk, but only a choice to incur costs of minimizing the imposed risks.

8.1.2 Avoidability of Risks and Risk Alternatives

The avoidability of risks involves the opportunity and freedom to select alternatives to risks facing the risk taker. The avoidability of risk must not be confused with the controllability of risk. Avoidability implies choice among options with different costs; controllability implies the means to affect the probability and magnitude of occurrence of a given risk.

When alternatives are available that have less risk than the primary risk imposed, the cost associated with these options may be higher, making them unattractive. In any case, the risk reduction must be balanced against costs in evaluating options.

For inequitable risks, the alternatives for reducing the risk often involve fleeing the source of the risk. This is balanced against the cost of not fleeing. For example, how many people have left the warm climate and style of living offered by the Los Angeles area as a result of the relatively high probability of major earthquakes in that part of southern California? Though other factors involve the valuation of the risk, the cost of changing one's status quo is often very high. For inequitable risks, the cost of alternatives must always be high, otherwise the risks would probably be equitable; that is, a favorable alternative might well be selected even without the imposition of risk.

Equitable risks also involve alternatives. For example, one often trades degree of mobility against risk in the choice to ride in a car, a bus, on a motorcycle, in a plane, or not leave home at all.

If there are no alternatives to a consequence (including control), the risk taker can do nothing but face the risk. The ability of man to rationalize in the face of unavoidable dire consequences is well documented in the case of war and ambulatory terminal disease. Thus rationalization often changes the degree of a consequence value, especially when probability of occurrence is high. When the probability of occurrence is low, such as in uncontrollable natural disasters, a person may ignore the consequences completely. This concept of probability thresholds is considered in subsequent chapters. However if there are alternatives, the risk taker may make tradeoffs and choices. Often, these tradeoffs have intangible parameters—for example, should I ride a motorcycle to work because it is exciting and inexpensive, or should I drive a large automobile with its increased safety factor? Such decisions are sometimes difficult for individuals to face.*

8.1.3 Imposition of Risks

The distinction used here to differentiate equitable and inequitable risks is a narrow definition based on whether the benefits are direct or indirect and the degree to which knowledge about the risks is available. The availability of individual choice is not considered in this definition, since the availability of risk alternatives is also a factor in individual choice. Neither equitable nor inequitable risk definitions or availability of risk alternatives can address the problem of individual choice alone. For this reason, a concept of "imposed" risks is necessary in discussing the interdependence of risk benefits, availability of alternatives, and individual choice.

Risks are either self-imposed by a risk agent (individual choice) or imposed on the risk agent from external sources. These types of risk imposition are termed endogenous and exogenous risk imposition, respectively.

8.1.3.1 Endogenous Risk Imposition

Choice is under control of the risk agent alone. The difference among individual choice situations depends on who receives the direct benefits of the activity that imposes the risk on the risk agent. These direct benefits are often fuzzy.

1. *Risk agent receives direct benefit.* An equitable risk is selected by the risk agent from available alternatives.

2. *Identifiable individuals receive direct benefits.* As opposed to the case of the risk agent who at best receives indirect benefits, this case covers the idea of "self-sacrifice for others" and "lifesaving by nonprofessionals." (Professionals are contained in the foregoing case, since they receive direct benefit in the form of remuneration for hazardous duty and the "excitement" factor of their jobs.) This is an inequitable risk that is self-imposed.

3. *Neither the risk agent nor identifiable individuals receive direct benefits.* The recipients of direct benefits are unidentified. This case covers the ideas of "service to others" and "patriotic sacrifices." The risk agent receives indirect benefits in the form of self-satisfaction of duty performed

* See Chapter 14 for one analysis of this risk tradeoff situation.

well and other intangible rewards. This is an inequitable risk that is self-imposed.

8.1.3.2 Exogenous Risk Imposition

Choice is not under control of the risk agent alone. Exogenous risks are imposed in four ways.

1. *Risks are imposed by fiat or unilateral action of others without direct recourse.* The drafting of individuals into the military is an example of risk imposed by fiat; a nonsmoker who is obliged to enter a smoke-filled room is representative of persons subjected to the unilateral action of others. In either case, the only recourse is to "flee" the situation (i.e., to remove one's self from the threat), an unacceptable alternative in most cases. A drunken driver is another case of unilateral action; here recourse is through preventive action, but not the direct action of the risk generator.
2. *Risks are imposed by a collective due process where the risk agent has no direct recourse as an individual.* Good examples of this type of risk imposition, are governmental actions, such as local boards raising property taxes.
3. *Risks are imposed by a due process where the risk agent has direct recourse as an individual.* Using the right of eminent domain or a revised tax assessment to take money or real property are good examples if the aggrieved individual has recourse to the courts or an appeal board. The appeal may be an expensive and unacceptable alternative, but it is available.
4. *Risks from natural causes and accepted cultural actions.* Risks from tornadoes and meteors are cases of the former, whereas risks of disease contagion from shaking hands in greeting is never questioned in this country. (Shaking of the right hand is considered an insult in some Arabic societies; this hand is used for personal hygiene functions.)

The definitions are not distinct, but they overlap, depending on degree of imposition and recourse; however these factors are identifiable as separate conditions for discussion.

8.1.4 Classification of Voluntary and Involuntary Risks

The three sets of factors just discussed provide the ingredients for a classification of risks into involuntary and voluntary categories. Four combined sets of factors (Table 8.2) furnish the gross basis for classification. As long as the concept of "informed consent" holds—that is, as long as information on risk is not "covertly withheld," and reasonable alternatives are availa-

Table 8.2 *Classification of Voluntary and Involuntary Risks*

Voluntary Risk	Involuntary Risk
1. Endogenous risk imposition with acceptable alternatives available *and* knowledge of the risk is not covertly withheld.	1. Endogenous risk imposition without acceptable alternatives *or* knowledge of the risk is covertly withheld (or both).
2. Exogenous risk imposition with acceptable alternatives available *and* the risk is equitable.	2. Exogenous risk imposition without acceptable alternatives *or* the risk is inequitable (or both).

ble—endogenously imposed risks and equitable, exogenously imposed risks are deemed voluntary. An acceptable alternative may be more or less desirable, but it must provide a reasonable alternative other than "fleeing."* All other risks are deemed involuntary whether self-imposed or exogenous.

Basically, a voluntary risk requires that acceptable alternatives to the risk agent be available, although these alternatives may be less desirable than the case of risk being imposed. Risks for which information is withheld purposefully are always involuntary. With these two basic conditions considered, self-imposed risks are always voluntary regardless of whether the benefits are directly obtained by the risk agent. Exogenously imposed risks are voluntary for all equitable cases. That is, the risk agent directly receives the benefits; however there are different degrees of acceptance based on the manner in which the risks are imposed. All other risks are involuntary.

The degree of acceptance of exogenous risks, whether voluntary or involuntary, depends on how the risks are imposed, as discussed in the previous section. Risks from natural causes or accepted cultural actions are basically involuntary, since direct benefits are seldom involved, but the risks are accepted as an act of God or an integral part of group behavior, respectively. These are generally more acceptable than risks imposed by others.

Conversely, risks imposed by fiat or unilateral action of others are resented most often and are probably least acceptable to risk agents. Even voluntary cases of the recipient obtaining the direct benefit are often distasteful to the risk agent whose free will is challenged.

* Fleeing as a response to a risk situation is usually unacceptable because it minimizes the risk by removing the risk agent from the source of risk and benefit. For example, to tell a coal miner to find less risky employment is unacceptable when no other work with equivalent remuneration is reasonably available to him in the immediate vicinity (hazard increment may be subtracted for equivalency).

Collective due process lies somewhere between the two extremes, with the degree of individual recourse determining the degree of acceptability. The more individual recourse (or free will), the more acceptable the imposition of risk, whether voluntary or involuntary.

8.2 DISCOUNTING IN TIME

Discounting in time occurs both in the past and in the future as shown by Linstone:*

> Apparently, discounting acts in both directions—future and past. A crisis about to happen or just experienced is discounted little, while events a generation in the future or in the past are discounted severely. The historical pattern of national wars suggests that a war is discounted completely in the space of about one generation.[1]

Such discounting may explain why young people take up smoking with a threat of lung cancer that will not develop for 20 to 30 years, or why young people feel that pensions for themselves are relatively unimportant. An example given by Linstone illustrates particularly the impact of discounting in time.†

> The striking impact of this discounting process on the part of individuals can be demonstrated by considering the world dynamics model that was created by Jay Forrester and Denis Meadows at MIT. . . . Consider an individual who was unconcerned about global pollution in 1950 and is still untroubled by current world population density and food supply. Normalizing these variable to 1950 and 1970, respectively, the Meadows "standard" run generated the pollution, population, and population/food production curves denoted in the illustration [Figure 8.1] by 0. Crises peak in 60 years for pollution, in 80 years for population, and in 90 years for food production. Application of a discount rate equal to, or greater than, 5 percent per year reduces the population and the pollution crises to minor significance—i.e., no dramatic worsening of the current situation is perceived by today's observer. It is not surprising, therefore, that cries of crises fall on deaf ears.[2]

The length of time one is subjected to a risk also seems to affect the valuation process in the form of discounting risk.[3]

A means for dealing with the problem is to assign to consequence values a discount rate, similar to financial discounting, over various periods of time. Note that it is the value of the consequence that is discounted, not the

* Reprinted by permission from reference 1. Copyright 1974 by the Institute of Electrical and Electronics Engineers, Inc.

† Reprinted by permission from reference 2. Copyright 1974 by the Institute of Electrical and Electronics Engineers, Inc.

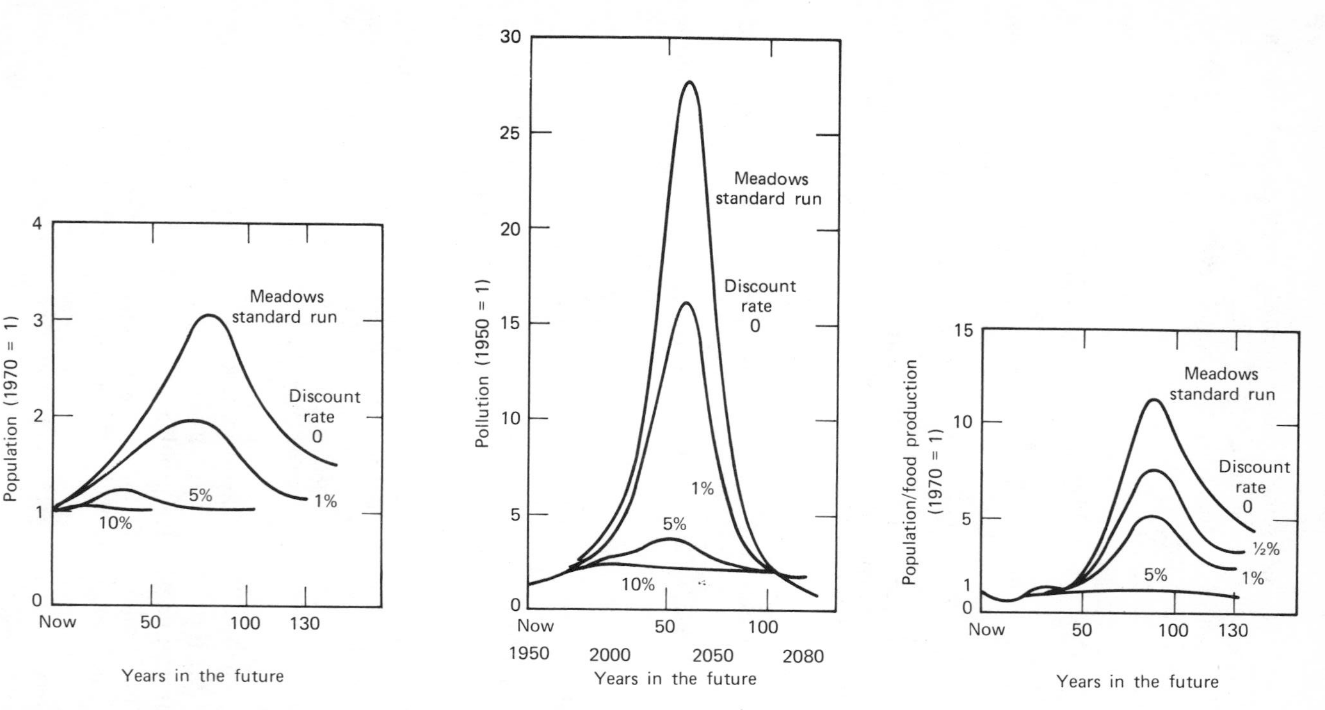

Figure 8.1. Discount phenomena. The striking impact of the discounting phenomenon on crisis perception is demonstrated by these curves, where future crises in population density, pollution, and food production are perceived as far less significant when discount rates as low as 5% are applied. The "0" curves are based on Figure 5 of *The Limits to Growth* by D. L. Meadows et al., New York, Universe Books, 1972 (p. 124). From "Planning: Toy or Tool," by Harold A. Linstone, Reprinted by permission from *IEEE Spectrum* Vol. 11, No. 4, *April* 1974, p. 45. Copyright (1973) by The Institute of Electrical and Electronic Engineers, Inc.

probability of occurrence, which remains unchanged. Perhaps the determination of this time discount rate is best accomplished by examination of societal behavior in terms of revealed preferences in the form of an "effective discount rate." However a number of factors can alter the effective discount rate.

Risks with direct benefits are discounted more heavily than risks with no direct benefits. As a result, discount rates for voluntary and involuntary risks differ, based on equity and also on whether they are endogenously or exogenously imposed.

The discount rate may be different for identified risk agents as opposed to statistical risks (see next section); more important, the actions taken by a risk agent may cause risks to his offspring and progeny, or to future generations of society. The discount rate may also be affected by whether the risks imposed are reversible or irreversible over the varying time periods.

The determination of effective discount rates is quite difficult because of the factors that affect the time discounting of risks:

- Voluntary vs. involuntary risk.
- Identifiable or statistical risk agent.
- Risk to risk agent, or progeny.
- Reversible and irreversible risks.

Such discount rates are measures of existing behavior and may not be directly applied, like economic discount rates.

In fact, there is some argument that one's values at the time of projection may change radically at an unspecified time in the future. For example, a childless couple may change their view of risks to progeny if they become parents. Furthermore, the world and society is not as "surprise free" as most forecasters would like to assume or hope. However people do make decisions involving future risks, and the effective discount rate is an attempt to measure what is done in coping, not what will occur.

8.3 SPATIAL DISTRIBUTION AND DISCOUNTING OF RISKS

Spatial distribution of risk involves the spreading of risk from individuals to others, either in society as a whole or to designated groups. Spatial distribution involves discounting of risk in three forms: (1) geographical distribution, (2) identification of risk agents, and (3) risk spreading. All these forms overlap to some extent, however; thus they are not necessarily independent in their effects.

8.3.1 Geographical Distribution of Risk

Spatial distribution and the discounting of risks can be distributed geographically as indicated by Linstone:

> Furthermore, discounting occurs in the space as in the time dimension. Most individuals are more concerned with events in their physical neighborhood than those occurring far away. Unfortunately, this very human space-time discounting phenomenon is poorly understood and constitutes a major reason for the ineffectiveness of long-range planning activities generally.[2]

Such distribution refers to geographic distance, but it cannot be completely separated from the problem of identification of risk agents. Furthermore, the ability of mass media and communications technology to effectively "shorten" distances and to identify particular risk agents has a major impact on the magnitude of spatial risk discounting. Possible forms for a spatial discounting function are given in Figure 8.2 for the purpose of illus-

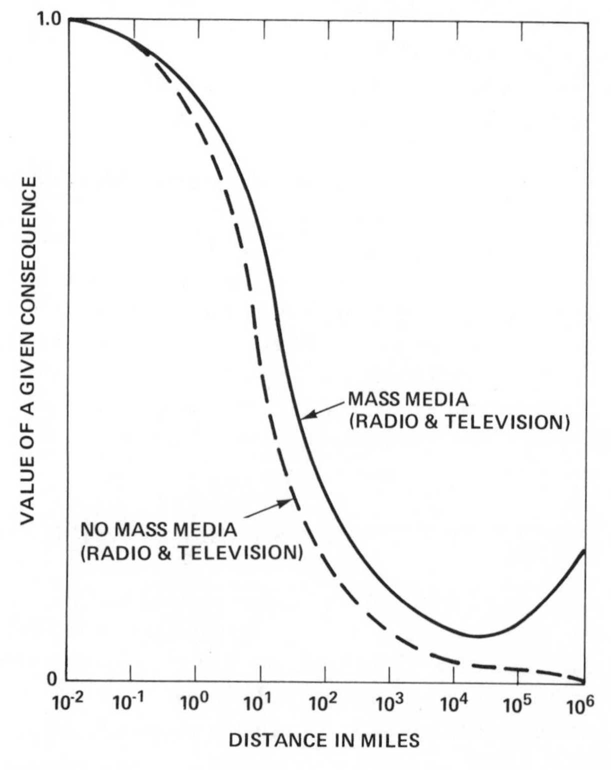

Figure 8.2. Some possible forms for spatial discounting functions.

trating the relationship, not to make any quantitative judgments. The dashed curve shows a slow decrease in the value of consequence as distance increases in one's immediate neighborhood. As the limits of the community are reached, the value falls off sharply, with a point of inflection at the effective community limit. Outside the community, value continues to fall off to some minimum level. This case includes newspapers and other means of providing neighborhood and community information, but no instant-coverage mass media such as radio and television. The solid curve (mass media present) widens the scope of interest, and value does not fall off as fast. An interesting phenomena at distances exceeding 11,000 miles involves astronauts who are specifically identifiable. As evidenced by the Apollo XIII abortive moon mission, the mass media, the special situation in space, and the identification of individuals had significant impact on society as a risk taker. The reversing of the direction of the curve illustrates this condition.

8.3.2 Identifiable Versus Statistical Risk Takers

If an individual or a group can be identified as the bearer of risk as opposed to statistical populations at risk, society tends to value this identifiable risk with increased concern, assuming equal probabilities and identical consequences, over statistical risk. For example, until the Coal Mine Health and Safety Act (1969) was given teeth in the form of the Mine Enforcement and Safety Administration (1973) many mine companies made little investment to protect miners as a group; but whenever a mine disaster occurs, millions are spent to rescue trapped miners, dead or alive.

If the risk taker expects to experience a consequence directly as an individual, he generally attributes a higher value to the consequence (positive or negative) than if he is only one of a group of people for which only one or a small number will experience the consequence.

In "The Evaluation of Life-Saving: A Survey,"[4] Linnerooth indicates that Thedie and Abraham were the first to distinguish an "identifiable" death from a "statistical" death.

> It is impossible to weight in the balance certain deaths and probable deaths, even if the latter are in greater numbers. The assessment we are going to make calls for anonymity and therefore, we can only deal with small risks, minor probabilities.[5]

Linnerooth[4] goes further:

> The more recent literature on the subject has also recognized this distinction. "Interest will not be centered in knowing the integral worth of every 'sparrow' but in knowing the value and cost of putting nets under 'falling sparrows'" (Carlson,

1963).[6] "It is not the worth of human life I shall discuss, but of 'life-saving,' of preventing death, and it is not a particular death, but a statistical death" (Shelling, 1967).[7] "It is never the case that a specific person, or a number of specific persons, can be designated in advance as being those who are certain to be killed if a particular project is undertaken" (Mishan, 1971).[8]

The difference between "individual" and "statistical" risks is thus well known qualitatively. Yet although little has been done to quantify the extent of the difference that is implied, the general shape can easily be illustrated (Figure 8.3). The value of a consequence for a single identifiable individual is shown as unity. If the consequence is to happen to one of a group of people of size *n*, the consequence value rapidly decreases and is limited by the total population *N*. The steepness and values for such a function are not defined.

The spatial distribution and the identification problem are interdependent, and Figure 8.4 combines both concepts for purposes of illustration. Nine groupings of risk agents are shown, with increasing distance and disinterest from an individual risk agent extending to the right. The nine bars indicate ranges of variation in a qualitative manner. The value is

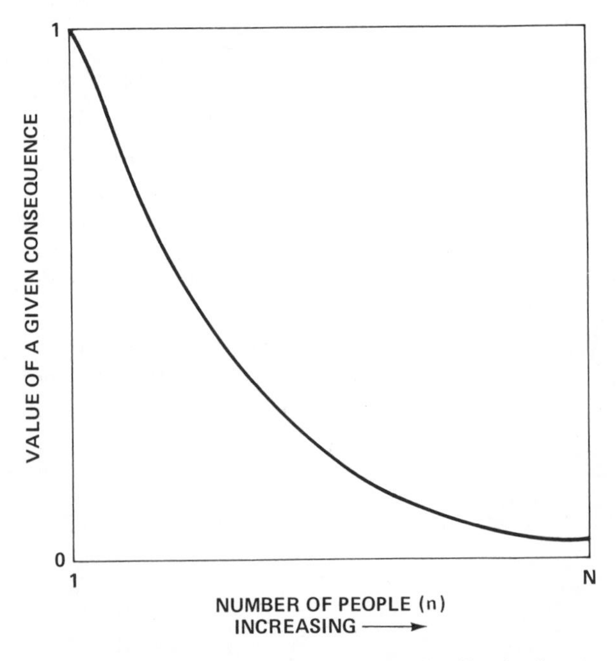

Figure 8.3. General shape of the risk agent identification function.

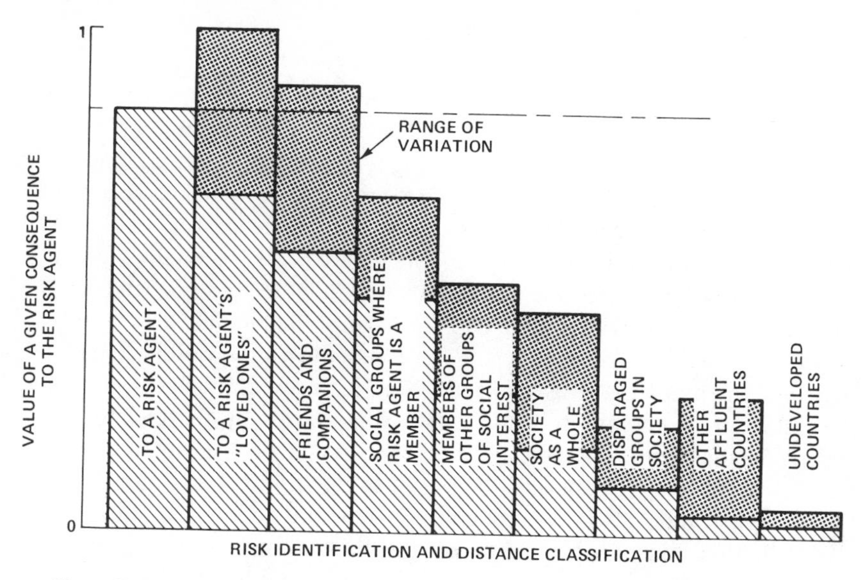

Figure 8.4. Possible effects of risk identification and distance on consequence value.

always to the risk agent, and when the risk is only to him, his value is set at unity. Risks to his "loved ones" (e.g., family) may range from less or greater than to himself. Though this depends on individual risk propensity, the shape of such a relationship tends to decrease as distance and inability to relate to people at risk increase.

8.3.3 Risk Spreading

Risk spreading involves the dependence of risk value on the relationship between magnitude of consequences and probability of occurrence. The uncertainty of a large consequence is often more disruptive to a risk aversive person than are many smaller more certain consequences. For example, a single $100 gamble may be more momentous to such a person than 10 wagers of $10 each. Although this concept is examined further in the discussion of probability, it is worth noting that the variation* of a sample decreases by the inverse of the size of sample for most probability distributions.

* The variance of the sampling distribution of a sample from most infinite statistical populations, which are distributed uniformly, normally, or are not particularly skewed, obey this principle.

8.4 CONTROLLABILITY OF RISKS

The ability to control risk has increased rapidly as technology has developed. Arguments for continued development of technology at high levels always involve factors such as a higher standard of living, longer life spans, and increased safety as well as growth for its own sake. Technology creates new risks but can also provide means to reduce existing risks as well as the new ones. In the absence of reliable data whether technology controls more risks than it creates is a matter of value judgment and depends on the orientation of the individual or group—pro- or antitechnology.

Fay[9] argues that there is a "right" to freedom from harmful technology. This fundamental "right" is pursued in the main by antitechnologists, and the ability of technology to reduce risk is supported by the technologists. The public has come increasingly to demand that both technological and social institutions operate at their respective limits to minimize risk. Technological innovations to reduce environmental contamination, product contamination and hazard, and accidental mishaps are commonplace. Old and new institutions are promoting occupational health and safety (OSHA) environmental protection (EPA), and consumer protection (consumer protection activities of the Federal Trade Commission) such that inequities in risk are ameliorated. These examples of federal efforts are matched at the state and local levels as well.

> There is an increasing demand for risk reduction. In part, it is generated from new forms of risk from increasingly complex technology. Our scientists and technologists have learned to compound substances unknown to nature. Against many of them, nature is virtually defenseless. There are no natural agents to attack and break them down. It is as if aborigines were suddenly attacked with machine-gun fire: their bows and arrows are of no avail. These substances, unknown to nature, owe their almost magical effectiveness precisely to nature's defenselessness—and that accounts for their dangerous ecological impact. It is only in the last twenty years or so that they have made their appearance in bulk. Because they have no natural enemies, they tend to accumulate, and the long-term consequences of this accumulation are in many cases known to be extremely dangerous, and in other cases totally unpredictable.*
>
> In other words, the changes of the last twenty-five years, both in the quantity and in the quality of man's industrial processes, have produced an entirely new situation—a situation resulting not from our failures but from we thought were our greatest successes. And this has come so suddenly that we hardly noticed the fact that we were very rapidly using up a certain kind of irreplaceable capital asset, namely the tolerance margins which benign nature always provides.[10]
>
> * The unknown risks and dangers associated with biological experimentation with DNA offer a good example [W.D.R.].

On the other hand, society is becoming increasingly serious in requiring that sophisticated technology be used to protect the population. Reduction of risk is in itself considered to be a benefit. A Mitre Corporation study[11] on a concept for categorizing benefits recognizes three general classes of benefit:

1. Materialistic (economic survival).
2. Physical protection and security.
3. Self-advancement (education, scientific, recreational, and aesthetic).

The physical protection and security class implies risk aversion. It is suggested that this benefit is as important as the other two.

In any case, the manner in which consequences are valued depends on the controllability of the consequence as well as the degree of control to be achieved. Clearly controllability also affects risk aversion.

Three aspects of controllability of risks are discussed here: (1) the perceived degree of control, (2) systemic control, and (3) crisis management.

8.4.1 Perceived Degree of Control

The perceived degree of control (as opposed to the "real degree of control") to avoid a risk consequence by a valuing agent is a major factor in determining consequence value. For example, the driver of an automobile or the pilot of an airplane is more likely to discount the risk consequence value than is a passenger. The driver or pilot feels that he is empowered to avoid an accident (event) through his skill and control. As long as the driver has this perception of personal control, (in reality he may be a poor, accident-prone driver) he feels that the objective statistics of accidents per mile per person do not really apply to him. Thus the degree of controllability, whether real or perceived, must be considered to be a major factor in the nature of a consequence. This difference between objective risk level and subjective risk perception has been recognized in the literature,[12, 13] and discussed in Chapter 3.

8.4.1.1 Group Behavior

Even though the objective risk of an undertaking may be well fixed, the subjective perception of the risk dominates. The three groups identified by Kates and described in Chapter 4 exemplify the different perceptions of controllability. The "Worry Beads School" sees society's ability to cope as limited; for the "Tip of the Iceberg School," control is too little and too late, and the "Count the Bodies School" believes that risks can be coped with.

The means to bring perceived degrees of control closer to objective levels call for education and information transfers at the nontechnical level. The ability to present complex technological issues in simple, straightforward language without resort to half-truths is, indeed, limited and severely handicapping those who try to implement new technological innovations. Even when the ability to present a lucid case exists, the credibility of the presenting group enters into consideration. The discussion in Chapter 2 on the squawk potential and credibility address this problem directly.

8.4.1.2 Individual Behavior

When a person gets behind the wheel of an automobile, a tradeoff is made voluntarily (perhaps not consciously, or the actual decision was made when one decided to be a driver): the risk of having an accident is balanced against the benefits of low cost, rapid mobility. In 1973 there were 55,800 fatalities and 2 million disabling injuries in the United States,[14] a death rate of 27.9 per 100,000 population. About 12,000 of the fatalities involved pedestrians or pedal cycles where it may be presumed that the driver was not personally at risk.

Does the average driver accept this risk on his own, or are other processes involved for which the driver discounts this level of risk? Any answer to this question must involve the psychological response of drivers; and in the absence of experiments to attempt to identify such processes, the author can only make observations based on experience. At least two processes seem to exist.

First, spatial distribution is involved in the "it won't happen to me" syndrome. Automobile risks are regarded as mere statistics until someone the driver knows personally is involved in an accident or the driver witnesses a serious accident. In these cases, the risks are brought home to the risk taker, and for some time thereafter he tends to drive more cautiously than usual.

Second, the driver perceives that he has some control over the risks because he has the ability to drive skillfully and cautiously, and he perceives that his reflexes will help him avoid serious situations. The perception of such control may be unrealistic, however, since 67.1% of all fatal accidents and 79.9% of all accidents in 1973, involved improper driving of some type.[14] Many have had the experience, as a passenger, of automatically pressing a nonexistent brake pedal. A good driver seems to need to feel he is in control of the situation.

Thus the risk factor considered is the perceived degree of control of the risk taker, not the actual degree of control, which may be quite different from the subjective perception. For example, the risk taker may be a poor, accident-prone driver who will not admit his shortcomings, even to himself.

Most readers have personal experience with driving a car. Motorcycling, swimming, skiing, and other sports provide other examples. When traveling on a commercial airplane, however, one looks to others—namely, the pilot and crew—for special expertise to minimize hazards. In this case positive systemic control is sought, not personal control.

8.4.2 Systemic Control in Risk Reduction

A society concerned about exposure to risks from new or ongoing activities of man or from natural causes, can act to reduce risk systemically. For example, the commercial airline system has posted a continual decrease in accidents and death rates, specifically and overall, and on a passenger per mile basis over the years. Flood control projects have saved many lives from naturally caused flood conditions. Though equipped to use his technology to consciously assure better safety goals, man does not always choose to make use of this capability. Society tends to accept certain risks from systems where vigilant attention to safety is provided in a real sense, as opposed to systems where safety is relatively uncontrolled. One method of measuring the degree of control is to see how well systems have been controlled over time in the form of learning curves, or at least the general trend in system safety.

Figure 8.5 plots the historic records of the U.S. aircraft industry from 1953 to 1973. The number of deaths from catastrophic accidents* per aircraft mile per year is given for a 21-year period in Figure 8.5*a*, and the total number of deaths from catastrophic accidents per year for the same period is shown in Figure 8.5*b*. Both sets of data show "learning curves" indicating that safety in commercial aircraft is influenced by a systemic approach to control. The efforts of the Federal Aviation Administration and the National Transportation Safety Board are effective in increasing safety.

Catastrophic accidents occurring in buildings and structures reveal a different pattern. Figure 8.6 shows the number of deaths per year for the same period as the aircraft industry. The trend line is the reverse of a "learning curve" and indicates that systematically the situation in preventing catastrophic accidents is uncontrolled, at least to the extent that other systems, such as the aircraft industry, are controlled.

* An accident is considered catastrophic if it meets one or more of the following criteria: 10 or more fatalities, 30 or more injuries, $3 million in damages.

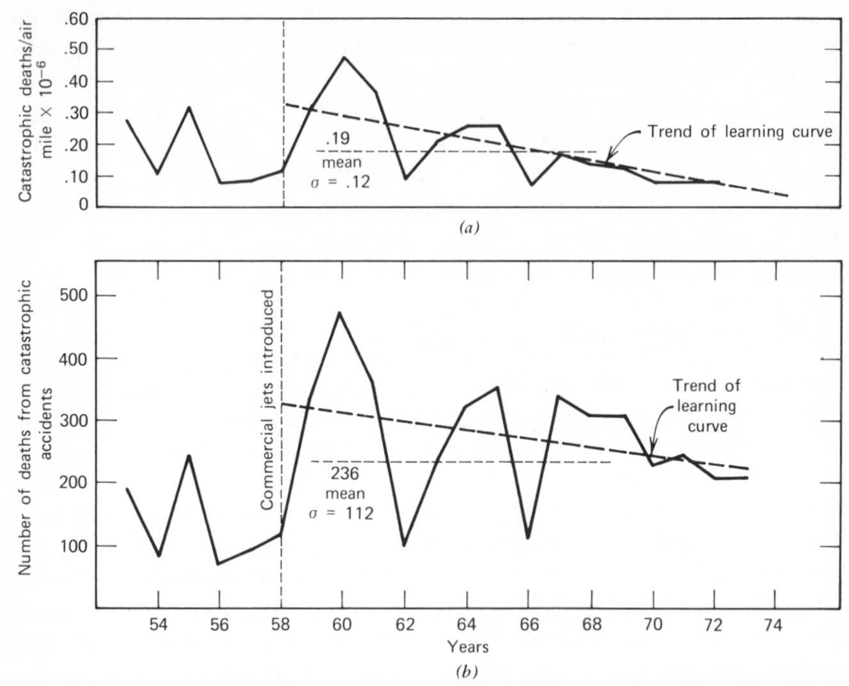

Figure 8.5. Learning curves for catastrophic accident deaths for U.S. commercial aircraft, (jet-prop), 1957–1973. (*a*) Deaths per aircraft mile. (*b*) total deaths per year. *Source:* Data from "The Consequences and Frequency of Selected Man-Oriented Accident Events," CONSAD Research Corp, EPA Contract Report No. 68-01-0492.

8.4.2.1 Systemic Control of Risk

More formally, systemic control of risk requires a plan and its implementation to control risks, which involves the following:

- A philosophy of controlling and minimizing risk as a major emphasis in the design and operation of the system involved.
- A means for regulation of the total system to assure maximum safety.
- A system design that includes the following:

    Quality assurance
    Redundancy for critical systems
    Training, licensing, and certification of personnel
    Inspection of equipment and operations
    Licensing and registration of operation
    Ongoing review of system performance to meet goals
    Enforcement and auditing system

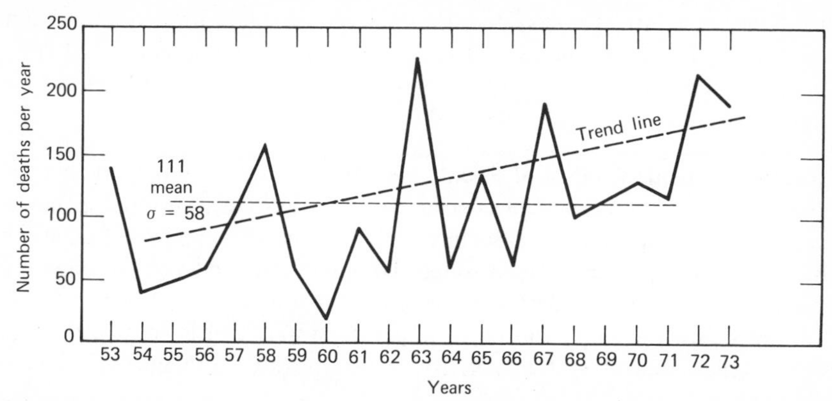

Figure 8.6. Absence of "learning curve" for catastrophic accident deaths in U.S. buildings and structures, 1953–1973. *Source:* Data from "The Consequences and Frequency of Selected Man-Oriented Accident Events." CONSAD Research Corporation, EPA Contract Report No. 68-01-0492, pp. 137–139.

For the purpose of definition, it is convenient to identify three classes of risk controllability on a systemic basis. These classes cover both man-oriented and natural risk systems.

Positive Systemic Control of Risk. For positive systemic control to be effective, over time as a system grows and expands, risks must increase at a proportionately lower rate, to ensure either that the absolute risk is decreasing over time or that the risk per unit of measure of system performance is decreasing over time (i.e., the system exhibits a "learning curve" for risk reduction over time). For example, commercial aircraft travel exhibits a decreasing accidental death rate on an absolute basis over time as well as a decrease of accidental deaths per million passenger miles flown. Based on the criteria used, positive systemic control may be either absolute or relative, depending on the criterion selected.* Absolute positive systemic control is generally more desirable than relative positive systemic control.

Systems that are designed with positive systemic control as a goal include aircraft travel, nuclear energy, space systems, and weapons and defense systems.

Level Systemic Control of Risk. Level systemic risk control implies that risks increase over time no faster than the system's rate of expansion, either

* Absolute control is the case of risks decreasing on a total industry basis regardless of the growth of the industry. Total deaths per year attributable to the aircraft industry is an absolute criterion. Relative control relates risk to an activity index. Fatalities per passenger mile flown is therefore a relative criterion. On this basis Figures 8.5*a* and 8.5*b* represent relative and absolute learning curves, respectively.

absolutely or relatively. A system that exhibits absolute level systemic control may at the same time exhibit relative positive systemic control of risks. In general, level systemic control involves systems whose risk behavior is characterized by a steady level of risk over time.

Negative Systemic Control of Risk. Negative systemic control involves the absence of a systemic control concept and/or a system whose risk behavior is characterized by an increase in risk over time (Figure 8.6). Essentially this condition represents uncontrolled risk on a systemic level, although it is possible that increased safety at a lower level is sought.

In all cases, the units chosen to illustrate systemic risk control must be realistic measures of system activity in that positive influence is taken to obtain safety.

Systemic control may be either demonstrated or proposed. Demonstrated systemic control requires that the system exist and that there be sufficient historic data to empirically demonstrate the level of control as in Figure 8.5. If systemic control is sought and is proposed, this is a desirable and necessary step toward demonstrated control.

8.4.2.2 Other Types of Control

Other forms of control involve less than the total system. These systems have many of the design features found in systemic design, but the overall approach is such that some important design features are lacking as well as the basic philosophic approach. For example, in the construction of buildings a variety of building codes specify construction design to meet strict fire regulations; but immediately on occupancy doors are blocked and safety systems not maintained or inspected; or the reverse may be true (i.e., design is bad but inspection strong). In other words, there is no total "systemic" approach to minimizing risk. Such control is termed, "specific control features."

Three types of specific control are noteworthy.

Control Through Specific Design Features. Safety is not achieved through systemic control, but through specific design features that help to assure safety in various parts of a system. These features are usually applied to parts of the system that have high probability of risk and/or large consequences. The installation of safety features in automobiles, the use of double-wall tankers for shipment of liquefied natural gas, and the use of dikes around inflammable liquid storage tanks are examples of specific design features to reduce risk.

Control by Inspection and Regulation. Positive control is to be achieved, not through specific designs but through operational systems that attempt to remove or minimize high risk situations primarily through the use of feedback of information for control decisions. The regulatory system for recall of defective autos for safety violations, inspection of buildings for code violations, and periodic state auto inspections are examples of proposed positive control by inspection and regulatory functions. Regardless of whether coupled with specific design features, control by inspection and regulation is an ongoing means to attempt to reduce risk in high risk situations.

Risk Management System. The National Aeronautics and Space Administration uses the term "risk management system" to combine control through specific design features and control by feedback mechanism involving inspection and regulation.

The Risk Management System (RMS) is a technique that uses systems engineering principles, operating personnel feedback, and management information and control systems to identify, evaluate, and reduce or control risks. Risks are identified by using engineering analyses and tests, and are evaluated by using the proper risk analysis technique. Through a system of management reviews and checklists, management is able to decide to eliminate or accept risks.

This technique is particularly effective in improving the management of risks in large, complex high-energy facilities. These improvements are needed for increased cooperation among industry, regulatory agencies, and the public.[15]

Such a system does control risk, but not at the systemic level itself. An RMS system may be an important ingredient in achieving systemic control, but it is only a tool for getting there.

8.4.2.3 Ranking of Control Factors

Table 8.3 arrays four sets of control factors as previously described in rank order of desirability. The four sets are

- Control approach
- Degree of control
- State of implementation
- Basis for control effectiveness

All four sets are needed to describe the effectiveness and resultant desirability of a control system. For example, the most desirable control approach would be demonstrated, absolute, positive systemic control, and the least desirable would be no control, the lowest ranking of all four sets. All other combinations fall in between. If cardinal values are assigned in each set, an interval scale can be obtained to provide a degree of controllability number for each combination. This is left for implementation in

Table 8.3 Control Factors and Ranking of Approaches to Control for Each Factor

Control Approach	Degree of Control	State of Implementation	Basis for Control Effectiveness
1. Systemic control	1. Positive	1. Demonstrated	1. Absolute
2. Risk management system			
3. Specific design features	2. Level	2. Proposed	2. Relative
4. Inspection and regulation	3. Negative		
5. No control scheme	4. Uncontrolled	3. No action	3. None

Part D, since a variety of value judgments are involved in assigning cardinal values.

8.4.3 Crisis Management

Crisis management is a term that has at least two meanings.

First, it can apply to a technique of management that pays little attention to problems until they reach crisis proportion, assuming that problems that do not create crises will go away. It is an extreme form of management by exception; there is little planning or little opportunity for planning, and action occurs only after a crisis. As discussed under methods of formal evaluation in Chapter 4, the activities of the legislative branch of government exemplify crisis management.

Second, crisis management can apply to means to alleviate man-made and natural catastrophes by managing available resources to minimize the consequences of such events after they occur. It is in the latter sense that crisis management is of interest in risk assessment. Kupperman, Wilcox, and Smith[16] have treated the second type of crisis management with considerable insight. Their discussion is reproduced in part in Appendix C for the reader's convenience.

Crisis management as discussed by Kupperman et al. has many of the same types of problem associated with risk management and assessment, at least in concept. More important, the degree to which a crisis management system exists, operates, and is successful directly affects the manner in which people value the consequences of crises. This is particularly true at the catastrophic risk level, since the existence of a crisis management system influences "group resiliency" directly and positively. It improves a group or community's ability to deal with catastrophic situations. An event that might overtax a community's ability to restore itself is made more palatable by spreading the risk across a broader base. Thus control of crises through crisis management techniques can reduce the value of catastrophic risk consequences.

REFERENCES

1. Harold A Linstone, "Planning: Toy or Tool?" *IEEE Spectrum*, Vol. 11, No. 4, April 1974, pp. 42–49. Copyright 1974 by The Institute of Electrical and Electronics Engineers, Inc.
2. *Ibid.*, p. 43.
3. Glenda Y. Nogami and Siegfried Streufort, "Time Effects on Perceived Risk Taking," Purdue University Technical Report No. 11, Lafayette, Ind., July 1973.
4. Joanne Linnerooth, "The Evaluation of Life-Saving: A Survey," RR-75. Laxenburg, Austria: International Institute of Applied Systems Analyses, March 1975, p. 24.
5. J. Thedie and C. Abraham, "Economic Aspects of Road Accidents," *Traffic Engineering and Control*, Vol. 2, No. 10, February 1961, p. 3.
6. J. W., Carlson, "Evaluation of Life Saving," Unpublished Ph.D. dissertation, Harvard University, April 1963. As quoted by Linnerooth.[4]
7. T. Shelling, "The Life You Save May Be Your Own," *Problems in Public Expenditure Analysis*, Samuel B. Chase (Ed.). Washington, D.C.: The Brookings Institution, (1968), p. 127.
8. E. D. Mishan, "Evaluation of Life and Limb: A Theoretical Approach," *Journal of Political Economy*, July–August 1971, p. 639.
9. James A. Fay, at the Engineering Foundation Conference on Risk Benefit Methodology and Application, Asilomar, Calif., September 21, 1975.
10. E. F. Schumacher, *Small Is Beautiful*. New York: Harper Colophon Books, Harper & Row, 1973, p. 17.
11. J. Watson, R. Kuehnel, and J. Golden, "A Preliminary Study of A Concept for Categorizing Benefits," MTR-6569. McLean, Va.: The MITRE Corporation, January 1974, under Environmental Protection Agency Contract 68-01-0490.
12. Siegfried Streufort and Eugene A. Taylor, "Objective Risk Levels and Subjective Risk Perception." Purdue University Technical Report No. 40, Lafayette, Ind., August 1971.
13. H. D. Van der Meer, "Decision Making: The Influence of Probability Preference, Variance Preferences, and Expected Value on Strategy in Gambling," *Acta Psychologica*, Vol. 21, 1963, pp. 231–259.
14. *Accident Facts*. Chicago: National Safety Council, 1973.
15. Letter from J. P. Claybourne, National Aeronautics and Space Administration (John F. Kennedy Space Center, Kennedy Space Center, Fl.), regarding a workshop on RMS.
16. Robert H. Kupperman, Richard H. Wilcox, and Harvey A. Smith, "Crisis Management: Some Opportunities," *Science*, Vol. 187, February 7, 1975, pp. 404–410.

9

Factors Involving the Nature of Consequences

The factors involving types of consequence discussed in the last chapter addressed the manner in which any consequence's value would be altered by the means in which the consequences are imposed, how they occur in time or space, and how they may be controlled. Here we consider the form of the consequence itself as well as the manner in which consequences are valued.

The value of a consequence is affected by the nature of the consequence, which can refer to the possible outcome (premature death, injury or illness, financial gain or loss, etc.) the magnitude of a consequence, the manner in which different consequences are viewed in different cultures, and several special kinds of risk that warrant separate discussion.

9.1 HIERARCHY OF CONSEQUENCES

Table 4.2 presented a hierarchal arrangement of consequences and consequence values, based on a quasi-Maslovian scale. Each major category of need is dominant over those below it as long as that need remains unfulfilled. Once it is fulfilled, the lower level needs become dominant, in turn, although higher level needs can preempt lower level needs at any given time. The hierarchy continues through exhaustible resources (e.g., survival and security factors), physical security, belonging, egocentric needs, and self-actualization. For each class of risk consequence, a number of illustrative factors and variables are shown. At one end of this scale, the value of a consequence is related to life and health. At the other end, it is related to the quality of life. Economic values in terms of money (and, perhaps, power as related to money) are important parameters in the middle ranges of the hierarchy. Essentially, the major societal problem is a continuous tradeoff of health and safety, versus the quality of life, versus money and power. These are indeed difficult terms to define, especially since it is arguable that given sufficient money and power, a person can improve the quality of his life and to some extent lengthen life and protect health.

9.1.1 Motivation and Risk

Maslow argues that survival is the dominant need; but when that need is basically fulfilled, the needs representing next higher level, security, become dominant, and so on. The need for fulfillment is a major human driving force and is the basis of motivation.

When this concept is approached in the light of risk values, it seems to work in reverse. The risk is always a threat to one of the needs. If the threat is to a need already fulfilled and no longer dominant, the threatened need becomes dominant again, with the added complication that all other fulfilled needs on a higher level are also threatened. For example, a wealthy, satisfied, egocentric person will not worry so much about illness and death (except for purchase of insurance) that such fears dominate his life. However once he becomes ill or is threatened with premature death, his values revert to the more primitive.

9.1.2 Problems of Measurement of Intangible Consequences

Many of the consequences involved at either end of the hierarchy are intangible. Thus the problems of measuring such values must be addressed.

9.1.2.1 Limitation of Monetary Scales for Tangible Consequence Values

When the consequences of an event are tangible and monetary, a monetary value scale is a directly measurable scale and is expressed in physical terms. That is, the "beans" can be counted, weighed, and manipulated. The case of manipulation often becomes a value in itself for ease of accounting, budgeting, taxation, and so on. As a result, the "real" value scale is often hidden or ignored. As mentioned previously, there is a nonlinear utility of money that represents a more meaningful scale than dollars. This scale may be different for different people, but Swalm[1] has generalized that the utility of money diminishes over a range of 4 orders of magnitude—2 orders of magnitude on each side of a sum of money the valuing agent is used to dealing with. Thus if one is accustomed to dealing with a sum of $10,000, sums outside the range of $100. to $1,000,000 have less meaning and value. A millionaire's range might be sums of $10,000 to $100,000,000.

In other words, the utility of money is discounted by people when they deal with sums of money with which they are unaccustomed or are uncomfortable. A major question that does not seem to have been approached visibly is the following: how much are we as a society willing to pay for "order" in the form of precise record keeping systems, as opposed to using the utility of money to individuals as a more meaningful value scale? Put differently, are many of the decisions made for the convenience of "bean counters" as opposed to people for which the decisions are to serve?

Using the utility of money as a cardinal value scale does not always ensure better valuations of tangible consequence values. There are many cardinal scales of value in which monetary values are not only meaningless but also erroneous. For example, one can count the number of cars entering our national parks and, by reference to the National Parks Service budget, develop a cardinal index of "dollars spent per car served." Yet such an index would be invalid if the object was to eliminate cars from overcrowded parks, since the index would be acting in the wrong direction.

Economists* often argue that people will purchase or pay according to their preference for particular items, and that the choices among intangible, as well as tangible values, will be made. This may be true, but buying activity may be the "result" of people being forced to make a choice among limited alternatives with a gross, inaccurate, and possibly misleading, economic measure of value. The problem of establishing preferences is one of measuring results; the problem of determining values or risk consequence is one of determining causal parameters when possible.

9.1.2.2 Scales for Intangible Consequence Values

The use of monetary values for intangible values, such as the value of a life to an individual, is marginal at best. Lederberg makes a strong case in this area:

> By any rational argument, the health of any individual is a priceless good. This does not set its value at mathematical infinity, so much as to point out that it is incommensurable with so-called strictly pecuniary evaluations. . . . Pecuniary estimates are hardly to be taken seriously except to suggest the scale of a cost benefit analysis. Many citizens may feel that they value their health and their lives more highly than does the multitude; and they wish to maintain the voluntary option to strike different bargains in areas that exercise their particular anxieties. It is one thing to advertise the merits of a transaction; it is another to impose it willy-nilly on the whole population.[2]

Two questions must be addressed for intangible value scales. First, what kinds of scale should be used for what kinds of value, and second, how may such scales be used effectively? The hierarchy of risk consequences presented in Table 4.2 permits us to deal with the first question. For the second question, Table 9.1 lists the value scales that might appropriately be used, expressed as primary scales and derived scales for both individuals and collective groups. Primary scales measure the value directly and are

* F. Y. Edgeworth, Irving Fisher, Vilfredo Pareto, for example.

objective; derived scales are interpretive and subjective. The scales are illustrative, and no claim is made to their appropriateness, exhaustiveness, or importance.

At either end of the hierarchy (i.e., premature death or self-actualization), the value scale becomes more intangible than in the middle. Thus monetary scales may be quite appropriate in the middle of the hierarchy, becoming less appropriate as one goes up or down the hierarchy. In fact, the use of a derived monetary scale for an intangible primary scale often distorts the whole consequence valuation by assuming a precision and accuracy in assigning value that is not warranted.

The problems of measurement of tangible and intangible parameters are covered further in Part C.

9.2 VARIATION IN CULTURAL VALUES

Different cultural systems have different sets of values. The wealthy, industrialized societies value health, wealth, and enjoyment of leisure time highly. The poor, undeveloped societies are concerned primarily with survival. In affluent societies, life is held dear while the opposite holds in the poor culture.

Differences in culture arise from the "symbolic dimension of social action . . . art, religion, ideology, science, law, morality, common sense."[3] These in turn affect the manner in which a society or members of a society value consequences and evaluate risks. Thus risks acceptable in one society may be unacceptable in others for any and all of the dimensions cited.

The Maslovian hierarchy of needs provides some basis for estimating dominant societal needs and values in that primitive cultures have values oriented around survival whereas more affluent cultures tend to be oriented toward higher level needs. Of course, in any culture there is a degree of variation of affluence across the culture—through castes, concentration of wealth and power, rising middle classes, and so on.

> Social learning includes each individual's arriving at a set of cognitive discriminations and expectations (meanings) associated with them—a set of standards for discriminating and interpreting—that he attributes generally to others in order to predict and interpret their behavior. These attributions do not perfectly coincide with the attributions each of the others makes to the same collectivity of others, but as long as the variance among individual conceptions of the generalized other does not interfere with the ability of the several individuals to accomplish their purposes with and through one another, they have the sense of sharing a common set of understandings, a common set of standards and mutual expectations—that something anthropologists call a culture.[4]

Table 9.1 *Hierarchy of Risk Consequences and Some Illustrative Scales*

	Individuals		Collectives	
	Primary	Derived	Primary	Derived
1. Premature death (actual death)	Perceived value of avoidance of: Immediate death Time-discounted death Statistically discounted death	Measured value of all three How much one would give to avoid these in terms of dollars, activities, and change of lifestyle	Average number of years foregone Number of deaths per million per year Cause of death	Cost to society Cost to average individuals Dollars spent per death avoided (or years saved)
2. Illness and disability (pain and suffering, loss of faculties and social capability that may or may not lead to death)	Avoidance of severe, lasting pain and suffering. Dehabilitation Death coupled with above measures	Amount paid for correction Amount paid for prevention Time and motion lost Earnings foregone Social activities foregone	Number of illnesses and injuries Purity of genetic pool	Lost productivity Cost of preventive and corrective health care
3. Survival factors	Adequacy of diet, shelter, and comfort Crowding factors	Calories/day Amount per period Portion of income for sustenance People/mile	Avg. diet, shelter, etc., levels Number and % with inadequate diet, shelter, etc. Birthrate/death rate	Cost of adequate survival factors Cost of avoiding deficiencies Resultant ill health and death
4. Exhaustible resources	Unavailability to meet current and future needs Cost to meet needs vs. substitution	Price–demand relationships Substitutability	Scarcity of resources Dependence on imports Price–demand relationships Substitution elasticity	Cost of waste and conservation National offense vs. industry needs Import–export balance

5. Security	Perception of rights encroached Crime incidence as an individual Sterility of government	Measures of these	Civil rights actions Number of rights threatened (e.g., privacy, fair hearing) Crime rate Inflation rate	Public opinion polls Value of the dollar Public safety dollars spent
6. Belonging—love	Perception of group Family and society membership Loneliness	Measures of these	Divorce rate Marriage rate Changing social mores	Number and kinds of clubs and activities Membership statistics
7. Egocentric	Perception of wealth, power, rank, fame, etc. Material wealth Freedom as an individual Self-esteem	Earnings and assets Position	Number and kinds of people who have satisifed their personal goals	Suicide rate
8. Self-actualization	Perception of the many factors of quality of life Self-satisfaction	Measures of quality of life	Collective measures of quality of life Collective satisfaction	Collective indices of the quality of life

Thus although all cultures are heterogeneous, any individual culture can be considered to be a whole set of behavior patterns that characterize the culture itself. On this basis, one value system can be differentiated from another within reasonable definition. This has been one of the major objectives of anthropology.

The intent of this study is to promote the realization that different societies value and evaluate risks differently based on cultural exposure. The projection of one culture's value as a means for estimating another culture's risk behavior is not realistic. Risk evaluation across cultural boundaries is not the same as considering the range of values within a culture.

9.2.1 Cultural Differences Across Societal Boundaries

9.2.1.1 Economic Status as a Basis for Cultural Grouping

Herman Kahn[5] has made some specific projections of cultural separation based on economic groupings for the year 2000. He identifies six economic groupings: visibly postindustrial, early postindustrial, mass consumption, mature industrial, transitional, and preindustrial. Kahn emphasizes the growing disparity among cultural groups based on their economic status and the resources available to those groups to develop a technological economy to move up the ladder toward an industrial state.

9.2.1.2 Anthropological Groupings

A more conventional anthropological approach is taken by Velimirovic,[6] who compares small-scale traditional societies with modern technological societies. The bridge between traditional and modern is technological progress and the desirability of and pressure for new technological advances. Velimirovic begins by examining technology and technological progress, to understand its impact on cultural change.

Kranzberg and Pursell[7] define technology as "man's efforts to cope with his physical environment—both that provided by nature and that created by man's own technological deeds . . . and his attempts to subdue or control that environment by means of his imagination and ingenuity in the use of available resources . . .".

Lock's[8] definition of technology is applicable for modern society rather than for a small-scale society of the present or of the past. He says that ". . . the role of technology is taken to be the application of scientific knowledge for the generation of improvements in, and benefits from, the societal subsystems, individually and collectively."

Technology is not new in our age; it is not the product of the Industrial Revolution of the 18th and 19th centuries. Technology was with mankind from the beginning, from the first manifestation of man's wish to master nature, to potentiate or replace his muscle, and to use the resources of his environment. Stone scrapers and spear points of early man are as much technological implements as the complicated machinery of modern societies. However, not only tools and machines can be identified with technology; it can be said to be "*man's rational and ordered attempt to control nature . . .*" including any theoretical and organizational attempt in the pursuit of that control (Kranzberg et al.[7]).[6]

In addition to the desire to control one's environment through technologic progress, is the desire to minimize physical effort and to develop power and prestige. Once contacts are made in a primitive society, a positive attitude toward technological progress results. Changes due to such progress lead to changes in the social fabric of the culture.

A social system is held together by the interaction of a group of people as long as a common body of knowledge, beliefs, skills, and learned behaviour manifested in their religion, economic and, political practices, arts, values and attitudes, is maintained for a number of generations. As a social system is a functional whole, it follows that all components of it are interdependent. This means that an alteration of one component automatically influences other parts and thus brings about a feedback in any or all of the other components. Culture change means "any change of cultural circumstances as far as it influences the structure or the function of the respective social entity."[9]

Culture change might come from inside a group or from the outside.[8, 9] Each technological advance is interwoven with psychosocial transformations.[6]

Velimirovic indicates that traditional small-scale societies are more exposed to risks from their natural environment than are modern societies. A wide range of resilience mechanisms, such as religious beliefs and taboos, may be mobilized to protect social unity in the traditional society, which is threatened by natural hazards it lacks the knowledge or resources to avoid or control. Conversely, modern societies are more exposed to risks from man-made environment. The most vulnerable society is one in transition, since in shedding its traditional social structure it is unprotected culturally, yet unable to establish technological control of its environment to any major extent.

"Societal values depend on the cultural pattern of a group. Stronger consensus on values prevail in traditional societies."[6] Both types of society frequently derive prestige and power from technological progress and are willing to take risks for social gratification. Nevertheless, some "risks from major technological innovations may be of global proportion for extended time periods."[6] The awareness of these new man-made risks may well be reshaping modern societal value systems.

9.2.2 Cultural Differences with a Society

Kahn[10] has related concern with basic societal issues to income and wealth class in United States society. His preliminary analysis shows considerable insight, and the flavor of this approach is illustrated in Table 9.2.

Modern society has a diversity of value systems within it, but mass communications media has made different groups aware of, and perhaps tolerant of, other value systems. Groups with wide value differences but common goals often band together, as discussed in Chapter 4, generating the squawk factor. Munch[11] indicates that marginal elements in a group (i.e., members of a particular group who are not well integrated) can serve as an important medium of change, since they may have less desire to maintain traditional patterns and values. Outstanding and marginal persons are influential agents in risk acceptance.

9.3 COMMON VERSUS CATASTROPHIC RISKS

Although events that commonly occur in society are often uncommon to an individual, in a variety of unanticipated events the consequences are large enough and occur infrequently enough to be called catastrophic. Spatial distribution affects the degree to which an event is designated as catastrophic or common (routine). Thus 50,000 fatalities per year from automobile crashes is considered a variable but recurring situation, and a single accident accompanied by a fatality is called routine. Conversely, the crash of a jet aircraft with more than 100 fatalities is labeled a catastrophe.

9.3.1 Disproportionate Attention to Catastrophe

Society gives to catastrophic risks attention that is disproportionate to that accorded to all other risks and means of death to which the public is subject. Why do catastrophic events cause so much concern?

One answer may be society's methods of communication and news media. Large events receive considerable coverage because larger events evoke larger headlines, more coverage, and the most likely reader or viewer interest. These events sell papers and invite video and radio coverage. More basically, it is human nature to be fascinated by catastrophes beyond one's control when they happen to others. The normal initial human reaction seems to be "Thank Heaven, it didn't happen to me or my family or community!" The second reaction of concerned people is "Now, what can I do to help?" Others seem to enjoy the excitement associated with the event, while still others just exercise morbid curiosity.

Table 9.2 *A Very Preliminary Analysis of Current U.S. "Classes"*

Issue				Some Basic Issues	
Class Description	A Useful but Misleading Nomenclature	Approximate Percentage of Households	More than 95% Fall Within Income Range	"Class" Status	Attitude to Economic and Social Status
Old wealth	Upper upper (UU)	1	>20,000	Ascribed	Ostensibly indifferent
New wealth	Lower upper (LU)	4	>40,000	Achieved	Great pressure to rise to U
Modern progressive middle class	Upper middle (UM)	20	7000–99,000	Ostensibly indifferent—often mostly ascribed but some achieved	Moderate pressure to rise—until recently almost no fear of falling
Middle America (i.e., traditional and conservative middle and modern working)	Lower middle (LM)	50	2000–20,000	Mostly achieved	Some pressure to rise—great presure not to fall
Traditional working and modern coping poor (e.g., "risers")	Upper lower (UL)	20	1000–10,000	Achieved	Often insecure
"Culture of poverty," noncoping poor, hard core poor, many "mentally incompetent," etc.	Lower lower (LL)	5	<5000	Mostly ascribed but some achieved	Almost always insecure

Source: Herman Kahn, *Basic Public Policy Issues,* Vol. II, HI-DFCC-1-I, Hudson Institute, Croton-on-Hudson, N.Y., December 1969.

Catastrophic events become historic events of considerable magnitude because they are often well documented by official and unofficial records and reviews. People still refer to the Johnstown Flood, and the record of human heroics, failings, and tragedy still evoke interest. Perhaps these catastrophes provide the ultimate human test of facing violent death in a world theater. One reacts by asking "What could I do in such a situation?"

In any case, there seems to be a "boomerang" effect in communication about risk, as indicated by Denenberg et al.[12] This is a "psychological defense mechanism which protects an individual against excessive fear."[13] When a communication becomes psychologically unbearable, the receiver will minimize or ignore it. This tendency is particularly observable in insurance sales, where emphasis on avoidability is more effective than emphasis on the horror of consequences.

Another aspect is the sense of community concern against threat, mentioned briefly in Chapter 4. Most communities have built institutions to protect them against threat and to cope with catastrophes when they have occurred. As long as risk consequences are within the resources and capability of the community to respond effectively, they can be accommodated. If consequences become so large that the community cannot control or cope with the problem, the members of the community who perceive the risks will assign higher value to them.

Starr[14] calls this effect "group resilience." It is the ability for any group to perceive the limits to which the group can respond; outside the limits they are helpless. Thus the value of the consequence is a highly nonlinear function of the size of the consequence (threat), with a breakpoint at the coping limit.

9.3.2 Size of Catastrophes

When does a "large event" become a "catastrophe"? Any particular definition will be arbitrary.

For accounting purposes in this study, a societal risk is called catastrophic if it meets one or more of the following criteria: (*a*) 10 or more fatalities, (*b*) 30 or more injuries, and (*c*) 3 million dollars in property losses. These unquestionably arbitrary figures appear to give a reasonable delineation.

James R. Coleman has pointed out some particular difficulties in using such an arbitrary definition:

The first question I feel deserves having further comment, is that of the arbitrary definition of a catastrophic accident. It would seem that whether the definition were

1, 5, 10, or 100 deaths must have a great deal of bearing on the interpretation attached to the historical data. A detailed discussion of the effect of separating what is assuredly a continuum of events into two arbitrary categories based on X deaths, Y injuries, or Z total cost is necessary. . . .

The second question that I think must be discussed is one related to this arbitrary categorization or catastrophic risk versus ordinary risk and its application to the evaluation of overall systems.[15]

Coleman amplifies these remarks by comparing the risks of many small nuclear power plants to those of one large plant. The actual risk from the former may be larger than the one from the single plant, since large accidents are only one part of the risk. The perception of the accident as a greater risk may subject one to greater objective risk.

Both these questions are relevant; however in the second question Coleman confuses the difference between objective and subjective risk. The perceived catastrophic consequence value determines risk aversive action, not the objective measure of the consequence. Public acceptance of more smaller reactors may very well increase the objective risks of premature death, ill health, or property damage, but the perception of averted catastrophic risk may well out-value the objective concern.

A realistic tradeoff would compare alternatives at the objective level *and* the subjective level as well. If the objective and subjective balances differ substantially, the subjective will dominate unless the subjective perception can be changed to align it more closely with the objective. As a result of this process of education and enlightenment, the new subjective balance will again dominate, but it will more closely describe the objective situation. Such changes of perception are difficult to obtain, and in any case the "educating institution" must have utmost credibility to carry off the persuasion.

9.3.3 Armageddon as a Consequence

In the past 30 years the new technologies and quantitative jumps in the size of technology have created man-made threats to total human extinction, at least at the conceptual level. Previously, Armageddon was "in the hands of God"; now it is also in the hands of man, and there are doubts regarding mankind's ability to constrain the misuse of technologies that might lead to such catastrophe. The development of nuclear weapons and delivery systems in the 1960s has resulted in megaton overkill capability for the United States and the USSR, and comparable capability is being developed in other industrial countries. Is Armageddon to be had by the touch of a button or the action of a misguided few, or can it occur through an inadvertent incident or accident.

The first part of the question is sociopolitical, and a nuclear nonproliferation treaty and activities to revise and strengthen this instrument are examples of efforts to minimize the "touch of the button" catastrophe.

Defenses against terrorists and malcontents calls for the establishment of security systems. At the military level, as a rule, only the quality and cost of security systems are addressed. In the civilian sector, however, in undertakings such as the enforcement of safeguards for commercial plutonium shipments, the quality and cost of security systems must be balanced against the encroachment of civil rights. The implications of such systems are under study by the Nuclear Regulatory Commission.[16]

For the second part of the question the chance of inadvertent incidents is subject to a degree of control in the design of reliable systems to meet acceptable performance criteria. As an example, in the mid-1960s the design and installation of the ground electronic system of the Minuteman II missile delivery system, being developed by the air force in conjunction with Sylvania Electronic Systems, directly took into account the problem of inadvertent launch of armed missiles. The author was personally responsible for this analysis, called "Functional Fault Analysis," which determined from independent component reliability and failure information and the overall system design, the availability of systems to launch missiles when ordered to do so; fault location was identified automatically to the lowest replaceable component level, and the ability of the system to prevent inadvertent launch of missiles was determined. The air force had established an acceptable level of inadvertent launch protection, and the system had to be redesigned several times before this requirement was satisfied.

The important point is that in the name of national security the air force set an acceptable level of risk for an inadvertent launch and possible Armageddon. The numbers used may well be classified yet, however the range of 10^{-16} to 10^{-20} event per year can suggest how low such an achievable, acceptable level (in model analysis form) was implemented. The calculated chance of a meteor large enough to kill everybody on earth is about 10^{-13} event per year.[17]

It remains to be determined whether such levels can be set for activities not involving national defense, in such problems as

- Random genetic splitting and recombination of viruses.
- Modifications of the ionosphere from chloro-hydrocarbons and supersonic air travel.
- Greenhouse effect, from carbon dioxide as a result of industrial and societal technology utilization of fossil fuels.

9.4 NATIONAL DEFENSE AS SEPARATE VALUE SYSTEM

It is quite evident from acts of heroism and patriotism during wartime that value systems involving national defense considerations are quite different from normal societal values. Sacrifice for an ideal, such as freedom or a political or religious doctrine, occurs on a plane that is well above the norm. Higher levels of acceptable risk are realized, often without question.

As a result, values of a society at war differ from those of societies at peace. Even during peacetime, however, military operations in the name of national defense are carried on, and a different value system is used. Justifiable or not, military systems impose higher risk levels than other societal systems. While arguing the desirability of operating nuclear power plants with a very low probability of a high consequence accident, one accepts the low probability of a nuclear holocaust from the inadvertent launch of a nuclear-armed missile sitting in a standby mode.

The distinction between voluntary and involuntary risk is blurred for military systems. People directly involved may be involuntary draftees, following orders involuntarily, whereas others have volunteered for the benefit of a career or for patriotic reasons. Civilian casulties of a military catastrophe might even be considered voluntary risk takers, since it can be argued that national defense benefits all citizens. In any case, national defense value systems must be considered in a different light from normal societal value systems.

9.5 NATURAL VERSUS MAN-ORIGINATED RISKS

Man has always accepted certain kinds of risk from natural causes and has attributed these to acts of God or at least acted as if they were beyond his immediate control. This is true for both ordinary and catastrophic risks, but it is most evident for large disasters and catastrophes.

9.5.1 Catastrophic Risks

Society faces avoidable and unavoidable catastrophes. Those which are avoidable generally stem from two conditions: (1) society is able to do something to prevent the catastrophe; or (2) individuals, by changing their exposure to potential risks, can reduce the chances of being involved.

To differentiate between avoidable and unavoidable catastrophes, two classifications can be considered: natural and man-made disasters. Natural disasters can be avoided only by choosing to live in a place that has a lower exposure to such disasters; man-made disasters are avoidable both by direct control and by choice of potential exposure.

9.5.2 Protected and Exposed Populations

The extent to which protected or exposed populations are willing to reject or accept higher levels of catastrophic risk depends on many subjective and situational factors. However several generalizations can be made.

First, society will accept nature as an adversary at levels of risk much higher than for man and man-made activities.* Natural catastrophes are thought to be beyond control by man; therefore they are accepted. One may argue that the populations at risk could move to safer areas. All populations cannot live in the low seismic activity, low flood potential, low weather event areas, however, since these are limited in number. Often too, one natural threat cancels another; thus many East Coast areas of low seismic activity are plagued by high hurricane and flood potential. More important, the way of life of many people has been developed over many generations based on the assumption of natural risk for their livelihood. The fishermen living on the sea coast and the operators of tourist industries are examples. They have learned to live with the risks imposed by nature; indeed, their way of life may be centered on these and associated risks.

Second, many risks are assumed voluntarily with reasonable knowledge of the risk. Passengers on airplanes, autos, trains, and ships are usually aware at least qualitatively of the risks assumed. However the convenience, mobility, and savings of time associated with these modes of travel offset the additional risks in the mind of the risk agent. That is, as a passenger, the risk agent is getting the benefit as well as the risk. But consider the possibility of an airplane crashing into an apartment house near an airport runway: the residents of the building are assuming risks, however small, without directly receiving the benefits of air travel. Thus these risks are accepted involuntarily.† The inescapable conclusion is that various groups in society willingly and knowingly accept relatively high risks to obtain particular benefits if the risks are assumed voluntarily. Conversely, those on whom risks are imposed are unwilling to accept these risks if they are man-made and avoidable, and if the risk taker does not directly receive the benefits.

9.5.3 Ordinary Risks

Although the foregoing statements about catastrophic risks are also true for ordinary risks, the perceived degree of control is an additional factor to be

* This will be demonstrated later.

† Those who were living in the area before the runway was constructed are inequitably treated. Subsequently, others may move in voluntarily to take advantage of reduced property values and rents.

considered. Ordinary natural and man-originated risks often imply a perceived degree of control not found in catastrophes. A person might feel that he could survive a washout of a bridge by stopping his car to observe the washout before endangering himself, but he may not be sure of his ability to survive a dam break upstream.

In the previous chapter the perceived degree of control was discussed in relation to objective-subjective differences. Here the perceived degree of control is seen to diminish as the size of a consequence becomes very large. It is this relationship that partly explains why individuals expose themselves voluntarily to natural risks they might otherwise avoid.

9.6 KNOWLEDGE AS A RISK

9.6.1 Risk from Experiments

Chapter 3 stated that event-outcome space does not directly involve risk unless an exposure pathway exists. In a sense this is not strictly true, and the question of whether an experiment imposes risk is a complex issue as well as a timely one, given present concern with freedom of inquiry. The freedom of inquiry issue addresses the problem of whether it is ethical to conduct research along certain lines or whether such investigations should be made illegal. DeWitt Stetten, Jr.[18] argues that conducting research represents a freedom to learn, teach, or to inquire and that these freedoms are, in fact, abridged from time to time, subject to the test of a real and present danger.*

Among the lines of research against which voices have recently been raised are the following:

1. What are the genetic contributions to intelligence?
2. What kinds of experiments may properly be performed on consenting adults? Minors? Fetuses? Prisoners?
3. May one screen infants for a variety of genetic defects, some with known, others with currently unknown clinical consequences?
4. Under what circumstances may one tamper with the genetic process, as by the introduction of foreign genetic material into the genome?
5. When may one meddle with human conception and pregnancy as by artificial insemination, abortion, cloning, *in vitro* fertilization, or the use of surrogate mothers?

In arriving at considered judgments on these and a number of other problems, it is

* Reprinted by permission from reference 18. Copyright 1975 by the American Association for the Advancement of Science.

suggested that we treat freedom of inquiry as we have learned to treat freedom of speech—that is, agree to abstain when there is a real and present danger. By this test, the fact that the problem may be difficult, or that its solution may prove politically embarrassing or unpopular, is insufficient ground for invoking constraint. Indeed, a science that shies away from a line of inquiry merely because the result may be difficult to manage is in a sorry state.[18]

John Arents responds by stating that Stetten "perpetuates a common confusion between two nearly unrelated questions: (i) Is a certain experiment dangerous? (ii) Is knowledge of certain kinds dangerous or undesirable?"[19]

This separation is particularly pertinent because the risk involved from exposure has concentrated on the idea that a particular experiment implies danger, not the concept that knowledge gained in an experiment might be dangerous. If there are risks associated with knowledge, all experiments to gain knowledge may involve risk.

Exposure to knowledge, as opposed to exposure to the conduct of an experiment, always exists to some degree. The manner in which such risk is valued is a complex ethical question. It involves establishing whether the existence of a "tool" or the "use of a tool" determines risk. For example, a knife is a tool for cutting and other useful tasks, but it may also be used to injure.

9.6.2 Views on Knowledge Suppression

Much of the controversy involving new knowledge as a potentially dangerous "tool" addresses the freedom of inquiry question and whether or not knowledge should be suppressed. Stetten expressed one view when he suggested suppression for a "real and present danger." Arent disagrees vehemently:* ". . . Everyone committed to science must utterly reject and oppose the doctrine that ignorance is better than knowledge, self-deception better than intellectual honesty, faith better than thought."[19]

Another point of view is expressed by Siekevitz:†

I would like to take exception to Stetten's editorial, for I believe he has misconstrued the whole tenor of the arguments for and against the advisability of conducting various lines of scientific research. The furor is not over scientific research in certain political and social settings. Stetten lists "Lines of research against which voices have recently been raised," but most of these items involve not a "research

* Reprinted by permission from reference 19. Copyright 1975 by the American Association for the Advancement of Science.

† Reprinted by permission from reference 20. Copyright 1975 by the American Association for the Advancement of Science.

line" but a social and political question . . . and the arguments in most cases are against the social policies which allow for the experimentation rather than the caliber of the science.[20]

Siekevitz argues further that the scientists involved in experiments cannot make value judgments about the consequences of knowledge and its value alone:

Experiments in artificial insemination, abortion methods, cloning, and *in vitro* fertilization again take place in a society which has set standards and codes of behavior for its members, and any tampering with these must take in the views of all members of society, not just scientists. . . .

. . . In contradistinction to Stetten's concern for a possible limiting of the freedom of inquiry in scientific research, I believe that nothing would be more disastrous for scientific research than for us to set ourselves up, not only as the sole judges of the scientific results of our research, but also as sole judges of the effects of this research upon society as a whole. I do not question the "ethics and humanity of scientists"; what concerns me is the attitude of many scientists that there is nothing more important in the world than what they consider is the God-given right for them to do research, and that no question of morality, of ethics, of social or even political judgments should interfere with this right. In a democratic society, no man, scientist or nonscientist, should stand alone upon this pinnacle.[20]

Shannon extends Stetten's "real and present danger" to unintended and future danger.*

Stetten suggests the criterion of a "real and present danger" as a way of judging whether constraint on research is needed. This is a clear and extremely useful criterion. But a criterion which may be more important is that of unintended and future danger. Granted that this criterion is not as precise or easy to handle as the "real and present danger" one; nevertheless in many areas of research it will be the long-term effects that are the most problematic. We are no longer in a position to be as cavalier about the future as we once were. Our knowledge of the ecosphere has shown that simple actions may have far-reaching and wide-ranging consequences. . . .[21]

Stetten summarizes these:†

. . . Arents finds that knowledge gained through science must take precedence over ethical judgments. Siekevitz would place research into a preexisting political and social setting. Shannon suggests that future dangers should be considered along with

* Reprinted by permission from reference 21. Copyright 1975 by the American Association for the Advancement of Science.

† Reprinted by permission from reference 22. Copyright 1975 by the American Association for the Advancement of Science.

present ones in evaluating research proposals, a view to which I subscribe. I had intended "present" as an antonym of "absent," not of "future."[22]

New knowledge in itself can result in unwanted consequences in the present or future. The manner in which these consequences are valued involves value judgments made by many participants in political and social settings. According to Stetten, "It appears self-evident to me that all rulings and decisions by panels and committees, and indeed, by the entire electorate are mere summations of personal judgments."[22]

9.6.3 Information Valuing Problems

It seems that some will opt for control of knowledge, others will reject the idea of control of "freedom of inquiry," and others will bridge the gap by looking for acceptable criteria for control on scientific and societal levels. The "right or wrong" of such control will be argued for years to come, but one fundamental aspect of knowledge makes such ethical arguments specious. Knowledge is a strange commodity; when one gives it away, one neither diminishes his own store of knowledge nor is he able to retract that given away (in the sense that information once exposed can be purposefully "forgotten" or erased.). Thus control of knowledge can never completely suppress information, only delay its spread. The spread of resultant "half-truths" may even be worse than the "truth."* An editorial by Bernard D. Davis in *Science* cogently makes this point in connection with social determination and behavioral genetics.†

The fusion of evolutionary theory with genetics has yielded several profound insights into the nature of man. We now know that most traits are determined by interaction between genes and the environment, rather than by either acting independently. Moreover, the traditional view of race, as a set of stereotypes with minor variations, has been invalidated by the knowledge that races differ statistically and not typologically in their genetic composition. Finally, the rapid evolution of our species implies wide genetic diversity, with respect to behavioral as well as to morphological and biochemical traits.

Unfortunately, the idea of genetic diversity has encountered a good deal of resistance. Some egalitarians fear that its recognition will discourage efforts to eliminate social causes of educational failure, misery, and crime. Accordingly, they equate any attention to genetic factors in human behavior with the primitive biological determinism of early eugenicists and race supremacists. But they are setting up a false dichotomy, and their exclusive attention to environmental factors leads them to an equally false social determinism.

* Quotation marks indicate the uncertainty of the nature of truth.

† Reprinted by permission from reference 23. Copyright 1975 by the American Association for the Advancement of Science.

To be sure, in behavioral genetics premature conclusions are all too tempting, and they can be socially dangerous. Moreover, even sound knowledge in this field, as in any other, can be used badly. Accordingly, some would set up lines of defense against acquisition of the knowledge, rather than against its misuse. This suggestion has wide appeal, for the public is already suspicious of genetics. It recognizes that earlier, pseudoscientific extrapolations from genetics to society were used to rationalize racism, with tragic consequences; and it has developed much anxiety over the allegedly imminent prospects of genetic manipulation in man. . . .[23]

The roles of the scientific community, the public at large, the media, and regulatory agencies in determining the value and probability of consequences deriving from new knowledge are not yet clear. If control of experiments and knowledge is pursued, it may be regulatory (by government) or voluntary (by participating scientists and their peer groups). In either case Pandora's box will eventually be opened. In many cases the imposition of controls can be delayed until risks are determined and methods for lessening them found. However the control of information involved in delay also slows the ability to find solutions.

Valuing such risks calls for relatively new approaches to risk behavior, and it is premature to expect any precedents to provide historical perspective. The arguments are still at the ethical level.

REFERENCES

1. Ralph O. Swalm, "Utility Theory Insights into Risk Taking," *Harvard Business Review,* November–December 1966, pp. 123–135.
2. Joshua Lederberg, "Squaring an Infinite Circle," *Bulletin of the Atomic Scientist,* Vol. 27, No. 7, September 1971, pp. 44–45.
3. Clifford Boertz, *The Interpretation of Cultures, Selected Essays.* New York: Basic Books, 1973, p. 30.
4. Ward H. Goodenough, "On Cultural Theory," *Science,* November 1, 1974.
5. Herman Kahn, "Summary Briefing on 'Non-Military Forces for Change' in the Seventies and Eighties." Croton-on-the Hudson, N.Y.: Hudson Institute, February 11, 1970.
6. Helga Velimirovic, "An Anthropological View of Risk Phenomena," IIASA Research Memorandum RM-75-55. Laxemburg, Austria: International Institute for Applied Systems Analysis, November 1975.
7. M. Kransberg and W. Pursell Jr. (Eds.), "The Emergence of Modern Industrial Society, Earliest Times to 1900," in *Technology in Western Civilization,* Vol. I. New York: Oxford University Press, 1967.
8. G. S. H. Lock, "The Role of Technology Assessment in Northern Canada," *Technology Assessment,* Vol. 2, No. 4, 1975, pp. 253–257.
9. R. R. Behrendt, "Einführung in die Entwicklungssoziologie." Manuscript. Berlin, Free University, 1966.
10. Herman Kahn, *Basic Public Policy Issues,* Vol. II, HI-DFCC-1-I. Hudson Institute, Croton-on-Hudson, N.Y. December 1969.

11. P. A. Munch, "Agents and Media of Change in a Maritime Community: Tristan da Cunha." Presented at the 34th Annual Meeting of the Society for Applied Anthropology, Amsterdam, March 21, 1975.
12. Herbert S. Denenberg, Robert D. Eilers, G. Wright Hoffman, Chester A. Kline, Joseph J. Melone, and H. Wayne Snider, *Risk and Insurance.* Englewood Cliffs, N.J.: Prentice-Hall, 1964, p. 62.
13. James A. Fay at the Engineering Foundation Conference on Risk Benefit Methodology and Application, Asilomar, Calif., September 21, 1975.
14. Chauncey Starr at the conference cited in note 13.
15. James A. Coleman, Assistant Director, Division of Environmental Health, Minnesota Department of Health, private communication to the author, May 9, 1975.
16. U.S. Nuclear Regulatory Commission, "NRC Studies Need for Security Agency," Release No. 75-220, September 9, 1975.
17. Extrapolated from meteor data reported in "Reactor Safety Study: An Assessment of Accident Risk in U.S. Commercial Nuclear Power Plants," WASH-1400 (NUREG-75/014). Washington, D.C.: Nuclear Regulatory Commission, October 1975, Executive Summary Report, p. 2.
18. DeWitt Stetten, Jr., "Freedom of Inquiry," *Science,* Vol. 189, September 19, 1975, p. 753.
19. John Arents, LETTERS, *Science,* Vol. 190, October 24, 1975, p. 326.
20. Philip Siekevitz, LETTERS, *Science,* Vol. 190, October 24, 1975, p. 326.
21. Thomas A. Shannon, LETTERS, *Science,* Vol. 190, October 24, 1975, p. 327.
22. DeWitt Stetten, Jr., LETTERS, *Science,* Vol. 190, October 24, 1975, p. 328.
23. Bernard B. Davis, "Social Determinism and Behavior Genetics," *Science,* Vol. 189, No. 4209, September 26, 1975, p. 1049.

10

Other Factors Affecting Consequence Values

A variety of factors affect the valuation of consequences not included under types or nature of consequences. These factors involve the magnitude of the probability of occurrence, situational factors, and the propensity of groups and individuals for taking risks.

10.1 FACTORS INVOLVING THE MAGNITUDE OF PROBABILITY OF OCCURRENCE

The magnitude of the probability of occurrence of a consequence, especially when very low or very close to unity, has considerable influence on the manner in which one values a given consequence. Very low probability levels for some consequences are often ignored completely, indicating the existence of thresholds of concern. Thus there is an interdependence between probability and value.

10.1.1 Low Probability Levels and Thresholds

According to Mole,

> One important concept is that of negligible risk, a risk which in practice is ignored, even by those well informed about it. It is greater than zero but, because it is ignored, does not require statement in numerical terms. It may not be acceptable, however, in all circumstances.[1]

As the magnitude of a probability for a given consequence becomes smaller, a particular magnitude is reached below which the risk taker ignores the probability of occurrence. This "probability threshold" is affected by many parameters, including the nature and magnitude of the consequence and the avoidability and controllability of the risk.

The threshold is most likely lower for high values of consequences than for low values. As a result, one might expect a lower threshold value for a consequence involving death than for one involving the loss of an inconsequential sum of money.

The threshold is also higher for uncontrollable consequences than for controllable ones. If one looks at the risk of death from major catastrophes in society, the risk threshold from natural disasters is generally in the order of 1 to 2 orders of magnitude higher than for man-made catastrophes.*

In any case, identification of probability threshold levels for specific consequences can be used to determine whether an event and subsequent consequence can be considered to be credible or incredible.

10.1.2 Spatial Distribution of Risks and High Probability of Risks

The probability of occurrence of a consequence to an individual can be expressed to that individual or as the result of an event that may occur affecting a group of risk takers of which he is a member. Although the actual probabilities may be the same to a given risk taker, the valuation of identical consequences can be different.

As an example, consider the case of a game of Russian roulette played with a revolver, one chamber of six being loaded with a live cartridge. Consider three different situations, all with equal probability and consequence description to a single risk taker. First, the risk taker holds the gun to his head and pulls the trigger once; second, each member of a group takes one turn (one of the group will be killed, and it is assumed that the risk taker does not know the results of previous trials within the group), and the risk taker does not know the identity of the five other members of the group; and third, a situation the same as the second, but each risk taker knows the identity of the other five. Is the value assigned to the consequences the same in each situation? There may be some question about differences among consequence values between the first situation and the others, the difference in consequence value between the second and the third becomes acute if one includes his immediate family in the third case group.

This discounting in probability space seems to occur only when the probabilities of occurrence are high enough to influence a risk taker meaningfully. For example, assuming a fair roulette game (random and excluding "0" and "00"), one would be indifferent to betting on red or black (even money) or a number (36 to 1) for 36 numbers, if one could play as many times as he wanted at $1 per play. For a single bet, the risk taker has a distinct choice to make, depending on gambling instincts, his utility of money, and propensity for taking risk. However people play indiscriminately at numbers games and lotteries where probabilities vary enormously (at very

* The validity of this statement is shown quantitatively in a subsequent chapter.

low levels) from game to game. The larger a payoff in a lottery, the more people it seems to attract, even though the odds of winning may have decreased by orders of magnitude.* One seems to be indifferent to one chance in 50,000 or one chance in a million for these stakes.

10.2 SITUATIONAL FACTORS

The value for a given consequence can vary as a result of new situational conditions. Such situations occur on a dynamic basis and actions resulting from them can be identified.

10.2.1 Surprise and Dissonant Behavior

The world is not "surprise free" as many forecasters and systems analysts would like us to believe (for the convenience of their methods). Surprises always do occur; and when they involve unexpected consequences, the motivational process of "dissonance" is introduced.

It is suggested that dissonance is aroused whenever an event occurs that disconfirms a strong expectancy. If a person expects a particular event (consequence) and instead a different event (consequence) occurs, he will experience dissonance. He would judge (the different event) to be less pleasant than if he had no previous expectancy.[2] Thus an unexpected consequence may have a higher value than if it was anticipated.

10.2.2 Lifesaving Systems

In some situations individuals and groups voluntarily disregard certain consequence values in the short term in order to act to reduce consequences to others. This is particularly true with respect to lifesaving. The situation is always voluntary, but it may be triggered by several different motivational conditions.

First, if those to be saved are known to the rescuer and are "loved ones," the rescuer may well value their lives over his own, as is reflected in Chapter 8 in the discussion on spatial distribution. Figure 8.4 illustrates that the value of a consequence to others can exceed the value a risk agent assigns to the consequence occurring to him.

Second, voluntary lifesaving action is undertaken by those who receive direct remuneration for their risk exposure, either as special hazard pay for

* This seems to be the practice of the Maryland State Lottery Board in extending the payoffs from $50,000 for several winners by adding a $1,000,000 winner.

taking unusual risks in lifesaving, or as salary, for those who are professionally employed as "lifesavers," such as lifeguards, firemen, and rescue squad members. Some may take such jobs because they lack skill, training, or opportunity for alternate employment, yet varying degrees of skill, professionalism, job satisfaction, and job security are also involved. Many are attracted by the excitement of these jobs.

Third, there are those who are motivated by self-sacrifice and/or a desire to serve others. This is particularly true of missionaries and other service type groups. The basis for sacrifice may be to fill psychological needs for self-martyrdom, an honest desire to serve others, or devotion to causes.

Finally, there is the "hero" who is compensated by earning the respect of his peers. The implication here is risk taken for glory as opposed to motives involving service to others.

10.3 PROPENSITY FOR RISK TAKING

The propensity for risk taking for both individuals and groups of people must be examined in relation to both valuation of consequences and evaluation of risks. To understand the concept and illustrate its use, the following explanation uses risk acceptance considerations and is related more to the concept of risk evaluation.

10.3.1 Risk Acceptance

The level of risk acceptance (i.e., the level of risk for which a risk taker decides to just accept a risk as opposed to rejecting it) depends on the probability of occurrence and the nature and magnitude of the value of the consequences that can occur. This is qualitatively illustrated in Figure 10.1 for a single valuing agent in a general sense. The abscissa shows an evenly spaced rank scale of consequence value in terms of the gross indication of the hierarchy of risk consequences of Table 4.2. A logarithmic scale of probability of occurrence appears on the ordinate. Changes in the spacing in the abscissa scale will alter the specific shape of the curve, but not the general downward slope to the right hand. In other words, consequences not involving death or pain have higher acceptable levels of probability.

The acceptable probability of occurrence of a specific consequence value is designated as a "risk acceptance level." The profile of the acceptability of the probability of occurrence for all consequences involved in a situation is designated a "risk acceptance utility function."

Figure 10.1 illustrates two alternate risk acceptance utility functions. The top curve represents the risk acceptance utility function for a risk taker with a lower "propensity for risk acceptance" than the original valuing agent.

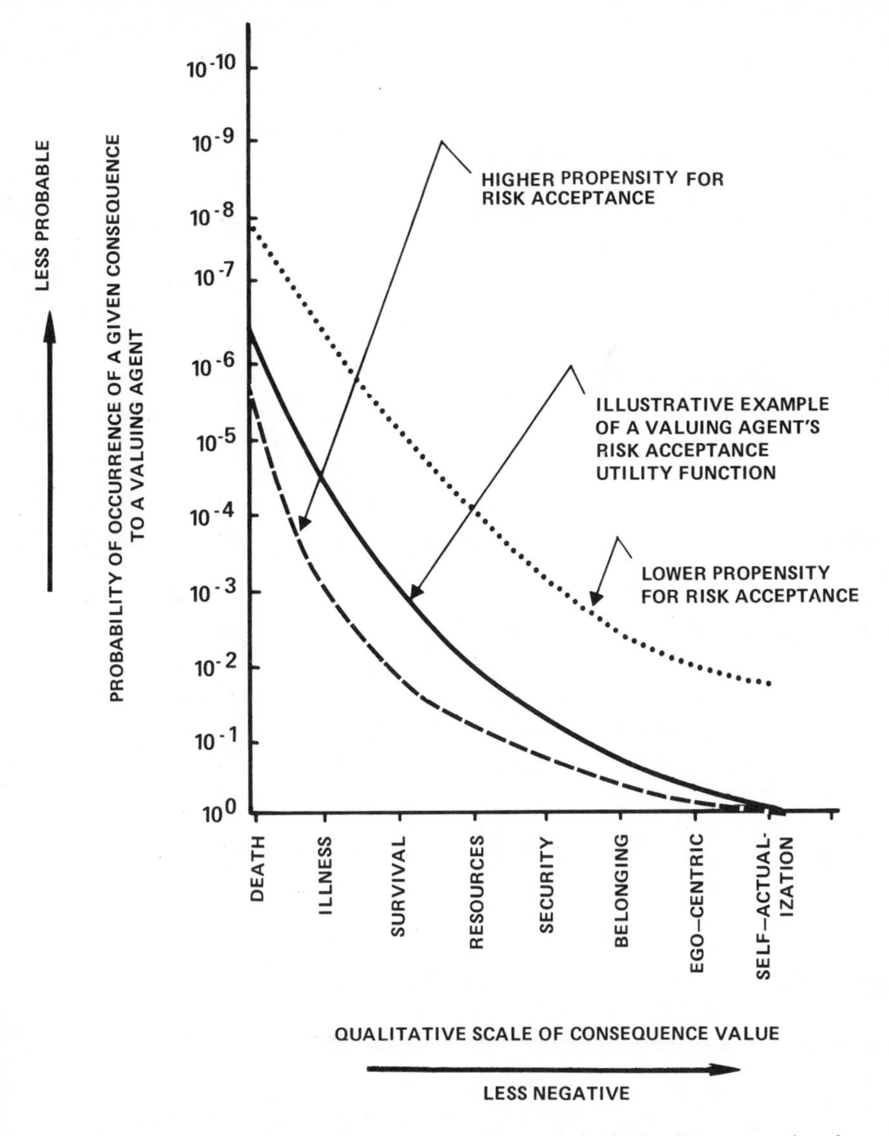

Figure 10.1. Risk acceptance utility functions for an individual valuing agent based on probability and value of consequences.

The "propensity for risk acceptance" is an individual subjective trait and is discussed subsequently in more detail. The bottom curve illustrates a higher propensity for risk acceptance; that is, the valuing agent is more likely to "take a chance" than the original valuing agent.

10.3.1.1 Risk Acceptance Notation

The following notation is used to denote risk acceptance. For a specified valuing agent, risk acceptance for a particular consequence C_{ij} is

$$R_{ij} = F_{ij}\,[\rho_i, \nu_{ij}] \tag{10.1}$$

where R_{ij} is the risk acceptance level and F_{ij} is a subjective operator. As such, R_{ij} is an acceptable probability of occurrence of the consequence specified. For a risk acceptance utility function R_j, we write

$$R_j = F_j\,[\rho_i, \nu_{ij}] \tag{10.2}$$

where F_j is a functional subjective operator on all consequences involved.

10.3.1.2 Uncertainty Considerations

The uncertainty in the assignment of probability can be expressed, at worst, as a uniform distribution around an error term b_i, or by a known distribution of the uncertainty, if this is available. The uncertainty in valuation can also be expressed in the same manner through use of an error term ϵ_{ij}, and the resultant uncertainty in risk acceptance probability is the form of an error term designated as r.

$$R_{ij} \pm r_{ij} = F_{ij}[\rho_{ij} \pm b_i, \nu_{ij} \pm \epsilon_{ij}] \tag{10.3}$$

$$R_j \pm r_j = F_j[\rho_i \pm b_i, \nu_{ij} \pm \epsilon_{ij}] \tag{10.4}$$

To minimize redundancy, discussion on risk acceptance functions includes risk acceptance levels unless specifically noted.

10.3.2 Individual and Group Propensity for Risks

Individual and group behaviors are different and result in different responses to consequence valuation. For example, group pressure on individuals in the group to conform to group standards or norms is a well-established pattern of group behavior.[3] The focus of interest is to identify the aspects of group behavior that are different from individual behavior in the propensity for risk acceptance.

One aspect of this difference can be addressed by examining the reaction of groups and individuals to pain, fear, and anxiety.*

* Reprinted by permission from reference 4. Copyright 1968 by Kaiser Aluminum and Chemical Corporation.

The fear of pain is probably more of a motivator in our daily lives than is pain itself. The fear of pain appears to be both instinctive and acquired. We may, for instance, flee a poisonous snake or a mad dog, even though we never actually experienced any pain from the apparition.

In fact, there does not even have to be a "real" object (cue) that could cause pain, in order for us to fear it and govern our responses and thus our motivated acts. There are people who fear other racial or ethnic groups, or even other ideologies and religions even though they never have experienced pain from them and perhaps, have never even had any sort of experience with them. Politics, as frequently practiced, is often the manipulation of the fear of pain.

However, pain, or the fear of pain, is not a sufficient explanation for the many anxieties and distresses that mark our daily lives. For one thing, infant animals and humans may react with fear even when they are not actually threatened, as in the case of a loud noise or the loss of support. For another, people who suffer from "congenital analgesia" (physiological inability to feel pain) nevertheless exhibit the same range of anxieties and distress symptoms that "normal" people do.

The older, simplistic view that anxiety states are derived from the fear of pain now appears too narrow a concept to account for much that we observe in human behavior. This is of special importance to the study of motivation because it is the creation of anxiety that makes punishment such a powerful motivator in our society.

These anxiety states do not necessarily have to be conscious. Vague senses of distress, despair, uneasiness, or of alienation may result from unconscious or subconscious tensions and forces. These may have an influence on both the direction and the intensity of the energy transaction that we have identified as a motivated act.

Anxiety behavior may, under some circumstances, be reverberatory, in that it begins with one system (a pain in my stomach) and spreads to others (maybe I have appendicitis) leading to motivated acts (putting on an ice pack, calling the doctor, lying down and curling up) that are inappropriate to the inner array of cues (it was only indigestion, after all).[4]

Irving Sarnoff then uses the concepts of fear and anxiety to explain differences in group and individual behavior:

The motives of fear and anxiety influence social affiliation behavior differently; the greater the fear aroused, the more the subjects choose to be together with others while they await actual contact with the fear-arousing object. Conversely, the greater the anxiety elicited the more the subjects choose to be alone while they await contact with the anxiety-arousing object.[5]

Thus group behavior is influenced by fear and individual behavior by anxiety. There are certainly many other factors involved.

Aspects of group behavior that affect the propensity for risk taking are discussed next. A subsequent chapter considers the individual propensity for risk.

10.4. GROUP PROPENSITY FOR RISK

Behavior of groups has been addressed from two major points of view that differ considerably. The first approach looks at the behavioral aspect of groups as distinct from the behavior of individuals. It emphasizes the interaction that gives rise to behavior patterns above those found in individuals and is based on the work of Homans, Maier, Bion, Burrow, Berne, and many others.[6-10] The second approach attempts to amalgamate individual preference structures into a group reference structure. It assumes a rational, economic set of individuals and uses analytical methods to synthesize group behavior. As addressed by Arrow[11] and Farris and Sage[12], the problem of group decisions is approached by way of decision theory structures.

10.4.1 Behavioral Characteristics of Groups

A group is an aggregation of individuals who associate for a common purpose.[13] Litterer[14] identifies four different types:

The Family. characterized by intense loyalty and satisfaction of members.

Friendship Cliques. Social interaction is the dominant purpose.

Task or Work Groups. Peers working toward accomplishing a common goal.

Command Group. Superior and subordinates working to achieve a common goal.

To this list the author must add one more type of group that is particularly important in risk behavior analysis.

Stress Reactive Group. A group with weak or strong affiliations, formed to cope with stress or threats to individuals.

This arises from Bion's basic assumptions that individuals under stress usually act in a manner that results in dependency, pairing, fight, or flight. They choose one of these means of release. Dependency can be aimed at other individuals or a group. Pairing involves joining to form a group for mutual protection or added strength to fight. This type of group is quite evident today in the large number of organizations set up to resist imposition of involuntary risk to the environment due to nuclear energy projects, urban renewal, highway encroachment, and so on. A group differs from an organization in that the group has only an informal structure, whereas an organization has both formal and informal structures.

An individual can be in several groups at the same time, and frequently these groups have overlapping members.

Groups have several principal features:

Goals or purposes. Group members share one or more goals, purposes, or objectives.[13]

Norms. Groups develop norms or informal roles of behavior that mold and quiet the behavior of members.[13]

Lifespan. A group has a lifespan related to its purpose.[13] When a group exists for any length of time, structure develops where individual members permanently fill different roles, including leaders and followers.[14]

Group Boundaries. As a result of roles and interaction within the group, informal boundaries develop and internal cliques can form. However, there is also an external boundary between the group and the outside world.[10] Conflicts arise across these boundaries.

Conflicts in groups arise from differences across the internal and external boundaries, competition for various roles, and forced conformity of individual members to group norms. Reaction to conflict produces different group behavior toward risk involving threats to the group. Before examining some specific cases, however, let us describe the analytical approaches to group behavior and decisions and consider their limitations.

10.4.2 Analytical Approaches to Group Decisions

Decision theory approaches to group decision making involve methods of amalgamating individual preference structures into group preference structures. The basic approach is encapsulated by Warfield.

> The idea of group decision making, such as goes on in legislative bodies and as practiced in earlier days in the town-hall meetings, is so firmly entrenched as a part of representative government that it comes as a surprise to many to find that there is no very satisfactory theory of precisely what is meant by a term such as . . . "the will of the majority."[15]

Thus decision theory approaches attempt to structure analytically such decisions. This method transcends classical means of expressing social preference (dictatorship, market mechanisms, voting, etc.). Kenneth J. Arrow as early as 1951 determined five conditions of desirability for methods of amalgamating individual preference structure into a group preference structure. Farris and Sage develop the concept of social welfare functions that must satisfy Arrow's set of five desirable conditions to be acceptable to society. Social welfare functions, such as majority rule, only hold for two alternatives; and some functions such as unanimity, are trivial. Utility combinations that supply weights to strength of preferences have been considered but provide little practical help when alternatives have multiple attributes.[12] Worth assessment models for individuals and groups hold some promise and Ferris and Sage[12] present a single group worth assessment model based on individual responses along with a complex structure that uses an iterative-interactive procedure for arriving at the group worth structure. Interactive computer terminals are suggested as a means of implementation and forcing a consensus. However, such methods have not been tested, and their theoretical and practical value has not been demonstrated.

Other approaches use "Pareto optimization" criteria and utility balancing among the group and utility exchange across the external group boundary.[16]

There are many who believe that such approaches are doomed before they begin.

> If decision theories offer a way of conceptualizing the relationship of values to behavior, they do so by incorporating certain facilitative assumptions about the nature of values. It is generally assumed, for example, that values are where you find them—preference hierarchies are assumed as given, and the problem of the genesis of values is simply avoided. The question of how values are translated into behavior is only one of the concerns of the motivational psychologist. The other question is that of genesis or development. How do evaluative dispositions come to be what they are? What are the antecedents for the development of human motives?[17]

This question of values and their nature is addressed in Part C. However the genesis and development of value and motivation still represent a major psychological question that has not been answered by decision methods.

10.4.3 Conflict Avoidance and Behavior Toward Risk Valuation

Conflicts arise from groups that affect individual behavior toward risk evaluation by virtue of group membership. Several particular cases are described here.

10.4.3.1 "Risky Shift"

The "risky shift" is a phenomenon recognized by behavioral psychologists[18, 19] and is cogently described by Trotter. It is directly associated with minimization of conflict among individuals involving conformance to group norms.

Conformity was once believed to be the major factor in making group decisions. People do tend to go along with the crowd. In recent years, however, social psychologists have reported on another phenomenon of "group-think." "Risky shift" is the term used to describe the tendency of certain groups to become more extreme or to take riskier positions in their judgments. The risky-shift phenomenon is explained by the pressure of an individual within a group to at least equal or preferably exceed the group average. The trend, therefore, is not to keep up with the Joneses but to surpass them. When this begins to happen within a group, says [Bertram H.] Raven [of UCLA], the effect is a movable or runaway norm that leads to more and more extreme positions.[19]

10.4.3.2 Cognitive Dissonance

Cognitive dissonance is a phenomenon that affects both individuals and groups; those who experience it attempt to minimize internal conflicts by rationalizing mistakes in decisions already made.

The theory of cognitive dissonance suggests that after making a choice (between cars, presidents, or whatever), people are subsequently motivated to believe that they have made the right choice and will commit themselves to that selection. Experiments have shown that people who purchase an automobile sometimes tend to seek out information (such as advertisements) that supports their choice. Even if the car is a lemon, there may be a tendency to overlook faults and continue to praise the car. By consciously or unconsciously disregarding contrary information, people avoid a mental or cognitive conflict.[20]

Cognitive dissonance described here is different from dissonant behavior previously discussed involving surprises. The implication here is more in line with Baruch Fischholt's concept of hindsight. He asks whether hindsight is better than foresight or just different, and his study concludes that outcome knowledge* is found to increase the perceived inevitability of the outcome reported:

Judges are largely unaware of the changes in their perceptions due to outcome knowledge. As a result, they believe that they and others had in foresight insights which they themselves only had as a result of outcome knowledge. Failure to appreciate the effects of outcome knowledge can seriously prejudice the evaluation of decisions made in the past and limit what is learned from experience.[21]

Since this effect occurs for individuals as well as for groups and involves the measurement of perceived behavior, we return to it in a subsequent section.

* As a result of hindsight.

10.4.3.3 Group Polarization

Whether informally or formally belonging to a group, "that when people are assessing their own opinions, they are more likely to be swayed by similar than by dissimilar people."[20]

In the former case they become members of a group informally by giving more credence to opinions of those with similar backgrounds. In the latter case they give more credence to opinions of their formal group. This allows a group bias to form that may ignore the outside environment to a great extent. It represents inbred thinking and reduction of internal conflict.

Many studies have demonstrated that in discussing risk-related issues, individuals tend to adjust their personal preferences in the direction of becoming more extreme in favoring a prevailing social norm (whether toward greater risk or greater caution). With such increased polarization comes increased confidence in one's opinion and increased commitment (or rigidity in the form of resistance to change in the opposite direction).[24-26] In a related group of studies, Castore and DeNinno[25, 26] have shown, using a series of extrapolations from Byrne's similarity-reinforcement-attraction model,[27] that individuals prefer to associate almost entirely with persons whose attitudes and world views are generally similar to their own. Thus the interest groups that develop on either side of an issue concerning technological innovation tend to be composed of like-minded persons with surprisingly low tolerance for opposing views. Continued interaction among such like-minded persons would generally move the group toward adopting a more extreme stand (pro or con, dependent on the prevalent within group norm).

*10.4.3.4 Group Homeostasis**

A group tends to be self-preserving and self-regulating. This concept leads to aspects well known in group behavior as fighting to preserve its entity from threat, group resistance to change, and so on. This is consistent with Berne's analysis of group survival discussed earlier. It means that threats to individual members that might in turn affect the group become group problems and may be augmented. In other words, the group may take on the problem of an individual if the problem itself may threaten group survival.

10.4.3.5 Group Identification

Individual members of a group tend to identify with a group and as a result are influenced by it.[28] Thus members of a group who are not directly

* See Glossary.

threatened may assume the consequences of a threat as their own. This is a corollary of homeostasis and involves acceptance of group norms.

10.4.3.6 Schism Reinforcement of Opposing Groups

Groups that have opposite views tend to polarize to extreme positions and reinforcement tends to widen the opposing views.[29, 30]

1. As such groups become more polarized in their attitudes, they will become more confident and rigid in their positions (consistent with the findings of Castore and Roberts).
2. Such groups will see fewer and fewer more moderate alternatives as acceptable courses of action.[27]
3. Once polarization has reached a moderately high level, interactions between groups at polar opposites on an issue tend only to reinforce the schism and, if anything, separate the protagonists further.

These are just a few of the more easily identifiable factors affecting the group propensity for risk. The overall propensity for risk involves individual behavior as well.

REFERENCES

1. R. H. Mole, "Accepting Risks for Other People," *Proceedings of the Royal Society's Medicine,* Vol. 69, February 1976, p. 108.
2. J. Merril Carlsmith, *Journal of Abnormal and Social Psychology,* 1963. Parenthetical comment by the author.
3. Harold Leavitt. *Managerial Psychology,* rev. ed. Chicago: University of Chicago Press, 1964, pp. 268–282.
4. "On Motivation," *Kaiser Aluminum News,* Vol. 26, No. 2, 1968, p. 29.
5. Irving Sarnoff, "Anxiety, Fear and Social Affiliation," *Kaiser Aluminum News,* Vol. 26, No. 2 1968.
6. George Homans, *The Human Group.* New York: Harcourt Brace Jovanovich 1950.
7. N. R. F. Maier, "Assets and Liabilities in Group Problem Solving: The Need for Integrative Function," *Psychology Review,* Vol. 74, No. 4, July 1967, pp. 346–358.
8. W. R. Bion, "Group Dynamics: A Re-review," *International Journal of Psychoanalysis,* Vol. 33, 1952, pp. 235–247.
9. Trigant Burrow, "The Basis of Group Analysis," *British Journal of Medical Psychology,* Vol. 8, 1928, pp. 198–206.
10. Eric Berne, *The Structure and Dynamics of Organizations and Groups.* Philadelphia: Lippincott, 1963.
11. Kenneth J. Arrow, *Social Choice and Individual Values,* 2nd ed. New Haven, Conn.: Yale University Press, 1963.

12. Donald R. Farris and Andrew P. Sage, "Introduction and Survey of Group Decision Making With Applications to Worth Assessment," *IEEE Transactions on Systems, Man, and Cybernetics,* Vol. SMC-5, No. 3, May 1975.
13. Martin K. Starr, *Management: A Modern Approach.* New York: Harcourt Brace Jovanovich, 1971, p. 628.
14. Joseph A. Litterer, *The Analysis of Organizations.* New York: Wiley, 1965, pp. 103–106.
15. J. N. Warfield, "Structuring Complex Systems." Columbus, Ohio: Battelle Memorial Institute, April 1974, pp. 3–4.
16. Irving H. LaValle, "A Solution Concept for Group Decision Problems," Tulane University Working Paper Series No. 64, August 1970.
17. K. E. Scheibe, "Five Views on Values and Technology," *IEEE Transactions on Systems, Man, and Cybernetics,* Vol. SMC-2 No. 5 November 1972.
18. Floyd L. Ruch, *Psychology and Life.* Glenview, Ill.: Scott, Foreman, 1967, p. 552.
19. K. L. Dion, R. S. Baron, and N. Miller, "Why Do Groups Make Riskier Decisions than Individuals?" In L. Berkowitz (Ed.), *Advances in Experimental Social Psychology,* Vol. 5, New York: Academic Press, 1970.
20. Robert J. Trotter, "Watergate: A Psychological Perspective," *Service News,* Vol. 106, July 1975, p. 378.
21. Baruch Fischholt, "Hindsight: Thinking Backward?" Hebrew University of Jerusalem, Oregon Research Institute, Office of Naval Research Contract N00D14-73-C-0438, 1974.
22. C. H. Castore, and J. C. Roberts, "Subjective Estimates of Relative Riskiness and Risk Taking Following A Group Discussion," *Organizational Behavior and Human Performance,* Vol. 7, 1972, pp. 107–120.
23. J. C. Roberts and C. H. Castore, "The Effects of Conformity, Information, and Confidence Upon Subjects' Willingness to Take Risky Decisions Following a Group Discussion," *Organizational Behavior and Human Performance,* Vol. 8, 1972, pp. 384–394.
24. J. C. Roberts and C. H. Castore, "Group Engendered Attitude Change," Purdue University, ONR Technical Report No. 16, June 1974. (Contract N00014-67-A-0226.)
25. C. H. Castore and J. D. DeNinno, "The Role of Relevance in the Choice of Comparison Others." Paper presented at the American Psychological Association Meeting in Honolulu, Hawaii, September 1972.
26. C. H. Castore and J. D. DeNinno, "The Role of Relevance in the Choice of Comparison Others. Some further findings," Purdue University, ONR Technical Report No. 4, March 1973. (Contract N00014-67-A-0226.)
27. D. Byrne, *The Attraction Paradigm.* New York: Academic Press, 1971.
28. C. H. Castore, "Diversity of Group Member Preferences and Commitment to Group Decisions, *Annals of New York Academy of Sciences,* No. 219, 1973, pp. 125–136.
29. C. E., Sherif, M. Sherif, and R. E. Nebergall, *Attitude and Attitude Change: The Social Judgment-Involvement Approach.* Philadelphia: Saunders, 1965.
30. C. Osgood, and P. H. Tannenbaum, "The Principle of Congruity in the Prediction of Attitude Change," *Psychological Review,* Vol. 52, 1955, pp. 42–55.

11

Individual Propensity to Take Risks

11.1 RISK PROPENSITY

It is well known that personality characteristics, as well as situational characteristics, affect the propensity of different people to take risks. The "accident-prone personality" is a well-known example.[1]

Different risk takers have different risk profiles, which are also relatively well established. For example, the air force conducted studies to determine the risk patterns of those who demonstrated a propensity to take high risks.[2]

Considerable work has also been done to investigate the propensity of businessmen for taking risks as recorded by Swalm.[3]

Most of the studies have been observations of patterns in real or simulated conditions. Here we attempt to provide some insight of the propensity for taking risks based on an examination of causal factors, using classical gambles as a point of departure.

11.2 CLASSIFICATION OF GAMBLES

Gamblers can be classified by the type and magnitude of the consequences of a gamble, whether tangibly or intangibly valued, and the "sum total" of the payoffs is the gamble. In games where the consequence values are cardinal, expected value is a measure of the sum of payoffs.

Consequence values may be classified as positively valued (+), indifferent (0), and negatively valued (−). These values may have magnitudes that can be considered minor or major. This classification involves the utility function of the gambler, is subjective, and is variable. Here, however, "minor" and "major" refer to the magnitude in perturbations of the life style of the gambler. Winning or losing a couple of hundred dollars would be minor, winning or losing a million would be major. A dented fender might be minor, and a serious injury major.

Table 11.1 *Form of Voluntary Gambles for Varying Consequences for a Naught-sum Payoff*

Negative Consequence \ Positive Consequence	(+) Minor positive (includes zero)	(++) Major positive (includes minor)
(−): Minor negative (includes zero)	"Fun" gamble	"Lottery" or big payoff
(− −): Major negative (includes minor	"Senseless" gamble (e.g., "death wish") and different perception of value of a payoff	"Life or death"; "up with a bang, down with a crash"

A "naught"* sum payoff indicates that the "percentages"† for the gambler and for the "house" (opponent) are equal. Consequences may be positive, indifferent, or negative, but the chance of winning or losing is equal when probabilities of occurrence are considered. If the "percentage" is positive for the gambler, a "positive" payoff results; conversely, a negative percentage implies a "negative" sum payoff.

The type and magnitude of the consequences and the payoff sum affect the propensity to gamble. The effect of consequences can be best seen when considered for a "naught-sum" payoff gamble, as represented in Table 11.1 in the form of a matrix of combinations of positive and negative, all negative, all positive. All zero sets of consequences for gambles are omitted, to ensure that only mixed combinations of positive and negative consequences are considered. Minor consequences include zero, and major consequences include minor ones as well. Four types of gambles are (1) "fun" gamble (+, −), gamble for amusement and excitement; (2) "lottery" (++, −),* little consequence if one loses, but possibility of big payoff exists; (3) "life or death" gamble (++, − −), win all or lose all are possibilities, and either win or lose changes one's way of life; and (4) "senseless" gamble (+, − −), what seems senseless to some may seem reasonable to others (e.g., compulsive

* The term "naught" sum is used to differentiate between this type of gamble and the "zero-sum" game of Von Neumann and Morgenstern. The latter implies no externalities in the sense that what one opponent wins, the other loses.

† The term "percentage" is based here in the manner of "house percentage" in casinos and is the long-term (many trials) realization of expected value.

* A (++) indicates a major positive consequence, (+) a minor positive consequence. The same nomenclature is used for major and minor negative consequences, (− −) and (−), respectively.

gamblers with a "death wish" and daredevils who feel that climbing a mountain because it is there is important).

However, when the sum of payoffs is taken into account, a broader characterization of gambles is possible (Table 11.2). As a result, five characterizations of gambles are shown with the form of gamble for each characterization.

11.3 DELAYED BENEFITS AND GRATIFICATION

Individuals react differently when benefits, payoffs, and gratifications are delayed while the costs and risks are assumed in the near term. For example, the costs of an education are assumed over the interval during which the education process takes place, but the benefits are not expected until the process is complete or in some time interval after completion.

Considerable psychological research has been done on the problem of delay of gratification. Experimental situations are devised such that a subject may accept a small reward now or a larger reward later. This research has established the existence of consistent individual differences in the capacity to delay gratification. Some individuals seem to make decisions in a larger framework of time than others.

11.4 VALUE OF STATUS QUO AS A MEASURE OF PROPENSITY FOR RISK TAKING

The value of the "status quo" (i.e., the value that is placed on an existing situation by a valuing agent at a given time) can have positive, negative, or no value.

The positive case involves satisfaction with a present state of affairs, such as being ahead in a contest and then playing defensively to protect the lead. Satisfaction or "happiness" with one's ways of life (i.e., having most of one's needs fulfilled) is a situation to be protected.

Negative value stems from dissatisfaction, including perceived anxiety that one's needs may not be fulfilled. Hoping for rain if one is behind in a ball game is an illustration of a situation with negative status quo value. Dissatisfaction with one's way of life or status is another example.

"No value" implies indifference to the present condition.

It is hypothesized here that the propensity for taking risk is inversely related to the value of one's status quo; the higher the positive value of the status quo, the lower the propensity to take risk. If one assumes a degree of perceived satisfaction with the status quo that ranges over positive to nega-

Table 11.2 ***Types of Voluntary Gamble***

Designation By Consequence Type	Values of Consequence	Payoff Sum	Most Likely Form of Gamble	Characterization of Gambler
All negative	−, 0	−	Chance of lesser evil (−, 0) Change better than status quo (− −, −)	"Nothing to lose"
Negative sum	−, 0, +	−	Fun gambler (+, −) Compulsive gambler (+ +, − −, +, − −) Lottery (+ +, −)	"Betting against the house"
Naught sum	−, 0, +	0	Fun gambler (+, −) Compulsive gambler (+ +, − −, +, − −) Lottery (+ +, − −) Daredevil (+ +, − −)	"Sportsman"
Positive sum	−, 0, +	+	Professional daredevil (+ +, − −) The "house" (+, −) "Thrill seekers" (+, − −)	"Sure thing bettor"
All positive	0, +	+	Can't lose (+, 0) (+ +, +)	"Something for nothing"

tive values on some ordinal or cardinal scale, this can be related to an ordinal or cardinal scale of propensity for taking risks from a low to high range. This relationship is shown in Figure 11.1 as an inverse relationship that may be a linear (dashed line) or a nonlinear (solid line) function concave upward.

To illustrate this hypothesis, the five characterizations of gambles discussed previously are shown with approximate ranges of coverage. The all-negative "nothing to lose" gambler has a very high propensity to take risk. The all-positive "something for nothing" gambler has a very low propensity for taking risks, since he has little to gain but a lot to lose. The positive sum condition involves a low propensity for risk, and a gambler in this situation will take risks with consequences that extend down to indifference with the status quo, but not to negative status quo conditions. The "naught" sum covers a wide range in the middle and overlaps the negative sum, which in turn overlaps the all-negative condition.

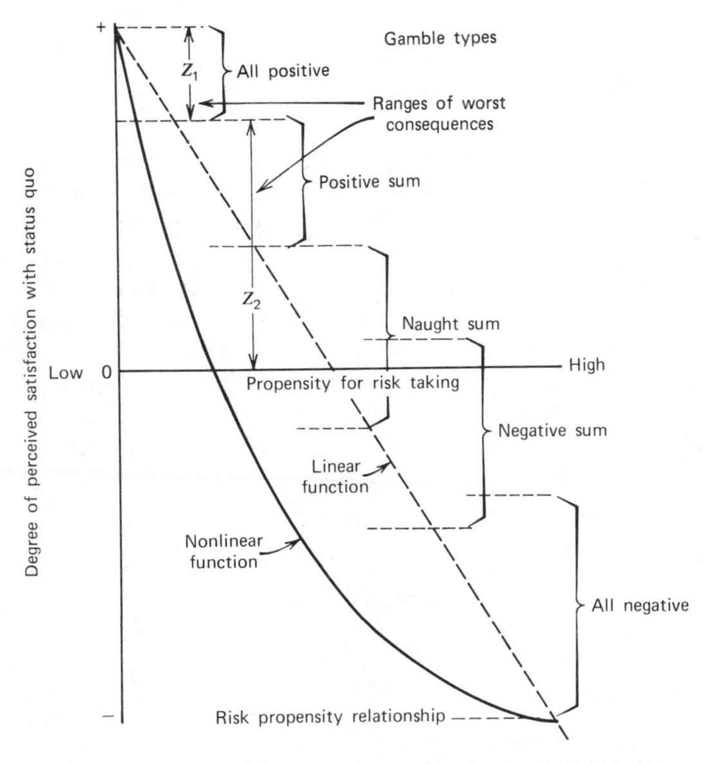

Figure 11.1. Qualitative presentation of the inverse functional relationship between perception of degree of satisfaction of the status quo and the propensity for taking risks along with ranges for different types of gamble.

Certain individual characteristics and values affect the propensity for risk independent of the status quo. This has been well documented by many investigators, including Swalm,[3] but much of the variability may be related to the condition of the status quo and how the gambler values it. As an example, it is often stated that a good gambler ignores or walks away from previous losses. However, few observers of gamblers would dispute that many gamblers take larger risks when they are losing, in an attempt to recoup. The loss has changed the status quo in a negative direction and the propensity for risk is increased, even though this may be done in spite of "good advice." It is also true in the opposite direction; that is, a winner often plunges with his winnings, since a loss of winnings will not change his original status quo. If he continues to be successful over a long period of time, he may adopt a new, higher valued status quo and become more conservative. That is, during the gamble the status quo changes slowly in the positive direction when winning, but more rapidly in the negative direction when losing.

11.5 INVOLUNTARY RISKS

As described in chapter 8 involuntary risks are risks inequitably imposed on a risk taker and from which he receives no direct benefit. These involuntary risks occur as a result of natural forces or by specific man-made actions or societal activities. Natural occurrences include catastrophes (earthquakes, tornadoes, floods, etc.) and normal activities (overexposure to sun, drought, lightning damage). Specific man-made actions involve such events as drunk drivers crossing to the wrong side of the road, which can be considered to be individual negligence, and encroachment of one's neighbor. Societal activities involve the acceptance of certain levels of societal risk, such as increased exposure to air pollution or radiation, by some or all parts of society to obtain some benefits (e.g., the production of electric power) for differing parts or all society.

11.5.1 Avoidability of Involuntary Risks

Involuntary risks may be avoidable by specific action of the risk taker on whom the risks are imposed. The person at risk may have no direct control on the events that impose the risk, but he may have direct control over his exposure to risk for those events. For example, individuals cannot control the onset of earthquakes, but they can choose to live in areas of low seismic activity.

When a risk agent perceives* the imposition of a new risk, he has two options to avoid this new risk. First, he may try to influence the manner in which the risk is imposed. For example, because of the construction of a nuclear power plant, neighboring populations might potentially be subjected to exposure to increased air pollution or radiation levels. Members of the population on whom these risks are involuntarily imposed can attempt to intervene, legally or otherwise, to either prevent the power plant from being located at that construction site or to assure that adequate controls are implemented to eliminate or at least minimize the potential risk. Society in the United States has provided institutions for recourse or control of imposition of risks on society as a whole through the jurisprudence system, and, more importantly, the executive branch of government in the form of regulatory agencies whose aim is to protect the health, welfare, and environment of society.

Second, a risk agent may seek to reduce his exposure to an imposed risk by removing himself from its effective range. In this sense, one can move from a high to a low altitude to avoid the risk from natural cosmic rays or move from the proposed site of the nuclear power plant to avoid assumption of any possible risks from its operation. Essentially, such avoidance action is individually motivated and there are few institutions, if any, involved.

In basic terms, the risk taker may fight (intervene), flee (avoid exposure), or accept the imposed risk. The motivation for taking action as opposed to acceptance is affected by a variety of factors. One approach to understanding the motivational relationships is to act as if an involuntary risk constitutes a threat to the status quo.

11.5.2 Involuntary Risks as Threats to the Status Quo

When a newly perceived risk is imposed on a risk taker, it is a direct threat to his status quo in terms of his way of life, his assets, and even his life and health. The anxiety response will be generally a function of the character and magnitude of the risk, the ability to avoid the risk at some cost, and the degree of satisfaction with the status quo of the individual. Despite an absence of formal data, one can tentatively examine the qualitative trends of these relationships. Such a qualitative approach is depicted in Figure 11.2, where surfaces for the thresholds of concern and fleeing action appear as functions of the degree of satisfaction with the status quo and the character and magnitude of the threat. For the character of the risk, the hierarchy of risk consequences (Table 4.2) is used. These eight classifica-

* Perception of a new risk involves both having a new risk imposed or becoming aware of an existing risk previously not identified.

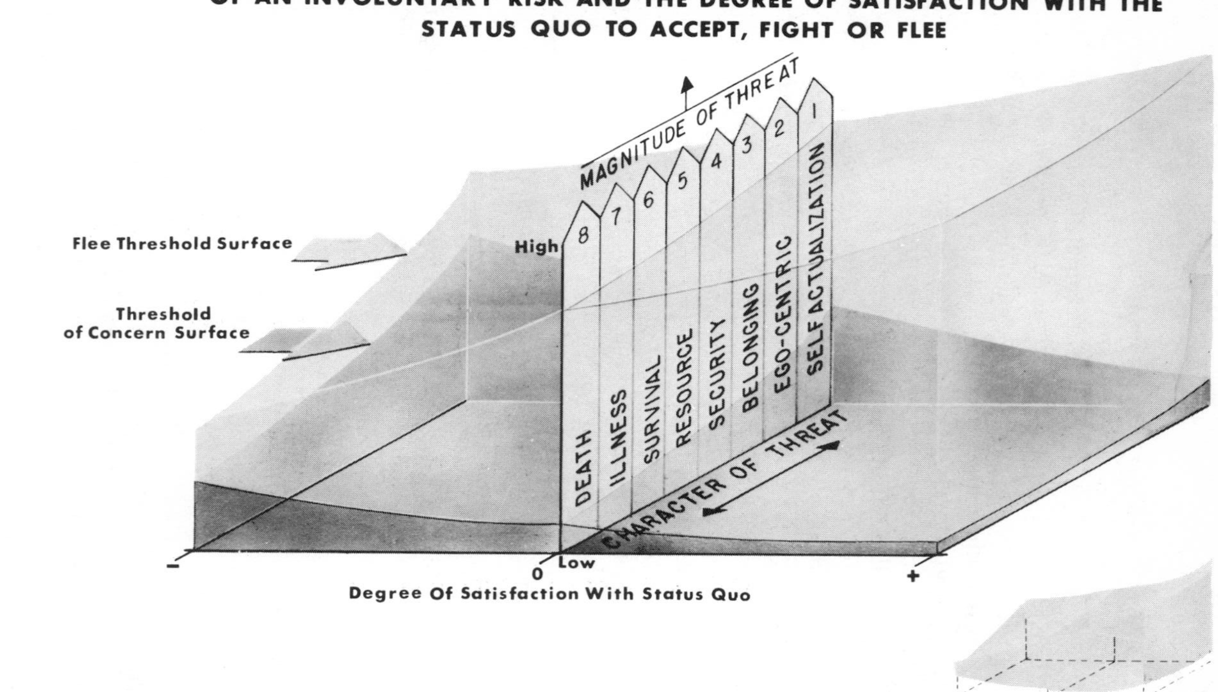
QUALITATIVE RELATIONSHIP AMONG THE CHARACTER AND MAGNITUDE OF AN INVOLUNTARY RISK AND THE DEGREE OF SATISFACTION WITH THE STATUS QUO TO ACCEPT, FIGHT OR FLEE
MAGNITUDE OF THREAT
8
7
6
5
4
3
2
1
DEATH
ILLNESS
SURVIVAL
RESOURCE
SECURITY
BELONGING
EGO-CENTRIC
SELF ACTUALIZATION
CHARACTER OF THREAT
High
Low
Flee Threshold Surface
Threshold of Concern Surface
–
0
+
Degree Of Satisfaction With Status Quo
Fight To Extent Possible Volume

Figure 11.2

tions provide eight profiles of magnitude of risk versus degree of satisfaction with the status quo which are qualitative, cardinal scales.

The lower surface is the threshold of concern. That is, the magnitude of the risk is sufficient for a risk taker to become actively concerned with risk avoidance. It would seem plausible that people who are happier with their status quo would be more sensitive to threats to the status quo than those who are unhappy with their present situation. Thus the curves decrease with increasing degree of satisfaction. Furthermore, threats to life and limb may concern people at lower levels of magnitude than threats to one's individualism or aesthetic value; hence the surface tips upward in these risk characteristics. The possibility of a discontinuous surface near the zero degree of satisfaction with the status quo must be considered. People who are positively situated may have sharper differences than people who are not.

The upper surface is another threshold: people will attempt to avoid an involuntary risk by retreating from it by fleeing, which necessarily means upsetting the status quo. People who are satisfied with their status quo will not concede changes as easily as people who are dissatisfied. The former would rather fight and intervene (represented by the volume between the surfaces) than change their status quo. This pattern becomes less distinct for threats to life and limb as opposed to threats at the other end of the hierarchy of risk consequences. As a result, the flee threshold surface increases from the front left corner in both directions.

The volume, as already indicated, illustrates the existence of an action between the two thresholds at which people are concerned and therefore attempt to fight the risk imposition, but are not motivated to flee. Between the two surfaces, this fight to the maximum extent possible action is greatest for people with a positive status quo to protect. If they cannot fight, they will accept the risk up to a certain level, then they will flee. Since fleeing represents a major change in status quo,* those presently satisfied must trade off the change due to fleeing versus change due to the threat.† In any case, the fight or flee concept is consistent with Bion's "basic assumptions."

* A decision to relocate inland to avoid hurricanes or to move to avoid high local levels of radiation (natural or man-made) involves disruption of interpersonal relationships, family, civil requirements, and so on. Changes in ease of mobility can, of course, have some effect on such decisions.

† There seem to be two different groups who tend to fight (intervene) the imposition of involuntary risks most vociferously. At one extreme we have the groups that are well satisfied and fight to preserve the status quo. Many environmental groups are in this category. At the other extreme are those who are dissatisfied and have nothing to lose. Students and youths against the draft and the Vietnam war joining with the "underprivileged" in an effort to upset the establishment in the late 1960s and early 1970s are examples of the latter. One might entitle these groups the "fat interveners" and the "lean interveners," respectively.

11.6 RISK PROPENSITY OF DIFFERENT VALUE GROUPS

There are many different value groups in society resulting from different cultures, different ethnic groups, groups of different heredity, and groups brought up in different environments, geographies, degrees of wealth, degrees of education, and so on.* The idea of the status quo as a measure of motivation to take risks provides a possible simplification as far as propensity for risks is to be considered. As a result, we can assume that there are three general classes of value groups: (1) those who are satisfied with their status quo, (2) those who are dissatisfied, and (3) those who are indifferent. The previous section described how these groups tend to react to involuntary risks of different types. This situation must be reexamined, however, when voluntary risks, which imply some motivation to undertake these risks to achieve particular goals, are concerned. It must be pointed out that increasing awareness of an individual's status quo in relation to the status quo of others can cause dissatisfaction with the former's status quo. That is, when a person comes to realize how badly he is situated in comparison to others, he might grow concerned about a particular condition. In some of the underdeveloped countries, for example, the rapid introduction of worldwide communication through television and news for large numbers of people is hastening the recognition that their original status quo is in great need of improvement. In other words, more people are becoming dissatisfied with the agrarian way of life and are substituting a desire for an urban manner of life, especially at very low income levels. At the upper level, the reverse may be true, because those who are quite affluent often tend to flee from the urban areas toward the suburbs and the remote parts of the country.

11.7 SUMMARY

The factors in risk evaluation discussed here have been treated qualitatively. The purpose has been to identify and characterize these factors. Probably other factors exist that have not been identified, but the author feels that the ones discussed are the major foci of concern.

For any given situation, all these factors may occur simultaneously to different degrees and interact. Apparently no studies have been carried out to determine the relative importance of these factors and how different groups in society react to them. The purpose here has been to identify the factors, to facilitate further investigation.

* These have been described more fully in Chapter 9 under the heading of cultural factors.

A subsequent part attempts to quantify these factors by observing the general behavior of society in reacting to risks involving different risk factors. By comparison, the effects of these factors can be investigated.

REFERENCES

1. H. W. Heinrich, *Industrial Accident Prevention,* 3rd ed. New York: McGraw-Hill, 1950, pp. 332–334.
2. E. P. Torrance and R. C. Ziller, *Risk and Life Experience: Development of a Scale for Measuring Risk-Taking Tendencies.* Research Report AFPTRC-TN-23, Armed Services Technical Information Agency Document No. 098926. Randolph Air Force Base, Texas: Air Force Personnel and Training Center, February 1957, pp. 5–7. Quoted in H. S. Denenberg, R. D. Eilers, J. J. Melone, and R. A. Zelten, *Risk and Insurance,* 2nd ed. Englewood Cliffs, N.J.: Prentice-Hall, 1974, p. 59.
3. Ralph O. Swalm, "Utility Theory Insights into Risk Taking," *Harvard Business Review,* November–December 1966, pp. 123–136.
4. W. R. Bion, "Group Dynamics: A Re-review," *International Journal of Psychoanalysis,* Vol. 33, 1952, pp. 235–247.

Part C

METHODOLOGICAL PROBLEMS AND APPROACHES IN THE QUANTIFICATION OF RISKS

The identification and qualitative description of risk factors in Part B is only a first step in understanding how individuals, groups, and society value and evaluate risks. It is necessary to attempt to quantify these processes. The quantities are imprecise, however, because of the subjective and often intangible nature of the factors and their effects on consequence valuation. Existing statistical and operational techniques have only limited usefulness for imprecise situations. Part C addresses the basic methodological problems in quantification for risk assessment and evaluates some possible approaches for providing pragmatic solutions.

Chapter 12 explores some generic problems in risk assessment. The initial portion of this chapter treats the basic epistemological processes that are useful for societal risk assessment and relates these to various attitudes and beliefs of those responsible for addressing risk evaluation. This analysis demonstrates the shortcomings of existing methods for assessment of intangibles and provides a basis for evaluation of alternative approaches to seek resolution. Two alternatives are examined: the rational-comprehensive methods of systems analysis, and the successive limited comparison method of Lindblom. The limitations of each are explored in terms of the information required to solve generic policy problems. A resultant compromise approach, using the better parts of each method, is proposed as a schema for a pragmatic (Singerian) resolution of the problem addressed.

Chapter 13 deals with specific problems related to measurement in the risk assessment process. The role of the risk evaluator and his accompanying bias are examined first. This is followed by an analysis of the meaning of value and utility, and application of these dimensions to the measurement of tangible and intangible values. Finally, a number of other measurement problems are examined such as the dynamics of specific situations and measures of value of a life.

Chapter 14 provides a brief overview of some approaches to the problem and the advantages and disadvantages of each. A generic approach to cost-benefit analysis is examined, and an error analysis is undertaken, to show the limitation of such approaches using a rational-comprehensive model. Two other methods for dealing directly with uncertainty in assignment of intangible value are described briefly: (1) decision making with uncertain utility functions, and (2) structured value analysis. The purpose is to indicate a few of the many approaches to the problem of evaluating subjective and intangible values that are not limited by the mechanistic nature of classical utility theory.

The casual reader may want to skip most of Part C, but several areas should be reviewed. Sections 12.3 and 12.5 provide insight to basic methodological approaches to problem solving. Section 13.1 explores the role of the risk evaluator, and Section 13.4.2 investigates measures of the value of a life. Section 14.1 examines classical cost-benefit analysis and its limitations.

12

Generic Problems in Risk Assessment

The need for societal risk assessment has become progressively more urgent as new and developing technologies thrust increasingly complex issues on a society more able to make individual choices than social choices. Thus the problem becomes partly one of analyzing social choice about technological development, and a considerable body of literature on this subject provides a means for discussing several of the basic problems associated with risk assessment. References to existing literature illustrate these problems, but no attempt is made to duplicate the sources extensively.

The problem of analyzing social choice is multidisciplinary, but a major dichotomy exists between the "scientific" approaches of the classical scientist, the economist, and the decision theorist, and the "behavioral" approaches of the political scientist, the social psychologist, and so on. One relies on structural models of what behavior ought to be and is inductive; the other relies on observation of behavior and is deductive.

The comprehensive model of decision making assumes that individual preferences are at least in principle knowable, known, and transitive. It is as if a referendum were held among 210 million Americans on a series of policy issues, with the decision rule "one man, one vote." For the political analyst this conception of social choice is accompanied by a number of problems.[1]

Political scientists' understanding of voting behavior is highly developed. Regression analyses and computer simulations based primarily on survey research can account for a very large proportion of the variance of electoral choice and can achieve a high degree of accuracy in prediction. Much of the improved power of these techniques has come from altering or abandoning assumptions that every citizen votes, that every citizen has a "complete set" of policy preferences, and that policy preferences are unidimensionally transitive.[2-6] In short, the political scientist's understanding of citizen policy preferences departs substantially from the comprehensive decision theorist's assumptions.[7, 8]

A second problem is that citizen preference is only the first step in a complex process of decision making. The relationship between societal preferences (even assuming they are knowable and known) and decisions made in political institutions is anything but automatically isomorphic.[9] Decisions in a pluralistic society are made in myriad institutions, public and private. Identifying the decision makers (individuals, groups, institutions) is at least as problematic as determining how a decision is reached.

The decisions facing individuals acting daily for themselves or as agents or evaluators for others are becoming increasingly dependent on factors involving tangible and intangible nonmonetary values. This holds in government, business, the academic world, and elsewhere, especially for decisions involving risk.

The new philosophies of modern management of Douglas MacGregor[10] and Abraham Maslow[11] are oriented toward human values, as opposed to the quantified scientific management philosophy of Frederick Taylor.[12] Nonmonetary goals, often conflicting with monetary goals,[13] must now be considered in decisions. Concern of government, management, employees, and the public for new factors (e.g., pollution and environmental degradation, social well-being and welfare) and satisficing to achieve personal objectives are examples of nonmonetary goals that are becoming increasingly accepted by society.

Simply to state the dichotomy between the scientific approach characterized by use of precise, tangible measures, and the behavioral approach that recognizes the need for imprecise, intangible measures when appropriate, is an oversimplification. To achieve better insight, the problem of social choice can be addressed at the epistemological level, the psychological level, and practical level of implementation.

12.1 EPISTEMOLOGICAL ASPECTS OF FORMAL INQUIRY

An evaluation of the basic limitations of models and structure, recently developed by Mitroff and Turoff,[14] also provides an epistemological understanding of how research in the area of social choice differs from previous work.

Five philosophical approaches to formal methods of inquiry are described based on the work of as many philosophers. These along with their salient characteristics are described below.*

Leibnizian Inquiry System. Truth is analytic, i.e., the truth content of a system is associated entirely with its formal content. A model of a system is a formal model

* Reprinted by permission from reference 14. Copyright 1973 by the Institute of Electrical and Electronics Engineers, Inc.

and the truth of the model is a measure in terms of its ability to offer a theoretical explanation of a wide range of general phenomena and in our ability as model-builders to state clearly the formal conditions under which the model holds. A corollary to this is that the truth of the model does not rest on any external considerations; in other words, the model is independent of the raw data of the external world.[14]

A prime example of Leibnizian inquiry is the field of operations research, in that the major energies of the profession have been directed almost exclusively toward constructing and exploring highly sophisticated models. The methods of decision theory and the approach to decision theory as exemplified by Thompson[8] and others are prime examples of Leibnizian inquiry systems.

Lockean Inquiry Systems. Truth is experiential, the truth content of a system is associated entirely with its empirical concept. The model of a system is an empirical model and its truth is measured in terms of our ability (1) to reduce every complex proposition to its simple empirical referents (simple observations) and (2) to ensure the validity of each of the simple referents by means of the wide spread, freely obtained agreement between different human observers.[14]

This philosophy underlies the major part of empirical science and is data oriented as opposed to the theory-oriented Leibnizian inquiry system. The Lockean inquiry system is unable to deal with subjective and intangible referents, since the means of agreement are not freely obtainable.

Kantian Inquiry System. Truth is synthetic; i.e., the truth content of a system is not located in either its theoretical or its empirical components, but in both. A corollary is that neither the data input nor the theory has priority. An important feature of the Kantian inquiry is that for any problem, one must build at least two alternate representations. Out of these alternatives, one will represent the problem better than the others. The defect of the Leibnizian and Lockean inquiry is that they only give one view of the problem.[14]

The alternative that gives the best mix of data and theory to solve the problem is desired. A balance between data acquisition and theory are used to provide operational answers. An operational system with subjective or imprecise data must at least be a Kantian inquiry type of system.

Hegelian Inquiry System. Truth is conflictual, i.e., the truth content of a system is the result of a highly complicated process that depends upon the existence of a plan and diametrically opposed counterplan. Their function is to engage each other in an unremitting debate over the "true" nature of the whole system, in order to draw forth a new plan that will hopefully reconcile (synthesize, encompass) the plan and counterplan. A corollary to this is that by itself data is totally meaningless and only becomes meaningful, i.e., information, by being coupled to the plan and counterplan.[14]

These systems are the epitome of conflictual, synthetic systems and are typified by getting opponents to agree about what they agree and disagree on, then finding a new system to settle the disagreements. Such systems are not well structured in the sense that formal system analysis models exist; the method is essential to the model. This method has been used extensively by Kahn,[15] who has even classified the kinds of agreement: (1) first-order agreement, agreement on substantive or normative issues (e.g., on values, assumptions, analyses or policies and/or predictions); (2) second-order agreement, agreement on what a first-order disagreement is about—the disagreement is made precise and the effect of the disagreement is analyzed; (3) third-order agreement; agreement on why second-order agreement cannot be reached; and (4) fourth-order agreement, an agreement that no useful degree of agreement can be reached.

Singerian Inquiry System. Truth is pragmatic, i.e., the truth content of a system is relative to the overall goals and objectives of the inquiry. A model of a system is teleological or explicitly goal-oriented in the sense that the truth of the model is measured with respect to its ability to articulate certain system objectives, to create several alternative means for securing these objectives, and finally at the "end" of the inquiry to specify new goals that remain to be accomplished by some future inquiry. Singerian inquirers thus never give final answers to any question, although at any point they seek to give a refined, specific response.[14]

Although the Singerian inquiry system is ill defined, it is the epitome of operational systems. The objective is to solve the problem at hand in some manner, and no aspect of the system has any fundamental priority over any of the other aspects. The Singerian system will become more meaningful when methods of implementation are discussed.

It is clear from these descriptions that the "scientific" approach is best described by Leibnizian system and the "behavioral" approach by the Lockean system. The existence of the remaining systems offers some hope that the dilemma of whether to use the scientific or behavioral approach can be discarded, and that new, consolidated approaches can be more successful in dealing with problems of social choice.

12.2 ATTITUDES AND BELIEFS IN ADDRESSING PROBLEMS AND VALUES

12.2.1 The Structure of Beliefs and Attitudes

The choice of epistemological methods of inquiry is often preselected through the basic attitudes and beliefs of the practitioners. Such beliefs are altered only with considerable effort, as Thomas Kuhn[16] demonstrates in

The Structure of Scientific Revolutions, by showing that the scientific community is constrained by methodology. At any given point in the development of science there is a set of paradigms that is accepted and changes slowly. Science and technology are carried out under a prevailing set of paradigms. These paradigms provide a freedom to conduct professional activities within the accepted set but constrain most practitioners to these boundaries.

An example of these phenomena is presented by the "grand designs" of the 1960s with emphasis on system analysis programs in government such as PPBS (planning, programming, and budgeting system) and social policy programs such as "community action" as the means to solve societal problems. Bartley, in reviewing changes from these, states:

> Not surprisingly, the early '70s are starting to see an exploration of this paradox, a reassessment of what part should be played in policy-making by both systems analysis and social science. Neither is being rejected, of course, and both may prove more helpful in the future than they have in the past. Yet, their practitioners are working themselves into a new modesty, a new understanding of the limitations of their methods.[17]

As an explanation, he further states that:

> This is an interesting development in its own right, but it is even more fascinating viewed against a background of political philosophy. For the root question in this reassessment is the power of rationality: Its theme is a new skepticism about man's ability to draw up rational prescriptions for changing society, or at least a new realism about the case of doing so.[17]

The result of the effort of system analysis had led Rivlin[18] to conclude that "so far the analysts have probably done more to reveal how difficult the problems and choices are than to make the decisions easier."

12.2.2 Classification of Value Paradigms

At any given time there is a plurality of sets of paradigms, some old and in the process of discard, some operational, and perhaps, with opposing views, and some new ones trying to be established. This is illustrated by perspectives on relationships that hold between values and technology developed by Scheibe.*

The Luddite. The basic premise of writers in this category is that technological development is inevitably and fundamentally dehumanizing and corrupting. In a

* Reprinted by permission from reference 19, pp. 576–577. Copyright 1972 by the Institute of Electrical and Electronics Engineers, Inc.

technologically developed society, man is forced to live in a way that is both unnatural and spiritually depraved. A common spectre is that of short-sighted little men, usually engineers and profiteering businessmen, who have taken over spaceship earth and are mindlessly extinguishing all human values. But there is hope. Charles Reich foresees a spontaneous emergence of a new, post technological mentality which will restore human authenticity. Theodore Roszak sees hope in the development of an antitechnological counterculture.

The Technocratic. B. F. Skinner asserts that technology is our strength and that if we want to survive we must play from strength. Technology is on the march and man must adapt to it. Science is accepted as universal ethic, not just a method of finding the truth. But the admixture of outmoded, traditional, quasi-religious ways of thinking and the scientific, sophisticated, correct way of thinking about man has produced the inefficient and potentially disastrous custom of "muddling through." We must clean up our thinking, design our futures, and control that which we can control; which, thanks to technology, is just about everything.

The Apocalyptic. This perspective has much in common with that of the Luddites. Both hold that man has created the means of his own destruction through the exercise of his rational powers. However the apocalyptic vision does not share the belief that technological development can be stopped or that man will spontaneously reject the insane world he has created and return to pastoral innocence. Scientists, who are still engaged in the pursuit of saving truths, are not likely to act as prophets of despair—it is incompatible with the requirements of their role. Instead, this view gains clearest expression from critics, such as Leslie Fiedler and Ihab Hassan, novelists and filmmakers, such as Kurt Vonnegut and Stanley Kubrick. Other writers, such as Paul Ehrlich and Alvin Toffler present visions of the future which seem almost as hopeless, though they may continue to express the belief that there is A Way Out. The one shred of hope presented in this perspective is that perhaps the apocalypse will act as a massive cultural electric-shock treatment. Possibly, when the dust settles, the remainder of mankind will live a long while before creating another massive disaster.

The Cautionary Moral Sermon. The most common practitioners of this art form are scientists themselves, who for one reason or another look up from their laboratory benches and are alarmed by what they see. The list of practitioners reads like an honor roll of science—René Dubos, Jacques Monod, George Wald, Linus Pauling, Garret Hardin, John Platt, J. Bronowski. The common theme is that scientists have been naive and unwittingly irresponsible in the pursuit of their calling. They have been on the glimmering path of truth and have trusted to politicians to run the world and to the social scientists to keep score and offer practical advice. Now it is clear that scientists have misplaced their trust. They must rekindle their humane values and must play a crucial role in the creation of a new and more benevolent world order. With Whitehead, scientists must recognize that "Mankind has raised the edifice of science, because they have judged it worthwhile." Science is value-laden in origin and effect, and it is up to scientists to redeem the trust humanity has placed in them by dedicating themselves to the highest human values.[19]

Scheibe adds a fifth perspective, his own, that of the "Curious, Hopeful and Sometimes Astonished Observer." In this stance Scheibe views the other perspectives as having little contribution to practicality and their presuppositions about human values are psychologically naive (i.e., they lack insight into values that direct human behavior). There must be a greater willingness to deal explicitly with human values in relation to science, an argument also made by Maslow.[20]

The merging of science, values, and policy to solve problems is the gist of these arguments as well as many other writers.[21] However, recognition of the methodologies for realizing this merger and their limitations occurred even before the 1960s in a classic article by Lindblom.

12.3 ALTERNATIVES FOR COPING

In his 1959 essay on the "science of muddling through,"[22] Lindblom sets out an alternative to the "rational" approach of systems analysis, decision making, policy formulation, planning, and so on. The alternative is called the method of "successive limited comparisons" as opposed to the "rational-comprehensive" method. (They are also referred to as the "branch" and "root" methods, respectively.) The classical rational-comprehensive approach attempts to establish structure, goals, and values at the outset, and different policy alternatives are evaluated against these measures to select the alternative that gives the highest measure of value.

> The hallmarks of these procedures, typical of the root approach, are clarity of objective, explicitness of evaluation, a high degree of comprehensiveness of overview, and, wherever possible, quantification of values for mathematical analysis. But these advanced procedures remain largely the appropriate techniques of relatively small-scale problem-solving where the total number of variables to be considered is small and value problems restricted.[22]

As problems addressed by the root method become more complex, the number of variables to be addressed becomes unwieldy and formal solutions meaningless, as demonstrated later in this Part when the subject of cost-benefit analysis is revisited.

The branch method, on the other hand, assumes at the outset that many critical values or objectives will bring up basic disagreement and conflicts among different parties. Lindblom states this succinctly:

> Administrators cannot escape these conflicts by ascertaining the majority's preference, for preferences have not been registered on most issues; indeed, there often are no preferences in the absence of public discussion sufficient to bring an issue to the attention of the electorate. Furthermore, there is a question of whether intensity of feeling should be considered as well as the number of persons preferring

each alternative. By the impossibility of doing otherwise, administrators often are reduced to deciding policy without clarifying objectives first.[22]

Lindblom maintains that in the absence of agreement on objectives, there is no test for the correctness of policy except agreement on a policy itself. Thus one must examine both the objectives and the policy simultaneously, or at least in an iterative fashion, using incremental steps at each juncture of the process.

In summary, two aspects of the process by which values are actually handled can be distinguished. The first is clear: evaluation and empirical analysis are intertwined; that is, one chooses among values and among policies at one and the same time. Put a little more elaborately, one simultaneously chooses a policy to attain certain objectives and chooses the objectives themselves. The second aspect is related but distinct: the administrator focuses his attention on marginal or incremental values. Whether he is aware of it or not, he does not find general formulations of objectives very helpful and in fact makes specific marginal or incremental comparisons.[22]

Both individual administrators and organizations seem to operate in this manner, and it has already been demonstrated that organizations have good logical and operational reasons to respond to their environment incrementally.[23] By using marginal differences among policies and comparing difference on a limited basis, and repeating these steps iteratively, feedback is provided at each step. This leads to simplification of the complex issues, achieved through limitation of policy comparisons to the policies that differ in relatively small degree from policies presently in effect. This reduces the number of alternatives and simplifies the investigation of alternatives. The approach may appear to ignore important possible consequences of possible policies, but it circumvents futile attempts to achieve a comprehensiveness beyond human capacity. Good sense it suggested as an alternative to comprehensiveness. The two methods are characterized side by side in Table 12.1.

The "branch" approach, a good example of a Singerian inquiry system, is the epitome of pragmatic problem solving. It has taken almost 15 years since Lindblom presented the "branch" system for society to perceive that the "root" system does not by itself solve problems. A good example is the comprehensive draft final environmental impact statement prepared by the Atomic Energy Commission on the liquid metal fast breeder reactor[24] just before the AEC's demise and the turnover of responsibility to the Energy Research and Development Administration. This effort was an extensive review of the whole industry, encompassing development and commercialization of the LMFBRs. The environmental statement was unable to answer overall policy questions involving the total program, but it did

Table 12.1 *Comparison of the Rational-Comprehensive and Successive Limited Comparison Methods*

Rational-Comprehensive (Root)	Successive Limited Comparisons (Branch)
1a. Clarification of values or objectives distinct from, and usually prerequisite to, empirical analysis of alternative policies.	1b. Selection of value goals and empirical analysis of the needed action are not distinct from one another but are closely intertwined.
2a. Policy formulation is therefore approached through means-end analysis: First the ends are isolated, then the means to achieve them are sought.	2b. Since means and ends are not distinct, means-end analysis is often inappropriate or limited.
3a. The test of a "good" policy is that it can be shown to be the most appropriate means to desired ends.	3b. The test of a "good" policy is typically that various analysts find themselves directly agreeing on a policy (without their agreeing that it is the most appropriate means to an agreed objective).
4a. Analysis is comprehensive; every important relevant factor is taken into account.	4b. Analysis is drastically limited: • Important possible outcomes are neglected. • Important alternative potential policies are neglected. • Important affected values are neglected.
5a. Theory is often heavily relied upon.	5b. A succession of comparisons greatly reduces or eliminates reliance on theory.

Source: Reprinted by permission from Charles E. Lindblom, "The Science of 'Muddling Through,'" *Public Administration Review,* Vol. 19, No. 2, Spring 1959.

provide focus on the important parameters and factors affecting such questions. The unknown factors* had such variability that precise answers were unachievable. Though a failure from a "cost-benefit" model standpoint,† the effort was nevertheless valuable because it yielded insight on the major issues. It further allowed decoupling of the problems of development and

* In this case, future energy demand and uranium supply.

† Although this study was required by the courts, a Lindblomian might well have anticipated the results. Whether the useful results were worth effort and expense is an entirely different question.

demonstration in the nearer term from the problems of commercialization in the long term.

The point to be made is that useful information can be provided by both the root and branch approaches, but neither guarantees success or even help. In the root method the cost of dealing with many complex variables and the imprecise results may well outweigh the usefulness of identifying critical factors. Society seems to operate in the branch method, but as numerous observers have pointed out, this is a major reason behind the increasing desperation of our plight.

Since both approaches have some merit, hybrid approaches that use the best aspects of both and restrict their disadvantages should be considered. Such mixed approaches are based on the types and content of the information needed to obtain results, or conversely, how uncertainty is reduced.

12.4 THE NATURE OF INFORMATION USEFUL IN SOLVING POLICY PROBLEMS

Warfield[25] has indicated that the tools of policy analysis and societal problem structuring are primarily tools of communication and presentation of information, not only to professional practitioners but to decision makers and the public at large. The basic elements of communication are (1) words and semantic notations, (2) graphics, and (3) mathematics of logic and structure and of content.

The two forms of mathematics, though not totally unrelated, are important to note. That of logic and structure deals with general approaches to problem solving, whereas that of content is concerned with solving specific problems. Much of the literature and effort in the root method is based on logic and structure to provide formal and informal means for solution. The policy maker, on the other hand, is interested in specifics and particular content, which are often accessible only through the branch method.

Each of the foregoing elements is used to some extent to reduce both descriptive and measurement uncertainty in the root and branch approaches alike.

12.4.1 Reduction of Descriptive Uncertainty

The reduction of descriptive uncertainty stems from innate cognitive ability to discriminate among different objects in an empirical sense and among different ideas in an abstract sense, and to describe and utilize these concepts in an understandable, abstract manner. Each individual is exposed to objects and ideas in his own unique way and subjectively evaluates them. Language provides a common means of communicating similar experiences among individuals by describing new experiences and definitions of these and historic experiences. The process of definition, analytically, involves identification of all variables pertinent to an activity in a mutually exclusive, collectively exhaustive manner (e.g., definition of all variables involving causative events, outcomes, pathways, and consequences in a risk situation).

In the root method, this entails many variables, each defined explicitly, with assurance that all variables are covered and properly defined. The problem is often so complex that purely semantic definition is inadequate. Graphic structures, such as morphologies, hierarchies, tree structures, flow diagrams, and other pictorial methods, are often used to supplement verbal description. Mathematical techniques incorporating methods of logic and structure provide formal methods of definition and description. Graph theory, formal logic, group and set theory, and matrix algebra are examples of these methods. They deal with definition and interconnection, as opposed to the numerical content involved. Mathematical notation is a manipulative shorthand that becomes increasingly abstract as the problem addressed grows in complexity. As one relies more on mathematical structures, more vigorous definition is possible, but at the expense of understanding that is inherent in the semantic structures. Only those schooled in the mathematical techniques can address and understand them, whereas all members of society can understand verbal or written definitions within the limits of the semantic skills available. That is, precise language definitions are required, but our language may be semantically limited in providing the needed level of discrimination clearly and unambiguously. The more complex the problem, the more difficult the elimination of ambiguity.

In the branch approach, only the variables that incrementally affect changes in policy are considered, and all other variables are excluded. This indeed simplifies the problem, perhaps oversimplifies it. The remaining variables then may be defined as previously. It may be easier semantically to define these variables, since both the definition of the variables and definition of policies may be at the same level of detail, such that one set of definitions augments the other, and vice versa.

12.4.2 Reduction of Measurement Uncertainty

Reduction of measurement uncertainty involves the assignment of specific measures of magnitude or interaction to defined variables. The concern is with the content of the structure, that is, the values assigned to variables and the expression of functional relationships among them. In the root approach, the structure (i.e., the system for solution) is established, then values are assigned to each independent variable in the structure. These values are derived empirically from actual data when objective measurements are possible or subjectively when only value judgments are available. The values are entered into the structure for different policy alternatives and the value of the outcomes compared. The outcome values are by definition measured on a scale of desirability toward achieving defined ends as expressed in entry *3a* of Table 12.1. Words and diagrams are useful in assigning ordinal (rank) values and in understanding how values are assigned for cardinal scales. However the mathematics of content is necessary for assigning numerical values and their manipulation.

The shortcomings of this approach become apparent as the number of variables increases. First, to each variable is assigned a measurement scale that may be ordinal, cardinal, or ratio.* Cardinal and ratio scales may have infinite precision,† but precision beyond a certain level may have little meaning to the problem at hand. To use a familiar example, it is seemingly meaningless to express a value by saying, "I like ham sandwiches 1.95843106 times better than cheese sandwiches." Thus the meaningful precision of the outcome value scale is limited.

Second, there is a limitation to the accuracy in assigning values to a given variable. This is especially significant when values are obtained empirically from direct observation, because of the limitations of the measurement systems (statistical errors) and the measurement observers (bias errors). Although even larger inaccuracies may arise in value judgment expression, limited accuracy is expected and acknowledged in this case. The inaccuracies in empirical measures are sometimes masked by the elegance of the data acquisition methodology.

The branch method, on the other hand, accepts imprecision and inaccuracy in both value assignment and policy alternatives at the outset. Being incremental the analysis compares both values and policies on a relative basis. The precision and accuracy need only discriminate among alternatives at each successive limited comparison. This goal may not always be realized and further evaluation may be required, but the pragmatic aspect of this approach is such that no more information is sought than is necessary to make a decision, at least in principle. Since only incremental changes in policy and incremental values are considered, one is always wary that significant factors may have been ignored and one has missed the forest because of the trees.

The shortcomings of reducing uncertainty in each approach lead to the development of a hybrid approach.

* Nominal scales are descriptive. See Glossary for definitions.

† Compare "precision" and "accuracy" in Glossary.

12.5 THE GROSS BALANCE METHOD

The hybrid approach is called the gross balance method.* It utilizes the root method as a prescreening technique to assure that all meaningful variables and alternatives have been considered and described in the structural sense; then gross estimates of initial values are used to exercise the structure, to determine the critical variables. As a result of this prescreening, further data acquisition is restricted to critical variables, which then serve in conjunction with the branch approach to evaluate policy alternatives. The decision may be made directly or iteratively depending on how agreement is reached on a policy. The precision and accuracy of the values assigned to critical variables need only to assure discrimination among alternative policies in the final iteration.

The steps in applying the gross balance method are as follows: (1) develop a theoretical, structural model that covers all significant possible considerations, (2) exercise the model using test values to determine which possibilities and variables are critical, (3) obtain further information on critical alternatives and variables if warranted, and (4) develop an analysis using critical variables and policy alternatives together to lead to an acceptable outcome.

12.5.1 Structural Modeling

The structural models envisioned in this step are similar to those in which the literature abounds; some of these are described in subsequent chapters. Here, however, the structure is supposed to provide insight into the problem under study (in the gross balance method); there is no expectation of a direct contribution to a solution. The structure simply ensures that all critical variables and alternatives have been considered.

The extent of resources devoted to development of such a model must be commensurate with the type of final decision to be made and the degree to which assurance that all factors have been considered adds to one's confidence in accepting the final decision. Such tradeoffs are not easily quantifiable themselves and will rely on the best judgment of those involved in the decision making process.†

The objective of the model is realized when it is exercised to isolate the critical parameters. Thus the model must be amenable to both completeness

* This approach combines the branch and root approaches, but the suggestion that it be called "broot force" was rejected because the opposite connotation is more closely appropriate to the method.

† The intent here is to assure that a "scientific and technical" examination has been made such that credibility of coverage is obtained, as opposed to elegance in method and means as a sop to credibility.

of coverage and ease of determining critical parameters. A good example of such a model is the cost-benefit computer model used in the LMFBR environmental impact statement.[24]

12.5.2 Exercising the Model

Available data can be used to set "initial" values for variables and interrelationships for model evaluation. When data are not available, best guesses and value judgments are used. The values are entered into the model to develop a "baseline." Next parametric studies are made by evaluating different alternatives for the baseline, then changing values from the baseline systematically as alternatives are evaluated.† The objective is to determine which variables are most significant in affecting policy choices.

The LMFBR study, for example, demonstrated that after exercising the model and performing extensive sensitivity analysis, three variables were the most critical in determining the cost-benefit balance. These were (1) the future demand for power, (2) the availability and quantity of uranium raw material resources, and (3) the growth of competing technologies. Since these variables all depend on future conditions to determine value, accurate assignment of value was not possible. Nevertheless, the critical parameters were isolated.

12.5.3 Need for Further Data

Once the critical parameters are identified, further data and information may be obtained on these particular parameters if it is desirable to do so. In some instance it may not be possible to acquire better data, however, or to acquire it in a timely manner, especially when projections into the future are needed.* In many cases, even if data are obtainable, the cost of obtaining the information may exceed its value in securing policy agreement. The pragmatic objective is credible agreement; and if gross measures are adequate to arrive at agreement, acquisition of additional data may be desirable but not mandatory.

12.5.4 Gross Balance Analysis

The critical parameters and the meaningful policy alternatives are at this stage available for analysis. The decision maker(s) examines the critical parameters and verifies that they are complete. Personal and political

† This technique is often called "sensitivity analysis" and is defined in the Glossary.

* Power demand estimates for the LMFBR are a good example.

parameters may be added if they have not been included in the structural analysis. For example, the political climate and the personal situation, integrity, and credibility of the decision maker are often important considerations.

Relative levels of importance may be assigned to different parameters by the decision maker(s), formally or informally. Policy alternatives are analyzed to determine how well they achieve amelioration of conflicting parameters. If no policy alternative clearly emerges as the most desirable, the analysis of interaction of alternate policies with each parameter is studied to attempt to synthesize a new alternative, perhaps a compromise featuring parts of existing alternatives. Conversely, weights of parameters may be altered and a clear policy agreement forced. This does not imply coercion, but a studied reevaluation of values and policies. The process of reevaluation in itself may help in reaching agreement.

Thus the policies and values are successively evaluated until a "position" is reached. This "position" may not be a solution, but the best that can be done in a given situation.

The gross balance method, as just outlined in four steps, is essentially the process that has presently evolved under the National Environmental Policy Act of 1969, known by the acronym NEPA. The preparation of environmental impact statements on new government undertakings is required to facilitate analysis of the cost and benefits and the possible alternative actions available. Such statements are issued in draft form for comments; then in final form after comments have been considered. The legal requirement is that such an analysis be made, not that it project a solution. The decision maker in the agency proposing the project makes the final decision among alternatives, and as long as his choice is not unreasonable in light of the impact statement, his choice may deviate from those recommended as best by the detailed analysis. He may take into account political, economic, and personal factors that the analysis in the impact statement could not possibly cover.

Thus the author has not invented the gross balance method but simply identified it, formalized it, and provided a stepwise structure for its implementation.

The advantages of the method are demonstrated later in several applications. The approach is epistemologically all-encompassing, however, in that the initial model development is essentially Leibnizian, the ability to constrain the final analysis to a few critical, observable parameters is Lockean, the multialternative structure is Kantian, and the successive comparisons of conflicting values as agreement is sought are Hegalian. Finally, the overall method is Singerian, since the goal is a pragmatic, limited solution.

In this analysis the gross balance method is primarily applied to risk decisions as part of larger societal decisions. As such, its use here is limited to aspects that are pertinent to risk evaluation.

REFERENCES

1. Excerpt from "Proposal for Analyzing Social Choice About Technological Development," submitted to U.S. Environmental Protection Agency by the Institute of Science, Technology, and Public Policy, Purdue University, Lafayette, Ind., 1974.
2. A. Campbell, P. E. Converse, W. E. Miller, and D. E. Stokes, *The American Voter.* New York: Wiley, 1960.
3. A. Campbell, P. E. Converse, W. E. Miller, and D. E. Stokes, *Elections and the Political Order.* New York: Wiley, 1966.
4. V. O. Key, Jr., *The Responsible Electorate.* Cambridge, Mass.: Belknap-Harvard, 1966.
5. I. de Sola Pool, R. P. Abelson, and S. L. Popkin, *Candidates, Issues and Strategies.* Cambridge, Mass.: MIT Press, 1964.
6. William R. Shaffer, *Computer Simulations of Voting Behavior.* New York: Oxford University Press, 1972.
7. Raymond S. Bauer and Kenneth J. Gergen (Eds.), *The Study of Policy Formation.* New York: Free Press, 1968.
8. Victor S. Thompson, *Decision Theory, Pure and Applied.* New York: General Learning Press, 1971.
9. Heather W. Johnston, "Representation and Aspects of Policy-Making in Four State Legislatures." Ann Arbor, Mich.: University Microfilms, 1972.
10. Douglas MacGregor, *The Human Side of Enterprise.* New York: McGraw-Hill, 1960.
11. Abraham Maslow, *Eupsychian Management.* Homewood, Ill.: Irwin, 1965; see also *Motivation and Personality* (New York: Harper & Row, 1954).
12. Frederick W. Taylor, *Scientific Management.* New York: Harper & Row, 1911.
13. Rensis Likert, *New Patterns of Management.* New York: McGraw-Hill, 1961.
14. Ian I. Mitroff and Murray Turoff. "The Whys Behind the Hows," *IEEE Spectrum,* Vol. 10 No. 3, March 1973, pp. 62–71.
15. Herman Kahn, *Basic Public Policy Issues,* HI-DFCC-1-I. Vol. 1. Croton-on-Hudson, N.Y.: Hudson Institute, December 1969.
16. Thomas S. Kuhn, *The Structure of Scientific Revolutions,* 2nd ed. Chicago: University of Chicago Press, 1970.
17. Robert L. Bartley, "On the Limits of Rationality," *Wall Street Journal,* September 10, 1971, p. 8.
18. Alice M. Rivlin, "Systematic Thinking for Social Action." Washington, D.C.: The Brookings Institution, 1971.
19. Karl E. Scheibe, "Five Views on Values and Technology," *IEEE Transactions on Systems, Man, and Cybernetics,* Vol. SMC-2, No. 5, November 1972.
20. Abraham H. Maslow, *The Psychology of Science.* New York: Harper & Row, 1966.
21. Kenneth Boulding, *The Impact of the Social Sciences.* New Brunswick, N.J.: Rutgers University Press, 1966.

22. Charles E. Lindblom, "The Science of Muddling Through," *Public Administration Review,* Vol. 19, No. 2, Spring 1959.

23. Donald N. Michael, "On the Social Psychology of Organization Resistances to Long-Range Social Planning." Paper presented at the IEEE Workshop on National Goals, April 1972.

24. Proposed Final Environmental Statement on the Liquid Metal Fast Breeder Reactor Program, WASH-1535. U.S. Atomic Energy Commission, January 16, 1975.

25. John N. Warfield, "An Assault on Complexity," Battelle Monograph No. 3. Columbus, Ohio: Battelle Memorial Institute, 1973, pp. 1–3.

13

Measurement Problems in Risk Assessment

The processes of risk estimation and risk evaluation involve some basic problems in measurement that must be examined in depth before the factors that directly influence the assessment of risk can be considered. These problems of measurement may be classified as several questions, as follows:

1. Who evaluates risk? Why? What biases are introduced?
2. What is meant by value and utility?
3. Can consequence values be assigned to both tangible and intangible consequences?
4. How do cultural, situational, and dynamic considerations affect consequence value assignment?
5. What factors affect the assignment of value?

Exploration of these questions gives insight into the problems of measurement and allows an evaluation of various methods to overcome these problems.

13.1 RISK EVALUATION BIAS

Chapter 4 presented the distinction between a "valuing agent" and a "risk evaluator." The understanding of this distinction and the problem of communicating values among people bring up a number of epistemological questions. A means for the author to communicate to the reader certain value concepts without causing a value conflict is an example of the communication problem. What, indeed, is the proper role of a "risk evaluator?"

All these problems are epistemological and are directly related to biases in valuing consequences.

13.1.1 The "Author" as a Risk Evaluator

It is important to realize that any judgments made by the author pertaining to actual assignment of values fall into one of two categories: (1) the value to the author as a valuing agent, or (2) the value assigned by the author in the role of a risk evaluator. The first category is, indeed, limited for this kind of study; and the author must take the prerogative of a risk evaluator to generalize about the subjectivity of assignment of value. In this role the author makes no claim to the validity of values assigned, however, but seeks to develop insight into the valuation process.

The valuation of consequences is a subjective process. Every reader may have a different reaction to a subjective condition. To take this problem into account, the person or group at risk is often represented by the reader and referred to in the first or third person personal sense. Ranges and alternatives are described so that the reader and the author will not find themselves locked in an argument about the subjective value of a situation but can agree, instead, that the methodology is at least valid or appropriate.

The objective is to be able to present concepts involving risk values to the reader, not necessarily to make value judgments on acceptability of different assumptions. The author and the reader are certain to differ on the latter, but this should not stand in the way of expressing conceptual agreements and disagreements.

13.1.2 The "Test Valuing Agent" Approach

To assess the effect of changes in conditions in the valuation of a consequence, a "test valuing agent" is required. This individual is neither real nor typical, but it provides a means to test the variability of value assignments. Both the author and the reader can assume the role of test valuing agent in a variety of conditions expressing ranges of differences in society. When possible, these ranges are examined. However the concept of a "test valuing agent" provides a means to simultaneously act as valuing agent and risk evaluator—the first to examine one's own value, and the second to estimate group and societal behavior in determining value.

13.1.3 Individual Experience Versus Social Experiments

Many of the concepts put forth in this book are the results of individual observations made by the author, or his interpretation and evaluation of societal behavior from data developed by others. The author has not carried out any social experiments as such; thus there is no formal proof of concepts. In this area of societal behavior and intangible parameters, generalizations for better conceptual understanding of the problems may be the most that is realistically possible.

13.1.4 Risk Evaluator Roles

A major question arises in the interpretation of the role of the risk evaluator. Since the risk evaluator is making intrinsic decisions for others, this action is valid under some conditions and invalid under others. Two particular conditions seem to justify a proper role for a risk evaluator: (1) a problem is so complex that special technical expertise is required to analyze and evaluate the situation, and (2) inequities caused by imposition of risks must be recognized and rectified.

A good example of the first condition is technological forecasting and assessment. The future impact of present actions is examined in terms of technological development to determine the positive and negative impact on society. The development of these forecasts and assessments is indeed complicated. When the technical phase is complete, however, the decisions to be made in determining whether the societal impact and the balancing of benefits and costs are acceptable are societal decisions, not technical ones.

Societal decisions are examples of the second condition if inequities exist in the balancing of cost and benefits. It is the role of government, both the executive and legislative branches, to assure that analyses are made on a technical basis and that the public is protected from unwarranted inequities. The Office of Technology Assessment was established in 1974 to handle the first condition, and the National Environmental Policy Act of 1969, requiring environmental impact statements for all projects with potential environmental impact, has achieved a high degree of analysis in the executive branch.

Regulatory agencies such as the Federal Trade Commission, the Environmental Protection Agency, the Food and Drug Administration, and the Occupational Health and Safety Administration, attempt to balance inequitable assignment of costs and risks. It would seem that those "restrictive regulatory" agencies play a proper role as risk evaluators as opposed to "permissive regulator," such as the Federal Power Commission, the Nuclear Regulatory Commission, and the Interstate Commerce Commission; their role is to promote and regulate an industry. This restrictive-permissive dichotomy is used to emphasize the two distinct roles that exist in regulatory activities. This distinction is often blurred in actuality, since many agencies have concurrent responsibility for both roles. The success of a dual function agency in disassociating the two roles increases the credibility for the restrictive function.

In any case, when one plays the role of risk evaluator, justification for the validity of such action must be available to the public and those affected if the action is to be credibly accepted.

13.2 VALUE AND UTILITY

We have values that guide our choices, cause us to act as we do, and influence the manner in which we look at risk and take risk aversive action. The genesis and operation of values are subjects of considerable controversy. For many decades behavioral scientists in their zeal to be scientific shied away from value concepts, which were thought to be too subjective, as well as unnecessary from a hypothetical viewpoint. However modern behavioral psychologists suggest that some human behavior follows from individual beliefs and values[1] that are related to cognition and motivation, respectively—major areas of psychological research. A person's action is conceived as depending both on what he believes (expects, knows, suspects) and what he values (wants, desires, prefers).[1] Beliefs are associated with epistemology, values with ethics.

Conversely, there are those who say that values are a by-product of action.[2] People infer values and attitudes from behavior and modify them to fit their behavior. Values are not a cause, but a justification of action taken. Values are thus criteria for judgment, not an innate property of objects.

Whether innate or criteria for judgment, values are the major determinants in risk behavior (as in any other behavior). If they are innate, it is possible to separate values and policy as prescribed by the rational-comprehensive method of analysis (root method). If they are only criteria for judgment, the values and policies are inseparable and are only treated by the method of successive limited comparison (branch method). As long as the gross balance method is available, the helpful aspects of both theories can be used pragmatically, and the nature of value derivation will be a philosophical problem that does not limit practical application.

Axiology is the study of value as opposed to the source of value. However another term, "utility," is used in decision analysis and is sometimes confused with the concept of value. It is important to recognize the difference between value and utility for purposes of analyzing risk behavior.

13.2.1 Value and Utility

The terms "value" and "utility" are often used interchangeably. However differences, primarily in the manner they are utilized, must be noted.

"Utility," which is often used in economic and statistical decision theory, denotes the rational behavior of people in satisfying their needs and wants. As shown by von Neumann and Morgenstern,[3] utility can be quantified, but always in terms of rational economic behavior. The concept of rational economic man pervades economic theory and generally limits its application to situations in which rational behavior is thought to exist.

The concept of value, used to express the satisfaction of man's desires and wants, is not contrained to the concept of rationality. Intangible factors, such as the fulfillment of emotional, aesthetic, and ethical needs, are also included. Axiologists such as Hartman[2] have attempted to define "goodness" and "badness" in terms other than economic concepts. Thus value attempts to measure total behavior, including aspects that are not necessarily rational.

For purposes here, utility is defined as a measure of desirability in terms of rational-economic behavior, and value is defined as a measure of desirability in terms of the totality of human behavior. Utility is then a special subset of value.

The perception of both utility and value is for a specific agent at a specific time. Each agent has his own set of utility factors and value factors that change with time and with situations. It is one thing to measure utility or value for an individual agent at a particular time for a particular situation, but to be able to handle large numbers of agents for which the perceptions of many individuals must all be taken into account is extremely difficult, although perhaps technically feasible. Practically, it is desirable to construct scales of utility or value against which the perception of many agents can be interpreted. These scales form a syntax for the communication of the perceived utility and value functions among individuals, and carry all the intrinsic errors involved in attempting to quantify intangibles.

Scheibe has identified the "individual-collectivity value paradox"[1] whereby individual prudence may inexorably produce collective disaster. For example, when each farmer demanded his right to have his cows graze on the common, there was overgrazing, to everyone's loss. "The point is that it is difficult to get an individual or a collective entity socialized to the interests and values of competitions, so that when *they* make decisions, they take *our* values into consideration."[1]

13.2.3 Value Groups

Within society, each individual has a number of wants that he wishes to satisfy. Based on these wants, the individual seeks to obtain maximum satisfaction at a given time.

According to von Neumann and Morgenstern,[5] these wants may be considered as variables, and the maximum of satisfaction can be expressed as the maximization of a numerical utility function. These variables or parameters can be tangible or intangible, and they can be enumerated for a given individual. All individuals in society have similar sets of variables, although the wants that are represented by these variables differ in magnitude. Von Neumann and Morgenstern[5] termed the set of variables for an individual a "partial set of variables." The partial sets of all participants in society constitute the total set of variables. The total number of variables is determined by the number of partial sets plus the number of variables in every partial set. For a society consisting of a single individual, the partial set of that individual and the total set coincide. In addition, the variables of any one partial set may be treated as a single variable that represents a scale of value for the particular individual involved and corresponds to this partial set.

Technically, it is possible to handle the large number of variables that exist within a total population, but economically this is impractical. Two simplifying assumptions allow us to reduce the problem a more manageable scale. The first is to select for each partial set a finite set of parameters, identical for all partial sets in the total set. That is, if n parameters are represented by $x_1, x_2, \ldots, x_n$ for one partial set, there will be the same set of parameters for any other partial set, although the magnitudes or values assigned to these variables will differ among sets. The second assumption is that some partial sets have magnitudes of values that are nearly identical. Partial sets with close identity may be grouped together and called value groups. As the allowable difference in exact identity among partial sets within a value group is increased, the number of value groups is decreased. Thus a tradeoff between the accuracy of representation of a partial set to an individual's set of values may be made against the number of value groups (i.e., partial sets that have to be considered in the problem).

These assumptions allow one to compromise the accuracy of the representation of individual values to achieve a situation that is manageable in a practical sense. Many different partial sets can be considered, which makes it possible to treat the values of major conflicting groups within society or the population as a whole, as distinguished from a single monolithic set of values for the total population involved.

If the actual values or the parameters identified could be determined for each individual, value groups could be established statistically. The allowed distance between the means of various groups as well as their variances could be used to determine the accuracy of the representation of the value groups and the number of value groups necessary. This approach would not be practical, however, because of number of samples that would be needed and the difficulty of measuring such values. Other less exact means for determining the composition and structure of value groups must be considered.

It is important to note that value groups as defined are not actually groups in the sense of policy and social groups covered in Chapter 10. Here value groups are collections of people with like values, but no expressed interaction. The interest of policy and social groups is to achieve an objective. A value group may cut across many policy and social groups; policy and social groups may contain a multiplicity of value groups.

13.2.3 Value Scales and Goals

13.2.3.1 Value Scales

A scale of measurement for a particular variable to which a scale of values is associated can be defined. This scale measures the degree of "goodness" or "badness" in achieving the concept of value involved. Hartman defines "goodness" axiomatically as follows: "A thing is good if it fulfills the definition of its concept."[4] In this manner, the ultimate "good" of a value scale may be defined, and its complement, "nongood," can be related to the other limit of the scale, namely, "badness."* Thus, a value scale can be associated with each variable to which a value is attached.

A value scale implies the existence of a goal. This implies that value scales should be ordinal and preferably cardinal in terms of measurement scales. Nominal scales may be made ordinal by ranking scale members preferentially. Transitivity and consideration of weakly and strongly ordered sets are important considerations in developing useful ordinal scales from nominal scales,[6] since the determination of preferences may be inexact. Conversion of an ordinal scale to a cardinal scale implies inexact degrees of preference,[6] and this impreciseness makes it very difficult to determine cardinal utility. However, this in no way invalidates the assignment of a value to a variable.

13.2.3.2 Goals

A goal is a point on the value scale for a particular variable for which achievement is desired and sought. Thus a goal is an assigned value on the value scale that may or may not correspond with the maximum of the value scale. The maximum value on a value scale may indeed be an ultimate level or ideal which is unobtainable. Therefore, a realistic goal can be set below this level, as illustrated in Figure 13.1.

13.2.3.3 Baselines

A baseline on a value scale is the level of value that exists at a given time, designating a starting point from which one works to achieve a goal in terms of the value for the particular variable involved.

* Essentially, if a thing achieves its goal (concept), it represents "goodness." If a thing has an evil goal and achieves it, the thing is a "good" evil thing. "Good" is fulfilling one's goal; "bad" is the converse.

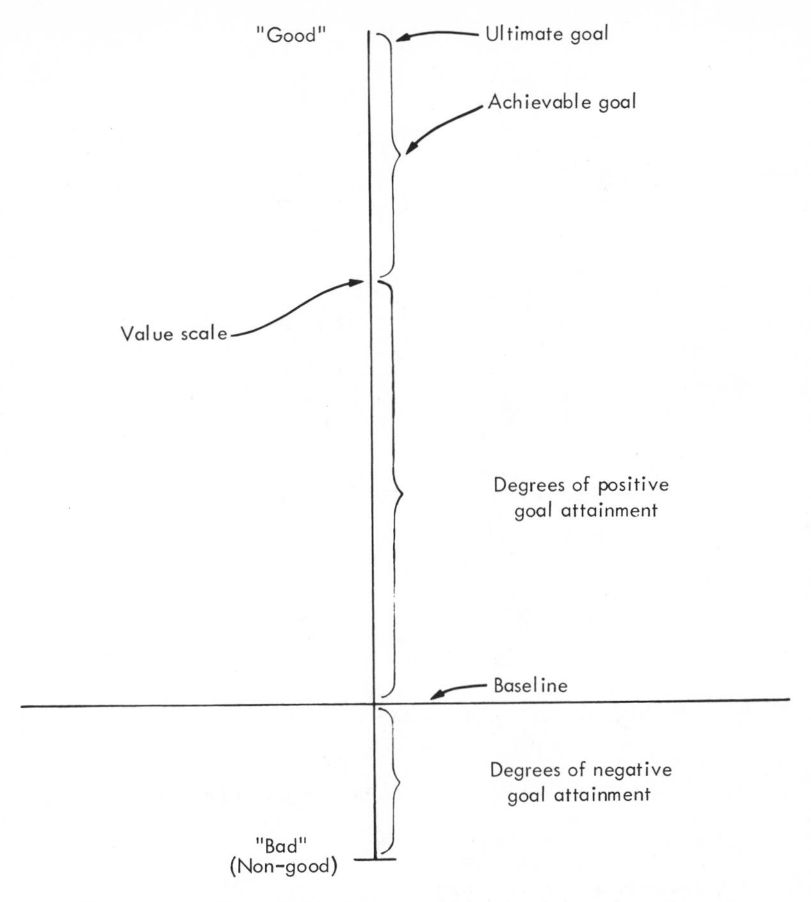

Figure 13.1. Values and goals for a variable related through a value scale.

13.2.3.4 Degree of Goal Attainment

Having set a goal and a baseline on a value scale, the actual measure of the degree of attainment of that goal is found to lie between the baseline and the goal.

13.2.4 Scales of Value and Utility

The assignment of utility or value scaled to particular parameters covers a wide variety of measures. In keeping with previous definitions, scales assocated with monetary values fall into the area of utility measurements. This split is illustrated in Figure 13.2.

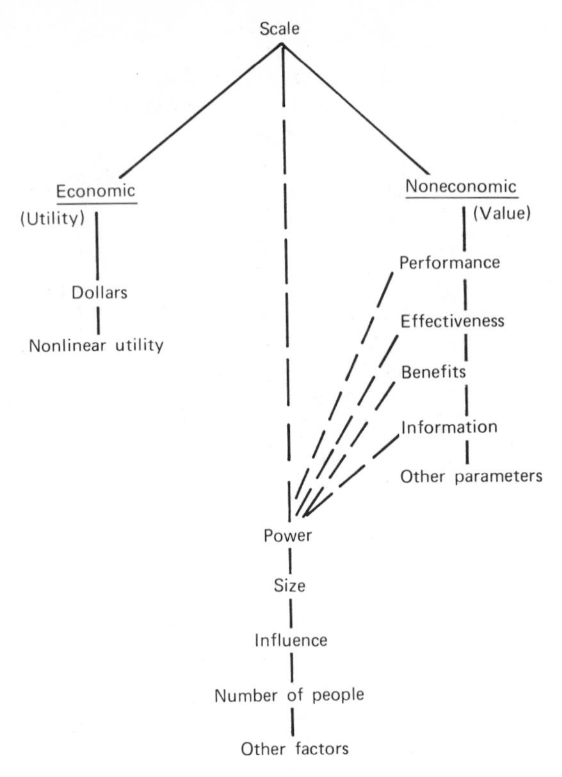

Figure 13.2. Some possible scale interpretations for economic and noneconomic parameters.

13.2.4.1 Interpretation of Utility

In economics, utility is most often defined in terms of dollars. However the meaning of dollars to an individual is dependent on the magnitude of dollars involved, as shown originally by Bernoulli,[7] as well as the risk to the user.[8] In these cases, utility is measured in dollars but then converted to a nonlinear scale of utility for the particular agent. These concepts are well covered in the literature and are not repeated here.

13.2.4.2 Nonmonetary Scales

A wide variety of nonmonetary scales is represented in Figure 13.2; however at least four areas are worth significant mention: value of performance, value of effectiveness, value of benefits, and value of information.

Value of Performance. The relative abilities of different systems to meet specified objectives can be evaluated against each other or the performance of a standard (or idealized) system. The context of performance implied here is that of a physical system, such as the relative performance of different automotive systems to provide transportation with maximum efficiency and minimal environmental impact. The quantities measured are usually extrinsic, but could be intrinsic, such as measures of aesthetic performance.

Value of Effectiveness. The relative evaluation of value systems against bounded, user-determined scales of effectiveness for the user's particular purposes involves the idea of effectiveness. The performance scale of value just discussed is peculiar to the system under evaluation. Effectiveness is peculiar to the wants and desires of a particular user or class of users in the manner that they want a given system to act. Therefore the effectiveness is bounded by the user's desires and wants.

Value of Benefits. The relative evaluation of different systems against an unbounded scale is made up of parameters that are not necessarily constrained to user goals, but are directed toward the total goals of one or more value groups. Here the user must not only account for his own wants and desires, but also for the requirements of others in society.

Value of Information. Information has value to the user or others in society. This information value can be measured in terms of performance, effectiveness, benefits, and other parameters; but once the value of information is ascertained, it becomes a useful scale. For example, cost-effective strategies for gathering data and information are immediately available when the marginal cost of obtaining new information is measured against the marginal value of the information obtained. When the cost of acquiring information equals the marginal value, further efforts to obtain information should be curtailed.

Other parameters that have noneconomic scales (aesthetic value, self-satisfaction, self-actualization, etc.) might be considered.

An Ultimate Value Scale. There are many who say that an ultimate value scale must be measured in terms of power, personal or otherwise. This is why Figure 13.2 shows a conversion of dollars and utility, as well as non-economic parameters, into power. Some of the possible components of power (size, influence, number of people, etc.) are also represented. The validity of this assumption is not debated here, but the implications of investigating the real meaning of power in terms other than monetary become immediately apparent.

13.2.4.3 Scale Measures

The measures of these scales may be nominal, ordinal, or cardinal. However value judgments are often made in the nominal and ordinal sense. In the first case, people are capable of making value judgments in terms of classifications.* The result is a nominal scale with a finite or infinite set of members. In the second case, people are able to order members of a set by preference to form an ordinal scale.† Value judgments may be made on a cardinal scale as well.‡

Cardinal numerical values may be assigned to values only on a cardinal scale. Therefore members of a nominal scale must be ranked to an ordinal scale, and members of an ordinal scale must be assigned values to form a cardinal scale. The valuation of scales to higher levels is in itself a value judgment and involves error and uncertainty.

13.3. PROBLEMS IN THE MEASUREMENT OF TANGIBLE AND INTANGIBLE VALUES

When assigning value to risk consequences, measurement difficulties intrinsic to dealings with value and utility cause problems in the establishment of value magnitudes. These difficulties arise from the need to assign cardinal magnitudes for tangible values and utilities that may be highly nonlinear, as well as for intangible values, which may be assigned only cardinal magnitudes with inherent imprecision.

A review of existing approaches to these problems and presentation of some new considerations is a prerequisite for dealing with valuation for those not familiar with the present state of the art.

13.3.1 Magnitude of Consequence Values

Von Neumann and Morgenstern[3] and Friedman and Savage[9] argue that measures of cardinal utility can be developed for all consequences by a series of "equivalent gambles" among ranked choices. Thus if A is preferred to B, and B is preferred to C, the cardinal utility of B can be determined by finding the equivalent gamble between A and C, for which a choice between the gamble and B represents indifference. Schlaifer[10] has emphasized that this gamble is made with infinite precision. Rowe[11] has demonstrated that such a gamble can be generated with finite precision only, because of the uncertainty and accuracy with which an individual can establish preferences. As a result, a dimension of uncertainty must be added to utility scales to cover uncertainty in cardinal assignment, especially for more intangible parameters.

* A martini drinker who believes he can identify the brand of gin he is tasting, is an example.

† An ordered set of consumer preferences is an example.

‡ An assessor who places a dollar value on a piece of property is an example.

Based on cardinal utility, whether precise or imprecise, the decision maker or risk taker attempts to maximize his expected utility. However Rowe demonstrates that with uncertain utility functions, adequate information is not always available to maximize utility, especially for intangible values. In this case the analysis breaks down and other factors must be considered. Some analysts insist that utility theory does not suffer from the problem of imprecise measures, only that the practitioners have failed to properly measure or express the utility functions. This is an absurdity at the level of practicality. Thus utility theory at its highest level of abstraction may be tautologically appropriate, but pragmatically it is worthless, since it provides no information to solve problems.

The nonlinear utility of monetary consequence with magnitude, the implication of treating intangible values, and the development of scales for tangible and intangible values are important factors involved in evaluating consequence magnitude.

13.3.2 Nonlinear Utility of Consequence Value with Magnitude

Consider a gamble that offers you a chance to flip a fair coin. If the flip comes up heads, the payoff is of magnitude M. If the flip comes up tails, the payoff is zero. The expected value is $\frac{1}{2}M$ for this single flip. You are offered this gamble for a price, but may make only one gamble for a specified magnitude.

In the first case, M is \$100. You are asked how much you will pay for the opportunity to gamble at the expected value of \$50. Anything less than a \$50 price is statistically in your favor, but there is a 50–50 chance of your losing whatever you put up. Perhaps \$25 to \$50 would be an acceptable price, depending on your propensity to take risk and the relative importance of losing \$25 to \$40 and winning \$100 (a gain of from \$60 to \$75). As a conservative gambler, you will not bet at the expected value level. The expected value of the overall gamble for a \$35 price is

$$EV = \tfrac{1}{2}(\$100) - \$35 = \$15 \tag{13.1}$$

Now consider the same gamble where the value of M is \$1. For this gamble, you would still like to pay between \$0.25 and \$0.40, but since

$$EV = \tfrac{1}{2}(\$1.00) - \$0.35 = \$0.15 \tag{13.2}$$

one may very well be willing to pay \$0.50 or even \$0.60 or more just for the sake of the gamble.

$$EV = \tfrac{1}{2}(\$1.00) - \$0.60 = -\$0.10 \tag{13.3}$$

In other words, the stakes are low enough to be relatively meaningless, and the value of the gamble or the act of the gamble exceeds the stakes.

Suppose now that M is \$1,000,000. You might not play even if \$100,000 were an acceptable price to the "house," since you could neither raise \$100,000 nor stand the 50–50 chance of its loss.

This concept of the nonlinear utility of money is schematized in Figure 13.3, where the abscissa is the magnitude of the payoff and the ordinate is the maximum price one must pay for the gamble. (Note that the ordinate is one-half the scale of the abscissa.) The curve shows that at low values, one might value the act of the gamble more than its payoff. A person becomes conservative at higher values, and at very high values one may even decide against gambling, as shown by the dashed line. The concept, well documented in the literature by writers as early as Bernoulli[7] and recently by Howard[12] and Swalm,[8] is based on the concept of diminishing marginal utility.

A variation of the gamble that substitutes a series of gambles with smaller payoffs for a single large gamble illustrates the spreading of risk

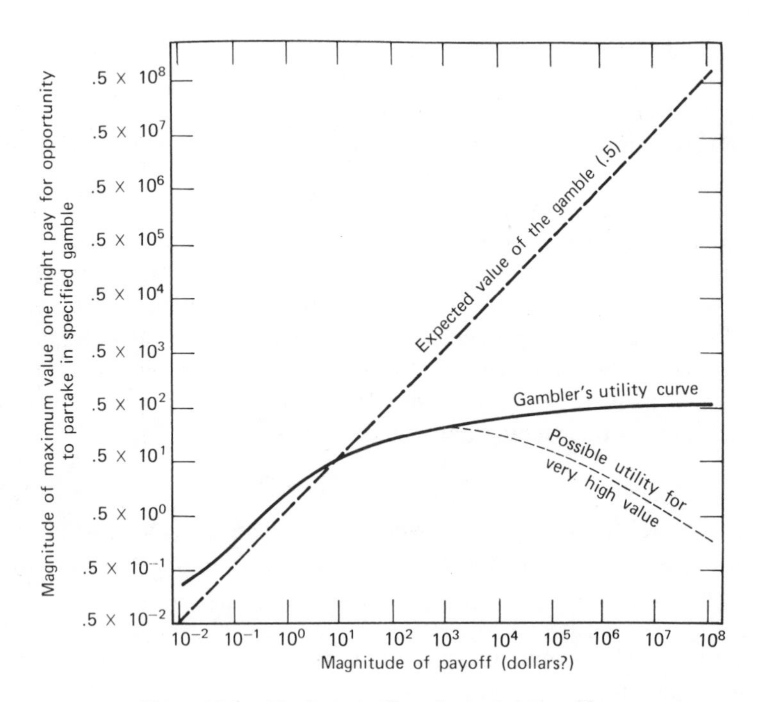

Figure 13.3. Nonlinear utility of monetary gambles.

and payoff over many trials. Consider one gamble of $100 versus 10 gambles of $10 each. In the first case, one might pay $40 for the single gamble and $4 each for 10 gambles. In the first case, the chance of losing $40 (or winning a total of $100) is one chance in two. In the second case, the chance of losing $40 (or winning $100) is one in 1024. A greater number of flips assures closer statistical conformity with the expected value of the gamble; therefore one may be willing to pay more for the opportunity of approaching the expected value as a limit by increasing the number of trials (or something less than the limit if one wants a "sure thing").

The smaller risks and payoffs that result from this series of gambles may make it more attractive to some and less attractive to others, depending on the risk taker's propensity to take risk. This concept, well documented by Swalm, is covered subsequently. Yet although the risks may be the same, there seems to be more concern for the problem of a crash of a large aircraft, such as a 747, and an equal number of fatalities from a number of crashes of smaller aircraft. An examination of this concept must include qualitative and intangible consequences, as well as quantitative and tangible ones.

13.3.3 Scales for Tangible and Intangible Consequences

A slight variation of the previous gamble illustrates the complete breakdown of expected value as a measure of risk when intangible values are assigned to a consequence. You are offered the gamble where the payoff M is for you to immediately kill yourself. If the outcome is tails, you live. Now you are asked how much money you will accept to take this gamble. In other words, there has been a move from a quantitative to a qualitative value. Most people would not even consider the gamble as proposed, but acceptance is highly dependent on particular situations. Consider as an example the case of a valuing agent who will soon die of cancer anyhow.

Gambles of this sort are taken all the time, except that the odds are more favorable, say a million to one or greater, against death. A highly hazardous occupation with premium pay is an example.

However the valuation of a consequence requires the development of at least an ordinal scale of value, if not a cardinal scale. Expected value is meaningful only with a cardinal scale, however; and the general tendency has been to use dollars or other monetary values as a basic cardinal scale. The serious limitations associated with the use of monetary values for values expect in limited economic situations have been described in Chapter 9 as factors affecting valuation of risk consequences.

13.3.4 Assigning Cardinal Values to Consequences

It is sometimes necessary to assign cardinal scales to intangible values. When this is done, it is important to retain the uncertainty in value assignment explicitly by assuring that the cardinal scale's precision is no greater than is meaningful to the valuing agent. When used, moreover, the inaccuracy of measurement must be retained, since our ability to be precise is inherently limited because of the meaninglessness of the mathematics of content when used to a level of precision beyond which information is not usefully communicated.

Thus attempts to assign cardinal values to a relatively intangible parameter encounter a subjective threshold of scale precision above which more precise values have no meaning to the valuer. That is, the valuer may rank his appreciation of three different colors for his office walls in the order of desirability as blue, green, gray. When using a cardinal scale of 0 to 1 to value appreciation (where blue is assigned 1, and gray 0), it is most likely meaningless to the valuer to assign a value to green of .375 (or any other value of three or more significant figures) when all that is really meaningful is his statement that green lies closer to gray in his estimation than blue. That is, the value of green lies in the bottom half of the scale, and the precision of the estimate is less than one part in three. There is no meaning to increased precision in that no more useful information is communicated to the valuer, and the range of precision defines the uncertainty of the estimate. Increased precision masks the uncertainty without an increase in subjective meaning, and it is improper to allow this masking, especially for sake of easy manipulation of mathematics.

13.4 OTHER PROBLEMS IN VALUE ASSIGNMENT

It is unquestionably difficult to measure values of risk consequences when the more intangible values (value of life or the quality of life, e.g.) are concerned. Several other factors must be considered in making such measurements and in understanding the level of uncertainty that exists for them. Also, as described in the next chapter, it is important to specify the uncertainty and to learn how to use it as a parameter.

13.4.1 Situation Dynamics

Values change as situations change, and situations change dynamically. For example, a person may place relatively little value on his health while in good health, but just after an accident or during a period of ill health, his value on good health may be his primary motivating force. Since changes in situations occur continuously, classes of situations usually have to be addressed, such as "normal," "abnormal," or "threat" situations.

It must further be recognized that it is not necessarily a new risk or a change in risk that occurs, but often only the alteration of knowledge or perception of risks represents a situational change. This involves a change in the knowledge base and is treated more fully in Section 16.1.6.

13.4.2 Measure of Value of a Life

Assigning value to premature death is a good example of attempting to attach value to an intangible situation. One method is to try to find a dollar value for a life.

13.4.2.1 Dollars and Value of a Life

Linnerooth has made a survey of various techniques for quantifying increase or decrease in human mortality.

> Four distinct approaches for quantifying human life values have been identified. The earliest, and currently most applied, is the "human capital approach" which values a life in accordance with its potential for future productivity. This approach has gained wide popularity because of the apparent ease in its quantification. A second approach, to quantify life values according to some implicit value, observable in societal acceptance of both public and private risks of death, has received some recent attention but very little application. A third approach, which has received some mention in the literature, is to relate "life values" either to the life-insurance purchases or to judicial awards for mortality cases. The last, and most recently developed approach, to value a life-saving or life-risking program according to the "risk" to which the population is exposed and the willingness of the population to pay to avoid this risk, has received much attention in the academic literature. However, contrary to the human capital approach, the criteria necessary for the "risk approach" are not easily quantifiable, and thus its applicability is still questioned.[13]

Human Capital Approach. The value of life is based on the premise that a man's worth to society depends on his productivity and as a productive unit is considerd human capital. Before discussing this approach, one must determine to whom the costs of premature death are attributable. Is one referring to cost to the individual at risk, cost to society, or cost to the individual's family? These costs are not the same and must be treated differently. In the "net value" approach, only the costs of society and the victim's family are considered. A "gross value" approach deals with all three.

There are major difficulties associated with both approaches. When considering cost to society, it must be recognized that the United States does not enjoy a full employment economy. If an individual working in a particular job dies or suffers a serious health effect, one must assume that he is expendable in that job and that there will be frictional* movement upward, replacing him and his earning power in a series of steps. The person best able to fil the position will move upward, leaving an opening that will be filled in the same manner as its occupant moves up, and so on, until an opening at the bottom of the chain occurs, permitting someone to be taken off the unemployment rolls to occupy the open position. Let us assume that the unemployed individual, who is now to be employed, was on welfare: thus there has been a reduction in public cost (i.e., one less recipient of welfare payments). There may be some adverse differences in earning power because of experience, but total salary involved cannot be considered as a cost to society. In this sense, the algebraic sum of wages lost because of frictional movement upward and reduction in welfare costs (a negative cost) must be used.

Estimating the "net loss" to the community in this way leads, for example, to the fantastic result that if we could have more road accidents, in which we succeeded in knocking down and killing old people we should reduce the "net loss." . . . It is indeed a sad commentary on the state of public conscience if we have to be persuaded that measures to reduce road accidents will pay, in some economic sense, before we will listen seriously.[14]

Rappoport states the following three grounds for rejecting both the net and gross value approach.

1. The lifetime-earnings measure is deterministic. Actually, there is no theoretical basis for using such a measure, in the first place. But if there were, it would *seem* to refer to a conceptual experiment of trading "a life" for money. In practice, we are usually evaluating small increases in the *probability* of death. Clearly, our answer must come to grips with risk attitudes; and it is well known that uncertain losses are (usually) subjectively valued at larger than the statistical expectation.

2. Lifetime-earnings take no account of leisure. If there were no saving or borrowing, consumption activity would be proportional to earnings and this problem would disappear. But the phenomenon of retirement places a severe burden on such a simplification. The human capital measure seems to undervalue old people's lives vis-à-vis working-age people.

3. The lifetime-earnings measure does not include externalities. Many people feel loss from the death of another even though they derive no financial benefit from that person.[15]

* "Frictional" is used in an economic sense to indicate a noninstantaneous process.

Implicit Societal Evaluation

Since society, through its political processes, does in fact make decisions on investment expenditures which occasionally increase or decrease the number of deaths, an implicit value of human life can be calculated. Such a method does not require any direct calculation of the loss of potential earnings or spending. Instead, it approaches the problem from a social point of view by estimating the expenditure society actually makes to save a life. If, for example, an arrangement is made that will increase safety, and save an estimated five lives, at a cost of $100,000, then the implicit value of a life is $20,000.[13]

Starr[16] has attempted to uncover historical risk versus benefit relations for the societal and individual acceptance of both voluntary and involuntary risk by estimating the implicit value individuals put on their own lives by accepting voluntary risks and the implicit value society attaches to social risks by preventing them. His selected risks include the Vietnam war, hunting, general aviation, motor vehicles, electric power, and natural disasters. The benefit is calculated by estimating money spent to participate in or to avoid the activity. It should be noted that the effort is to assign dollar values to risk, as opposed to the more acceptable approach of using historical risk to determine risk preferences. Mishan criticizes this procedure for decisions that have been implicitly made in democratic societies. On an implicit value placed on life by the political process, he comments as follows:

> [The] justification appears somewhat circular . . . the idea of deriving quantitative values from the political process is clearly contrary to the idea of deriving them from an independent economic criterion. . . .

In other words, unless it is assumed that societal decisions on life-death tradeoffs have been made in the past according to some notion of optimization, such decisions should not be an input into current decision-making.[17]

Insurance Premiums and Court-Decided Compensation. It is suggested that the amount of life insurance one is willing to purchase is related to the value one places on his life and the probability of being killed by some specific condition or activity. First, the insurance premiums do not represent risk aversion, since the probabilities of death are unchanged by the insurance purchase.* Second, protection purchased by insurance is for one's dependents and beneficiaries, not for one's own protection. Thus the amount of insurance a man takes out reflects only his concern for his dependents; it is not a measure of the value he sets on his own life.

* Insurance companies sponsor safety programs and can affect probabilities and subsequent premium rates, but individuals' purchases of insurance have no impact on probabilities, barring suicide.

Court-decided compensation or jury awards have been used extensively as estimates of the disbenefit from "pain and suffering" but seldom have served as an estimate for the disbenefit of "loss of life." As in insurance premium cases, court decisions which are after the fact do not affect the probability of risk to the individual. Also, the awards are to dependents, not to the person killed post hoc.

In spite of the difficulties, the Federal Aviation Administration has recently suggested this approach. In a draft order to develop a benefit/cost method for selecting and ranking airport traffic control towers, the Federal Aviation Administration recommends a figure of $300,000 and $390,000 for an air carrier fatality and general aviation fatality, respectively. These figures are calculated from the CAB non-Warsaw Pact accident payments for the period 1966 to 1970 and extrapolated to 1974.[13]

The Risk Approach. A more meaningful measure that often can be explicit is the amount of money society and the infrastructure is willing to pay to prevent a premature death. This can be observed by actually measuring what society pays for safety and antipollution measures. This is a derived measure. It should also be recognized that there is a significant difference in how society regards a statistical premature death as opposed to an individually identifiable case.

Discounting in space (i.e., spreading the risk from an individual to a statistical member group) reduces the value society places on premature death. Society will spend millions to save an infant who has fallen into a well, but will spend at least an order of magnitude less to prevent premature death on a statistical basis. Studies of compensation, willingness to pay, and so on, indicate that society is willing to spend between $100,000 and $500,000 to avert a single premature death on a statistical basis.[18-21]

The evaluation of a public program which increases (decreases) human mortality, ignoring the statistical expectation of lives saved (lost) and concentrating only on the reduction (increase) in the "risk" of death, is termed the "risk approach." Appropriate within the context of a cost/benefit analysis (sometimes referred to as benefit/risk analysis), it eliminates, on the part of the decision-maker, the necessity to define an absolute "value of life." The appropriate benefit (or cost) is not the value of a life saved but becomes the value of reducing the probability (usually small) of loss of life. The assumption is that preferences for risk reduction are not linear.[13]

13.4.2.2 Non-Dollar Measures

There are some noneconomic measures that may be used to express the value of a life. For a given situation, the risks of life shortening may be balanced against life extending benefits directly. A case in point is the use of X-rays for medical diagnosis and therapy; properly applied, such treatment can extend life, but it involves radiation exposure that can increase somatic and genetic risks.

Different cultures place different values on life. When one is barely surviving in an undeveloped nation, life is "cheap." This also may be true for different value groups in society. If so, degree of satisfaction with the status quo may be a gross means of indicating different magnitudes of life value.

Finally, direct, weighted cardinal measures of value in the form of indices may provide useful measures as long as one is aware of the range of uncertainty and lack of precision of such techniques.[11, 22] Since the use of dollars as an index often masks inherent imprecision, such noneconomic indices may well be preferred.

Chapter 14 presents an overview of the various techniques and methods that have been developed to address the problem of measurement of imprecise and intangible values.

REFERENCES

1. K. E. Scheibe, "Five Views on Values and Technology," *IEEE Transactions on Systems, Man and Cybernetics* Vol SML-2, No. 5, November, 1972.
2. Milton Rokeach, *The Nature of Human Values.* New York: Collier-Macmillan, 1973.
3. John von Neumann and Oskar Morgenstern, *Theory of Games and Economic Behavior.* Princeton, N.J.: Princeton University Press, 1953.
4. Robert S. Hartman. *The Structure of Value: Foundations of Scientific Axiology.* Carbondale: Southern Illinois University Press, 1967.
5. von Neumann and Morgenstern, *Theory of Games and Behavior,* p. 10. All aspects of this paragraph are from this reference.
6. Peter C. Fishburn, "Utility Theory with Inexact Preferences and Degrees of Preference," *Syntheses* Vol. 21, (1970), pp. 204–221.
7. Daniel Bernoulli, "Specimen Theoriae Novae de Mensura Sortis," 1738, as referenced by David W. Miller and Martin K. Starr, *The Structure of Human Decision.* Englewood Cliffs, N.J.: Prentice-Hall, 1967.
8. Ralph O. Swalm, "Utility Theory—Insights into Risk Taking," *Harvard Business Review,* November–December 1966, pp. 132–136.
9. Milton Friedman and L. J. Savage, *Journal of Political Economy,* Vol. 56, 1948, pp. 279–304.
10. Robert Schlaifer, *Analysis of Decisions Under Uncertainty.* New York: McGraw-Hill, 1969.
11. W. D. Rowe, "Decision Making with Uncertain Utility Functions," Doctoral thesis, American University, Washington, D.C., 1973, pp. 27–33.
12. R. A. Howard, "Decision Analysis: Applied Decision Theory," *Proceedings of the Fourth International Conference on Operational Research,* Vol. SSC4, No. 3, September 1968, pp. 211–219.

13. Joanne Linnerooth, "The Evaluation of Life-Saving: A Survey," IIASA Draft Report, Laxenbourg, Austria: International Institute for Applied Systems Analysis, March 1975.
14. Ely Devons, *Essays in Economics.* London: Allen & Urwin, 1961, p. 108 (from Linnerooth).
15. Edward Rappoport, "Economic Analysis of Life and Death Decision Making," in *Applying Cost-Benefit Concepts to Projects which Alter Human Mortality,* J. Hirschleifer, T. Bergstrom, E. Rappoport, UCLA-ENG -7478. Los Angeles: UCLA School of Engineering and Applied Science, November 1974, pp. 2–3 of Appendix 2.
16. Chouncy Starr, "Social Benefit Versus Technological Risk," *Science,* Vol. 165, September 1969, pp. 1232–1238.
17. E. J. Mishan, "Evaluation of Life and Limb: A Theoretical Approach," *Journal of Political Economy,* July–August 1971 (from Linnerooth).
18. *Insurance Facts,* 1966 ed. New York: Insurance Information Institute.
19. H. J. Otway, "The Quantification of Social Values," LA 4860-MS. Presented at the Symposium on Risk versus Benefit, Solution or Dream, LA 4860-MS, Los Alamos, N.M.: February 1972.
20. R. Wilson, "Tax the Integrated Pollution Exposure," *Science,* Vol. 178, October 1972.
21. J. Coates, "Calculating the Social Costs of Automobile Pollution—An Exercise." Presented at the Symposium on Risk vs. Benefit, Los Alamos, N. M. November 1971.
22. W. D. Rowe, "The Application of Structural Value Analysis to Models Using Value Judgments as a Data Source," Technical Report M70-14. McLean, Va.: The MITRE Corporation, 1970.

14

Methods for Evaluation of Imprecise and Intangible Values

A great deal of effort has been expended in the development of methods for formulating decisions involving social parameters and values. Many of these techniques are quantitatively oriented: decision theory, probability and statistical theory, operations analysis, utility and preference theory, are examples. An exhaustive review of these methods is beyond the scope of this book.* However certain key methods that deal with the uncertainties associated with evaluating imprecise and intangible values require some attention. They involve problems in dealing with uncertainty.

As a first step, classical cost-benefit analysis is reviewed to identify its shortcomings and the limitations of this approach through examination of various uncertainties inherent in the methodology. Subsequently, two methods of treating uncertainty in values and utility are addressed. The first involves treatment of imprecise utility functions in decision theory approaches, and the second features a method of structuring nonlinear value functions and value weights with direct means of determining the effect of uncertainty.

14.1 A REVIEW OF COST-BENEFIT ANALYSIS AND ITS INHERENT UNCERTAINTIES

There are many versions and many aspects of what is called cost-benefit analysis, including what is meant by "cost" and by "benefits." The nomen-

* For those interested in pursuing the array of methods available in this area, the author recommends review of a summary of methods prepared by the MITRE Corporation entitled "A Preliminary Study of a Concept for Categorizing Benefits."[1] In developing this concept, the authors initially surveyed available methods, determining their applicability and their limitations. This survey provides a point of departure for review of the broad scope of methods available.

clature of Chapter 5, where the concepts of direct and indirect gains and losses were introduced, is used and amplified here.

A cost-benefit (loss-gain) analysis is a two-step process. The first step consists of a broad, gross comparison of gains and losses, direct and indirect, to determine whether the undertaking for which the analysis is made is such that direct gains outweigh direct losses and indirect gains and losses can be balanced or inequities reasonably ameliorated. The second step involves a more detailed analysis to determine whether inequities have been ameliorated adequately. Wilson shows four steps:

> First, we must be sure that we understand the benefit and the risk and that the former outweighs the latter.
>
> Second, we must be sure we have chosen the method of achieving the benefit with the least risk.
>
> Third, we must be sure we are spending enough money to reduce the risk further.
>
> Fourth, we go back (iterate) and recheck our numbers with new perspective from the preliminary calculations.[2]

Primary emphasis here is on the latter three steps of Wilson, equivalent to amelioration of inequities.

14.1.1 An Economic Model of Cost-Benefit Analysis

14.1.1.1 Model Structure

Direct expenditures to minimize risks and other indirect costs or to obtain specific benefits involve commitments for dollar outlays. These expenditures are direct losses and are costs in the classical economic sense. Indirect losses, including risks, are reduced by expenditures resulting in direct losses. When the indirect losses are all risks (involving probability of occurrence of consequences as opposed to relatively certain occurrences), a cost-effectiveness of risk reduction curve may be developed (Figure 5.2).* Superimposed over this curve, a curve for achieving the cost effectiveness of obtaining direct and indirect gains (benefits) also can be drawn, but this requires a scale different from that used for losses (indirect costs). This latter curve is

* The term "cost-risk-benefit analysis" is often seen in the literature. The following equivalence with gains and losses is used here:

Costs. Direct losses (economic).

Risks. Indirect losses (economic and otherwise).

Benefits. Gains, both direct and indirect.

convex upward, since the steps to obtain benefits can be ordered by ratio of gain to direct cost. Both curves appear in Figure 14.1.

Economic theory indicates that assuming the scales for indirect losses and gains are identical, balancing the two curves at the margin will provide an economically optimum condition. This means that when the slopes of the two curves (their first derivatives) are equal, another dollar spent to achieve benefits will be no more efficient than a dollar spent to reduce risk.

14.1.1.2 Model Limitations

The assumption that the two scales are identical seldom holds in practice and is the exception rather than the rule. Attempts to find weights to assign to the scales to equate them involves considerable uncertainty. There are a variety of indirect losses and direct and indirect benefits. Some losses may be equated with some benefits at the level of individual items. The problem then is how to aggregate the individual items to achieve an overall balance, or, conversely, how to aggregate against each scale and then make an overall balance. This is a problem of method of aggregation.

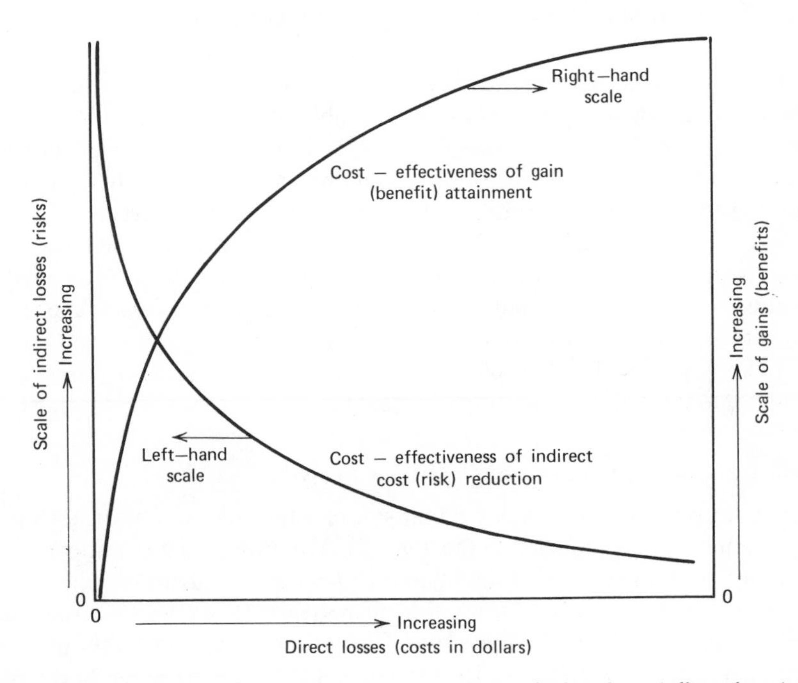

Figure 14.1. Gain and indirect loss curves as a parametric function of direct loss (cost) expenditures.

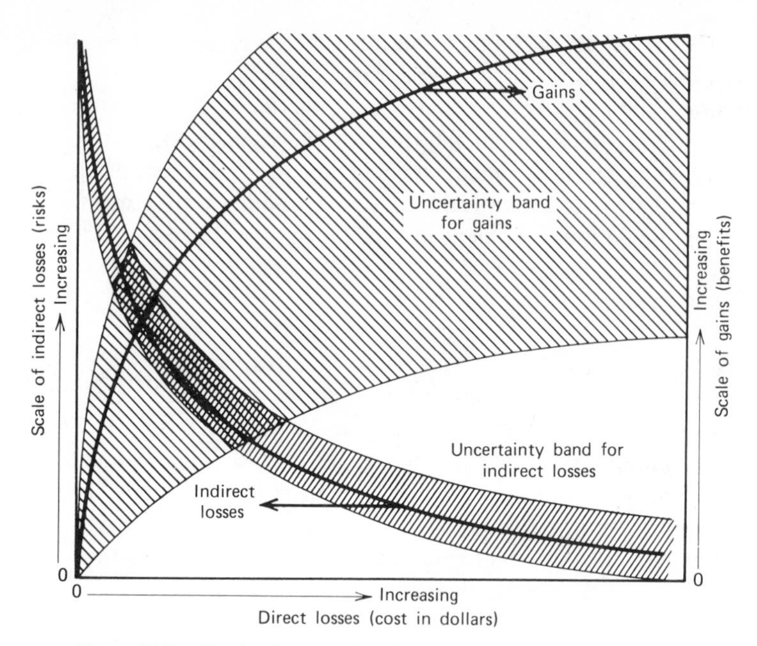

Figure 14.2. Bands of uncertainty for gain and indirect loss curves.

The uncertainty in measuring each parameter is another limitation. Direct losses involve estimates that probably entail the least uncertainty. Determination of risks involves uncertainty in knowledge of exposure-risk relationships, and uncertainties in specifying intangible benefits are often much greater than uncertainties in risk estimates. Even when gain and indirect loss scales are identical, the uncertainties in measurement are so large that meaningful analysis probably is not obtained. This is illustrated in Figure 14.2, where the bands of uncertainty for indirect losses and gains indicate a relative basis of knowledge of each.

A thorough examination of how these uncertainties arise requires some insight into the detailed structure of a gain-loss analysis.

14.1.2 Detailed Aspects of Gain-Loss Analysis

A gain-loss analysis involves comparison of gains and losses for each type of parameter to be included in the analysis. Moreover, each parameter must consider both direct and indirect gains and losses and must identify different value groups where those who experience losses are not the same value group who receive gains. This involves a taxonomy of gain and loss parameters with identical scales for particular parameters for both gains and losses.

14.1.2.1 Taxonomic Scales for Gains and Losses

The hierarchical scale of consequence value used in previous chapters could be a starting point for a taxonomic scale. To this scale let us add strictly monetary gains and losses as one entry. On this basis a first-order taxonomy of a scale for both gains and losses is made.

- Premature death
- Illness and disability
- Survival factors
- Exhaustible resources
- Security
- Monetary status
- Belonging and love
- Egocentric factors
- Quality of life
- Self-actualization

Each of these Maslovian needs may be broken down into further detail. For example:

- Illness and disability

    Immediate pain and suffering
    Latent illness
    Blindness
    Etc.

- Monetary Status

    Short-term gain (loss)
    Midterm gain (loss)
    Long-term gain (loss)
    Capital investment

Each of these can be broken down further, as required to develop scales for which data can be acquired.

14.1.2.2 Value Group Specification

A number of different value groups in society receive the gains and losses for each integer entry. In fact, there are an infinite number of such groups, but there is a small subset that may be particularly meaningful, such as:

- The group proposing the activity
- The investors
- The operators and workers
- Neighbors of proposed activity
- Society as a whole
- Low income, elderly females

14.1.2.3 Direct and Indirect Effects

Another level of separation must be considered. Each entry must be separated into two classes—one for direct effect and one for indirect effect. Definitions for these were developed in Chapter 5.

14.1.2.4 Complexity of the Taxonomic Scales

Each scale, one for cost and one for benefits, is made up of taxonomic gain-loss factors, value group factors, and indirect-direct entries for each of the foregoing entries. On this basis, the two taxonomic scales, one for gain and one for loss, may be made identical. The result is, indeed, complex and quite large, depending on the level of detail required to cover all needed entries. What to include and what level of detail is required are basic value judgments determining the scope of the analysis.

14.1.2.5 Scale Factors

For each taxonomic entry, one or more measurement scales must be developed to describe the magnitude of the gain or loss experienced by each factor. The scales may be nominal, ordinal, cardinal, or ratio* (Table 9.1). Multiple scales may be necessary to differentiate between individual and collective gains and losses, and more than one parameter may be needed to express the magnitude of a given factor.

Each taxonomic entry then has one or more parameters and associated measurement scales for both gain and loss entries. The scales must be identical in each case for gain and loss.

14.1.2.6 Probabilistic Considerations

Measurement of Probability of Occurrence of Gains and Losses. For each of the above-mentioned taxonomic entries whose magnitude is expressed on some scale, there is a probability of occurrence of the event that leads to the resultant gain or loss magnitude. The probability may be statistical (i.e., based on some random function) or based on a specified process, certain or

* See Glossary.

uncertain. Thus each entry has some frequency of occurrence that may be specified. Each taxonomic scale entry has at least two entries to describe it, one involving magnitude of gain or loss and one the probability of occurrence. This is illustrated in Figure 14.3 in the form of a probability distribution function involving probability of occurrence. This is different from measurement uncertainty, which exists for both probabilistic and certain events.

Measurement Uncertainty. The use of a functional relationship involves errors in the evaluation of the function which derive from the precision of measurement of meaningful values and probabilities, and accuracy of the measurements made. The uncertainty in parameter measurement, involving both the precision and accuracy of the measurement process is designated here as the "measurement uncertainty." As represented in Figure 14.3 for a certain result, this parameter may be described as a probability distribution of measurement error.

Estimator Uncertainty. The utilization of complex functions for comparison of gains and losses is often unwieldy, especially when the functions for each are based on different functional relationships. As a result, single value estimators to replace the functions are convenient to use with some increase in uncertainty. Central value estimates of dispersion are good

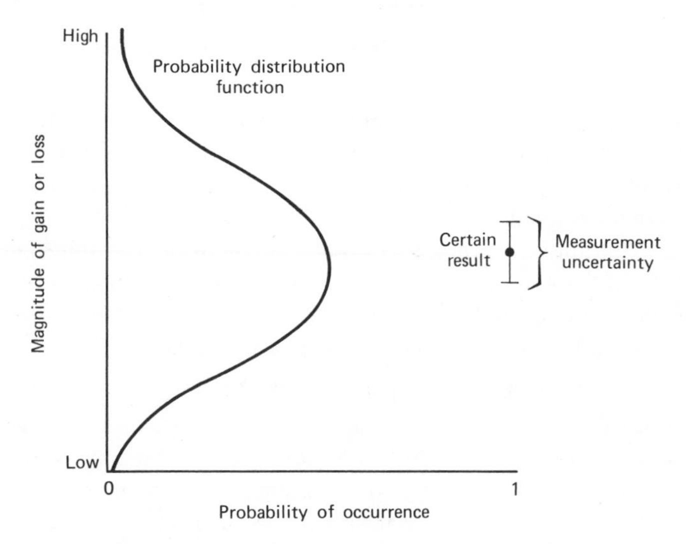

Figure 14.3. Probability of occurrence distribution function and measurement uncertainty for "certain" results.

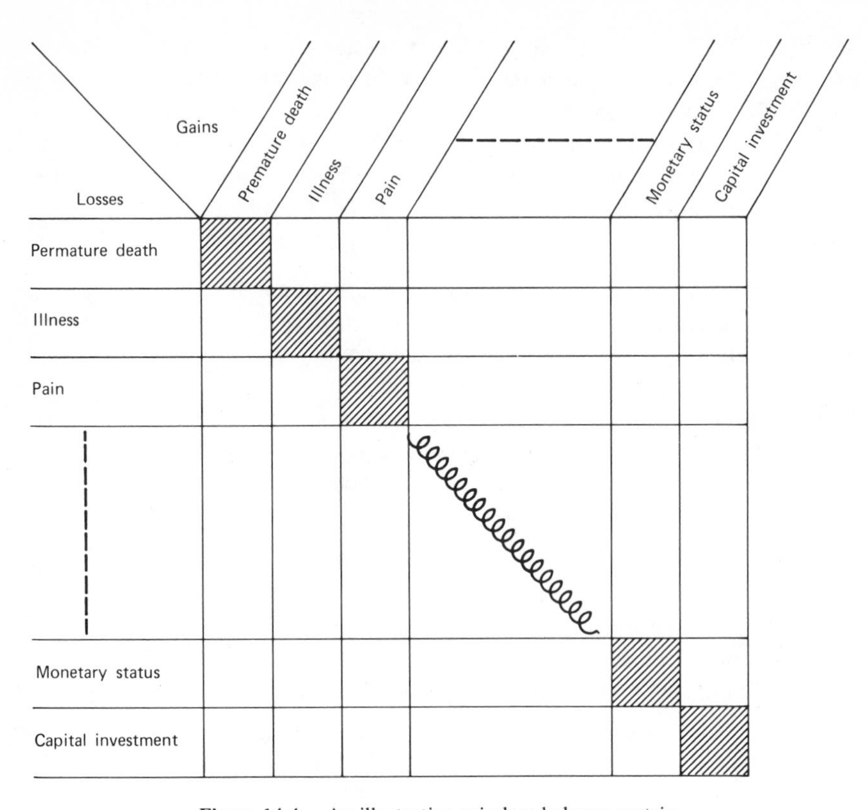

Figure 14.4. An illustrative gain-loss balance matrix.

examples of estimators for probability-value distributions. Means, medians, modes, standard deviation, and other methods of describing dispersion are illustrative. For time-dependent functions, various smoothing and cyclic techniques are available.

These estimators provide convenience in using the information at the expense of increased uncertainty. This increase in uncertainty is usually definable explicitly and may be analyzed by conventional methods. This type of uncertainty is designated "estimator uncertainty."

14.1.2.7 Comparison of Gain and Loss Scales

A proposed activity for which gains and losses are to be evaluated presupposes levels of gains and loss for each entry in the taxonomic scales for gains and losses, respectively. If for a specific entry the scales for gain and loss are identical, a gain is a negative loss, and vice versa. However this is seldom the case, as illustrated by the comparison matrix of gains and losses in Figure 14.4. The gain and loss taxonomic scales supply column and row entries, respectively. Scales are identical only on the diagonal of the matrix for taxonomic entries. Within the same headings, off-diagonals for different value groups have identical scales. Balances made on these squares can be evaluated directly on a gain-loss basis. However all off-diagonal intersections require that a functional relationship be established between gain and loss scales. Since these relationships are symmetrical with respect to the diagonals, only one-half (upper or lower) need be considered. The relationships represent either measures or value judgments about the relative merits of two different scales. For example, how much does a direct gain in economic security offset an indirect loss in chronic illness? A balance occurs at each entry point on the diagonals and one-half of the off-diagonals. Direct balances can be made on diagonals, but only indirect balances through a scale relationship can be made on the off-diagonals.

14.1.2.8 Aggregation of Balances

Each intersection on the matrix represents a gain-loss balance. From the overall gain-loss balance, we infer some aggregation of the individual balances. A variety of choices are possible for aggregation to make higher level balances. These aggregations are necessary if decisions cannot be made on the unaggregated matrix, that is, the balances are not all in favor of gains or all in favor of losses.

Aggregation by Value Group. The gain-loss balance for each value group can be ascertained by aggregating like entries.

Aggregation by Parameter. Gain-loss balance for each integer taxonomic entry can be made across all value groups, yielding an overview of gain-loss balances by each parameter.

Aggregation by Direct-Indirect Entries. Aggregation may proceed by grouping direct and indirect gain-loss balances first. Thus it is possible to begin with the separation of equitable and inequitable balances.

Aggregation by Grouping. Each of the three foregoing methods of aggregation is subject to higher levels of aggregation by grouping on the basis of taxonomic scale structure. For example, all illness subentries, including pain, would be aggregated.

Compound Aggregation. Aggregation can proceed to any level by sequentially applying all methods in any combination. Differences in measurement scales, and the ability to aggregate without conversion to some common scale, may cause the results to differ depending on the sequence of aggregation chosen, and the level of aggregation in each step, even when aggregation is on an equal basis.

Common Scales. Translation into common scales is one method that can be used to aggregate across measurement scales of different types. The array of scales discussed in Chapter 13 (effectiveness, value, dollars, etc.) gives typical examples. The scales may be arbitrary, and the translation function is often a value judgment made subjectively.

Relative Weights. Within common scales, different parameters or value groups may have differing degrees of relative importance in the scheme of those making the gain-loss balance. For example, different value groups might be weighted on the basis of population, political clout, or importance to the balancer. Different parameters would have different levels of importance on a common scale during transformation. These value judgments are particularly subject to individual valuation and interpretation.

14.1.3 Precision of Comparisons

The generation of scales, methods of aggregation, and the ability to accurately measure gain-loss values involve relatively low levels of precision in most cases. As a result, any comparisons made, from the lowest to highest level of aggregation, have finite precision and limited accuracy. Thus gain-loss comparisons are not meaningful if the precision of the comparison exceeds the accumulated imprecision of scale choice, measurement abilities, and aggregation.

To estimate the level of meaningful precision of comparisons, it is necessary to examine all measurements, scale transformations, and value judgments to determine their precision, levels of uncertainty, and the ability to measure such parameters.

14.1.3.1. Examination of Judgments

Table 14.1 lists the ten major kinds of value judgment, along with an expression of source of uncertainty, the method implementing the judgment and three types of resolution of the judgment.

Type A. Value judgments involving the choice of what to include and what not to include and the method of inclusion.

Type B. Value judgments involving the determination of how to measure parameters.

Type C. Value judgments involving individual valuation in terms of forcing nominal or ordinal scales into cardinal scales.

These definitions are based on the ability of the judgment to meaningfully express the gain or loss function as opposed to the ability to measure the parameter once the judgment has been made. On this basis the precision of the judgment as opposed to the accuracy of specific values is considered.

Table 14.1 *Judgments Involved in Gain-Loss Balances and Precision Estimates*

Judgment	Uncertainty	Method of Implementation	Degree of Resolution	Precision
Selection of taxonomic scale parameters	Degree of inclusiveness and completeness	Individual choice and mutual agreement	Type A	.90
Determination of taxonomic parameter measurement scales	Suitability, measurability, meaning	Capability to measure	Type B	.95
Selection of value groups	Degree of inclusiveness and completeness	Individual choice and mutual agreement	Type A	.90
Definition of direct and indirect gains and losses	Preciseness of definition	Mutual agreement	Type A	.90
Selection of scale transformation for off-diagonals	Suitability, measurability, meaning, acceptability	Individual choice, mutual agreement, capability to measure	Types A and B	.85
Level of aggregation	Suitability, meaning, need for comparison	Mutual agreement, achievement of level of comparison	Type A	.90
Method of aggregation	Suitability, need	Mutual agreement	Type A	.90
Sequence of aggregation	Suitability, need	Mutual agreement	Type A	.90
Assignment of common scales	Need, suitability	Need, mutual agreement, nominal-to-cardinal scale conversion	Type C	.90
Assignment of relative weights	Subjective evaluation	Ordinal-to-cardinal scale conversion	Types A and C	.81

14.1.3.2 A Cursory Error Analysis

A cursory error analysis, using the 10 value judgments in Table 14.1, gives some insight into the precision of gain-loss balance. Assume that each value judgment has a precision, depending on its type, as follows:

Type A. One part in 10 expressed in fractional form as .90.
Type B. One part in 20 expressed in fractional form as .95.
Type C. One part in 10 expressed in fractional form as .90.

The multiplicative resultant precision is used for compound cases. Furthermore, since each judgment depends on preceding judgments, the overall precision is a multiplicative relationship among the constituent value judgments. Central measures of dispersion involve tradeoffs between the utility of information and accuracy of presentation and primarily reflect problems of accuracy rather than precision. The constituent values appear in fractional form in the last column of Table 14.1.

The result of the multiplicative relationship among the 10 fractions is 0.3 or an overall gain-loss balance precision of one part in three. As a result, the gain-loss balance can indicate only three conditions: favorable, unfavorable, and no gain or loss, or perhaps, high, medium, and low.

Regardless of the actual precision used in making detailed value judgments for such a complex gain-loss balance, the precision of the result is finite, and the resultant gain-loss balance is relatively imprecise. Detailed analysis of cost and benefits by formal methodologies can offer insight into the many factors involved, pointing out specific inequities (cases of indirect losses that outweigh indirect gains for specific value groups). Decisions made by these methods may not be any more precise than judgments reached through gross methods of analysis that focus on the primary issues and average over many parameters of lesser concern. Detail does not imply better precision. Furthermore, the costs of attempting to implement detailed, formal methodologies may be better allocated by focusing on key issues that are the swing factors, and understanding these more fully.

14.1.4 Conclusions

Detailed cost-benefit analyses cannot provide precise answers to questions involving societal and political value judgments. The belief that a cost-benefit methodology can ever provide a means whereby data are "cranked in" and an answer "pops out" the other end is unsupportable. At best, detailed cost-benefit analyses are tools to furnish certain insights that may aid in decision making.

The process of going through a detailed cost-benefit analysis is a means of identifying the crucial and most sensitive factors affecting a decision, supplying insight into their interrelationships and enabling the decision maker to focus on the key decision factors. The value of such information to the decision maker, of course, must exceed the cost of making the detailed analysis. The value-cost judgment must be made for each specific case and often cannot be evaluated until after the analysis is complete. Even then the value of the results is basically a value judgment itself.

Once having focused on the critical decision parameters and key issues, the social and political questions (as well as technical and economic ones) can be addressed on a gross, but meaningful basis. The factors considered do not have to be very much more precise than the least precise factor involved. This means that a detailed economic analysis of high precision may be unnecessary, since the driving parameters in the decision may be much less precise. Thus the economic analysis may be relatively imprecise without affecting the decisional outcome. For example, if the key decision is which group to favor in an inequitable situation, to say that two groups will cost be $0.31 and $6.52 per person, respectively, is no more meaningful than saying that one group is favored over another by a factor of about 20 to 1, with a high cost of about $7 per person.

Cost-benefit analysis is thus a tool for gaining insight into decision-making parameters. It is not a means to make decisions.

14.2 DECISION-MAKING WITH UNCERTAIN UTILITY FUNCTIONS

Decision analysis techniques, though holding much promise in theory, have generally been inadequate in aiding decision makers in arriving at operational decisions involving value judgments in the form of intangible and subjective parameters. A case in point was cited in Chapter 5 in regard to the work of Barringer et al.[3] in comparing energy generation systems for coal and nuclear power. Nevertheless, the concept is an attractive one, and the author has attempted to extend the technique to allow its use with problems in which precise determination of utility functions is neither possible nor meaningful. A brief treatment is provided as an overview.*

14.2.1 Limitations of Classical Decision Theory

Formal decision-making methods based on cardinal utility, such as utility and decision theory, are applicable at present only for decisions in which the parameters are directly quantifiable, and are therefore inadequate when

* Reference 4 provides a more detailed treatment of this method.

some or all of the decision parameters involve value judgments, intangible measures, and nonmonetary measures not directly expressed in cardinal scales. It may even be argued that since almost all societal decisions involve nonmonetary parameters, not directly quantifiable, decision theory as it exists today has rather limited application to these problems. In fact, much of the work in management science and operations research, despite its heuristic value, has been aimed at the convenience of the theoretician and has resulted in elegant theories and methods having little practical use. This is well illustrated in the context of statistical decision theory by Schlaifer: "The actual analysis of a decision problem becomes a routine application of arithmetic if the decision-maker first decides on his probability for every event branch and his preference for every terminal value in his decision diagram."[5]

Based on the assumption that probabilities and preferences can be determined to infinite accuracy and precision on a cardinal scale, a total theoretical structure,[5] using Bayesian logic, has been developed for decision making. However this structure has had limited operational use, since the basic assumptions seem to be artificially derived and operationally insufficient.

14.2.1.1 Limitations of Expected Utility for Risk Evaluation

The probability of occurrence of every outcome of a decision and the preference for every terminal value of an outcome (i.e., a utility assignment for each outcome) are multiplied to derive a measure of "expected utility." Decision theory is based on the premise that higher expected utility is more desirable than lower values. Expected utility is calculated, assuming that the probability of occurrence and the utility assignment are independent of each other.

This assumption is not true for risk evaluation, since the magnitude of a consequence value is often a function of the probability of occurrence, and possibly the magnitude of the consequence can affect the assignment of probability values (not necessarily the probability itself). If probability and utility are not independent, the concept of expected utility has no meaning. It can be used only where independence is assured.

Since expected utility is a statistical measure of central tendency, it does little to explain future behavior except when a large number of trials are involved, as discussed in Section 13.3.1.

14.2.1.2 Infinite Precision in Utility Assignment

The development of a cardinal value assignment for an outcome, based on the von Neumann-Morgenstern method of an equivalent gamble* has been

* See Section 13.3.1.

the point of departure for most subsequent work in the field by others such as Churchman and Ackoff[6] and Luce and Raiffa.[7]

The major exposition of the use of this method is that of Schlaifer.[5] In all cases, this gamble is made to infinite precision. Schlaifer states this in terms of a certainty equivalent for a terminal value in the decision tree, as follows:

> Although a decision-maker can always decide on some one, definite value for a given certainty equivalent for the uncertain terminal value represented by a given terminal fork, he will often find it difficult to do so. Often he will feel that he might just as well select any one value in some fairly wide range as any other; and for this reason the reader may be tempted to think of the certainty equivalent actually assessed by the decision-maker as being a kind of "estimate" of this "true" certainty equivalent and as being therefore subject to some kind of "range of error."
>
> Notions of this sort amount, however, to a complete misunderstanding of the meaning of a certainty equivalent.
>
> There exists no such thing as a "true though unknown" certainty equivalent and, therefore, there exists no such thing as an "estimate" of a true certainty equivalent. A certainty equivalent is purely and simply a *decision**; and it follows that no matter how hard it may be for a decision-maker to decide on one particular value among all possible values for a given certainty equivalent, the value he decides on is to be thought of as exact and accurate to an infinite number of decimal places.[5]

Taken at face value, the argument begs the whole question of the decision in that one must decide on a certainty equivalent before coming to a decision operationally on the problem. The real decision then is the valuing of the certainty equivalent, since the remaining "decision" is simply the computation of expected utility.

Operational problems with nonmonetary parameters involve uncertainty in utility assignment. Therefore statistical decision methods that do not consider uncertainty in utility have little applicability for analysis of operational problems of this type.

14.2.2 Imprecise Utility Functions

Extension of statistical decision theory by use of imprecise utility functions is one means of overcoming the inadequacy stated.

14.2.2.1 Degree of Precision in Valuation of Utility Functions

The key question in addressing utility functions with subjective measures is the degree of precision possible in valuing utility functions from nominal and ordinal preference scales. The question is academic unless some estimate of the degree of precision attainable is determined empirically.

* Emphasis is Schlaifer's.

A series of experiments was conducted to elucidate the empirical aspects of the problem. The experiments attempted to test the hypothesis that ranked preferences are valued meaningfully only to finite levels of precisions. Some effort to determine the level of meaningful precision was also undertaken. The details and results of the experiments are given in reference 4.

Valuation of the ordinal scales resulted in finite levels of precision for all cases attempted. However the degree of precision attained is affected by several factors:

- Method used in valuation
- Choice of scale limits
- Number of members of an ordinal scale to be valued
- Internal reference of test subject
- Learning factors in repeated trials

These experimental results support the conclusion that intrinsic measures of both objective and subjective scales are imprecise.

14.2.2.2 Compound Utility Functions

A value judgment can be made for each value that a decision maker has for a decision. That is, although one utility function is required for a decision problem, a decision maker may have a number of different, sometimes conflicting, value scales that affect the decision. Each of these value scales ought to be quantified to some extent, then combined with the others to form a single utility function for a decision. The uncertainty in obtaining a value judgment for each value scale must be preserved in the combined utility scale. A combined utility scale of this type is called a compound utility function. Consideration must be given to the quantification of single value scales and the compounding of these scales as well.

14.2.2.3 Uncertainty in Compound Utility Functions

Maximum uncertainty is expressed by a uniform distribution of possible points on the utility function, where a given value of utility u lies on the utility scale. It is a range of equiprobable points. If the uncertainty is intrinsic to the scale itself, the uncertainty is in terms of the precision of the scale. If the uncertainty involves the ability to assign a certain level of utility to a preference u, the uncertainty is in terms of accuracy. Both types of uncertainty can be present.

Precision. The uncertainty in precision can be represented by an error term e; thus the uniform distribution of error ranges around the central value u.

$$(u - e) \leq u \leq (u + e) \tag{14.1}$$

The compound utility function with finite precision U is of the form:

$$U = f(u_1 \pm e_1, u_2 \pm e_2, \ldots, u_k \pm e_k) \tag{14.2}$$

Accuracy. The accuracy g of assigning a particular value u on the utility scale defines the uncertainty in accuracy of measurement in the form

$$(u - g) \leq u \leq (u + g) \tag{14.3}$$

The imprecise, inaccurate utility value is

$$u = u \pm e \pm g \tag{14.4}$$

The difference in precision and accuracy can be illustrated by reference to Figure 14.5. A utility scale with between limits of zero and unity is shown

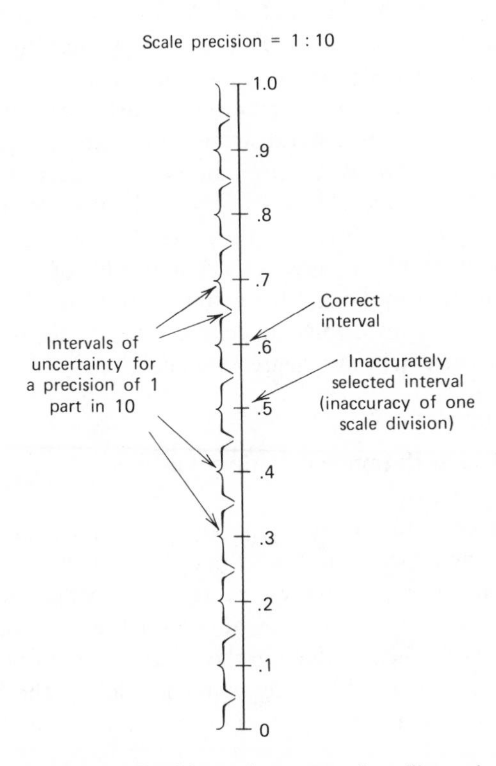

Figure 14.5. Precision and accuracy of a utility scale.

with a precision of one part in 10. The brackets indicate the assignment of a utility value u to the interval between .45 and .55, when it correctly should be in the interval between .55 and .65 and represents an inaccuracy of one scale division.

Accuracy lies only in the degree of belief of a single decision maker in assigning a value. Since his degree of belief is intrinsic, as is his utility function, his accuracy cannot be questioned when he assigns a utility value. As a result, accuracy is meaningless for this case. Indecision about the increment to which a value should be assigned is a matter of precision and/or indifference, not accuracy. If one is indifferent about which interval the value is placed at, the scale may be too precise. If he wishes to distinguish assignment of value within a given interval, a higher level of precision is required.

When more than one individual is involved in determining u, accuracy becomes important. Here inaccuracy represents the departure of the degree of belief of some members of the group from others. Value groups as discussed in Chapter 13, as opposed to individuals, must be considered.

If all values of e or of g are equally probable, the uncertainty is represented by a uniform distribution. A uniform distribution of error around a central value seems to be the most straightforward means of representing an uncertainty interval. If the most likely value in the uncertainty range is discernible or believed to exist by the decision maker, a normal distribution [or some other distribution] of uncertainty is possible.

The assignment of values to probability of occurrences is infinitely precise, since by definition probability is a cardinal scale. However it may not be meaningfully precise for some purposes. For one flip of the coin that has a probability of .50001 to .49999, the unfair odds of coming up heads is not too important. It is important for a 100×10^6 trials. The assignment of values to probabilities does involve inaccuracy. This inaccuracy results from deficiencies in measurement and degree of belief.

14.2.3 Decisions with Imprecise Utility Functions

Decision making with imprecise utility functions employs the same technique of determining expected utility as in ordinary decision making, except that the uncertainty in the resultant expected utility values is preserved. The central value of expected utility is the product of the best estimate of probability and the central value of the uncertainty range of the utility assignment. The extreme values are found by taking the product of the lowest value of probability and utility and the same for the highest values.*

* This may also be accomplished for precision considerations alone by rounding off the expected utility to the same number of significant digits as the utility assignment.

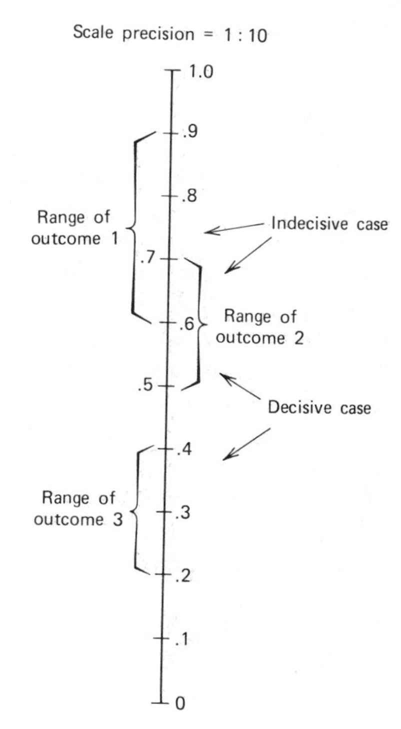

Figure 14.6. Examples of decisive and indecisive expected utility decisions.

The limits of range of a uniform distribution or the second, third, and so on, standard deviations of a normal distribution are used for this purpose.

When the expected utility of an outcome is higher than that of another, *without overlap of expected utility ranges*, the first outcome is selected as preferable to the second, and the decision condition is *decisive*. If there is overlap of error ranges, the decision is called *indecisive*, and no information on preferences is available. The indecisive condition can be refined by reexamining error limits resulting from inaccuracy, but not from imprecision. These conditions are illustrated in Figure 14.6 for three outcomes. The separation between outcomes 1 and 2 results in an indecisive case. The separation for outcomes 1 and 3, or 2 and 3 results in decisive cases.

Some decisions are truly indecisive. This should not be surprising. It is important to know if and why such conditions exist. If the decision must be made, the result must be based on other parameters outside the formal decision structure. Examples of decisions addressed by this technique can be found in reference 4.

The use of uncertain utility functions assures that decisions made are meaningful, not *charades* brought on by attention to method elegance as opposed to problem limitations.

14.3 STRUCTURED VALUE ANALYSIS

Several years ago (while the author was employed by the MITRE Corporation), the author developed a technique for formally arriving at value estimates, including means for treating nonlinear value functions, for providing on-line sensitivity analysis to evaluate the criticality and effect of different value choices, and for error analysis. A number of aspects of structured value analysis tell us what can and cannot be done with detailed formal techniques, and the present discussion serves this purpose.

Structured value analysis was first developed to evaluate the cost and technical performance of weather data gathering systems.[8] Applications were extended to the cost effectiveness of air quality monitoring systems[9] and cost-benefit analysis of computer service systems.[10] Since then, the methodology has been extensively used and has been extended into areas for evaluation of technology-oriented systems[11, 12] and into social areas involving less tangible measures.[13]

14.3.1 Structured Value Analysis Decision Space

The basic decision space for structured value analysis (Figure 14.7) is three dimensional if time dependency of cost and value variables are taken into

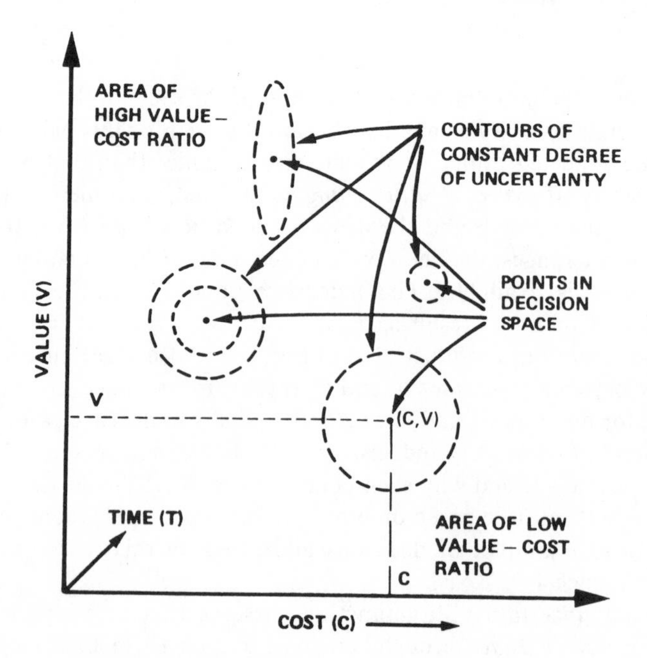

Figure 14.7. Structured value analysis decision space.

account. For a given point in time, the decision space is made up of all cost-value points (c, v) and their computed contours of uncertainty in both cost and value. The upper left-hand corner is the area of high value–low cost and the lower-right hand corner the area of low value–high cost. Decision points in the upper-left corner are preferred over those in the lower right. The decision space may be that for a single parameter value scale and cost scale, or it may be the composite aggregation of many value and cost scales.

To develop each parameter's value and cost scale and the means of aggregating these into single scales, a "value set" for a given decision agent must be "structured." Costs are economic parameters in the cases referenced and are aggregated conventionally, but there is a different value set for each decision agent (assuming they have different values). For each "candidate system" to be evaluated against a value set, a set of specific values of (c, v) is generated for each parameter. This is called a candidate "data set." Each candidate data set is then evaluated against the value set (or sets if different value systems are to be considered), and the one (within limits of uncertainty) that has the best value-cost ratio is selected for consideration.*

14.3.2 Parameter Value Functions and Weights

The overall value set is an aggregation of parameter values developed through examination of the value function of each parameter and weights of importance among parameters. The value scales for all parameters must be the same in each case, although a parameter examined may have any scale. The value function converts the parameter scale into a value scale. The value scale may be performance values, effectiveness values, or benefit values. Table 14.2 defines the differences along with cost considerations.

The value scale is arbitrarily given a value between 0 and 1, and the parameter scale is a functional relationship between the parameter measurement scale x and the value scale. The resultant functional relationships $F(x)$ are the value functions. Aggregation of parameter values is either multiplicative (factors) or additive with weights (addends).† Two examples of aggregation appear in Figure 14.8. Data sets consisting of x values for each parameter are evaluated against the aggregated value set.

14.3.2 An Illustrative Example

A simple and rather trivial example is given to illustrate the implementation of structured value analysis.

* A decision cannot be made on this ratio alone, since a value-cost ratio has an infinite number of possibilities such as those lying on the 45° line representing a c/v ratio of unity. When this occurs, cost limitations, utility of costs, and acceptable levels of value enter into the decision.

† The weights for a set of addends must add to unity to assure normalization.

Table 14.2 *Scales of Structured Value Analysis*

Scale	Definition
Cost-performance	The relative performance of candidate systems evaluated against the performance of standard (or idealized) systems as performance scale vs. total cost of implementing each candidate system.
Cost-effectiveness	The relative evaluation of candidate systems against a bounded, user-determined scale of effectiveness for his purposes vs. total cost of each candidate system.
Cost-benefit	The relative evaluation of candidate systems against an unbounded scale made up of parameters that are not necessarily constrained to user goals vs. total cost of each candidate system.

The system chosen is for the author, for driving home from work. Only two model parameters are used: time to drive home and risk in driving home. Three candidate systems, an automobile, a motorcycle, and a bus, are to be considered. The data for the three candidate systems are presented in Table 14.3.

The value set is derived by considering each model parameter. The value functions for time to drive home and for the risk parameter, along with the rationale for each, appear in Figures 14.9 and 14.10, respectively. A weight

Figure 14.8 Methods of aggregating structure value analysis parameters.

Mode 1: one set of addends, one set of factors

$$V_1 = \prod_{j=1}^{n} F_j(x_j) \sum_{i=1}^{m} A_i F_i(x_i)$$

where

$$\sum_{i=1}^{m} A_i = 1$$

Mode 2: two sets of addends

$$V_2 = \sum_{j=1}^{n} A_j F_j(x_j) \sum_{i=1}^{m} A_i F_i(x_i)$$

where

$$\sum_{i=1}^{m} A_i = 1 \quad \text{and} \quad \sum_{j=1}^{n} A_j = 1$$

where V = value
A = weight
F = value function
x = input value
i and j = indices

Table 14.3 *Sample Problem: Three Candidate Systems for Driving Home*[a]

System	Time (minutes)	Risk per 100,000 Miles	Cost per Year
Motorcycle	12	200	$250
Automobile	45	50	$750
Bus	90	3	$200

[a] In addition, ±10% variations in cost and time to drive home have been carried to describe the uncertainty that may occur in the data.

of .6 has been selected for time (quality of life) and .4 for risk (quantity of life). The value set is defined by:

$$V = .6\ F_T\ (\text{time}) + .4\ F_R\ (\text{risk}) \tag{14.5}$$

The results of the calculation are shown in Figure 14.11, along with the ranges of uncertainty for the three systems.

It is obvious that a choice cannot be made between the motorcycle and

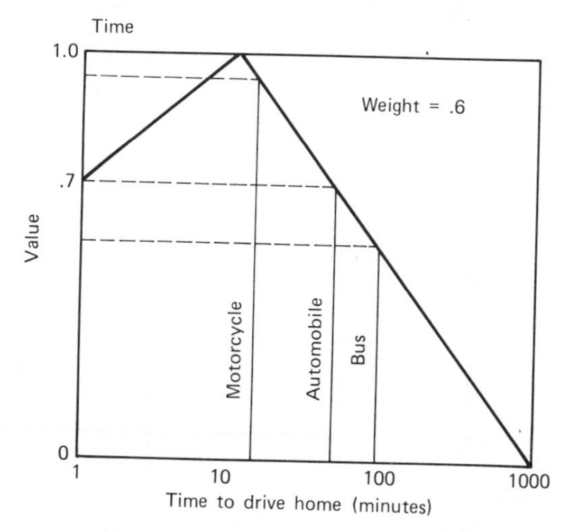

Figure 14.9. Time value function. *Measurement Scale.* 1 to 1000 minutes, logarithmic scale. *Value Judgment.* Getting home from work in about 10 minutes would be perfect, hence a value of 1.0 is assigned to this point. Acceptability falls off rapidly as time increases, in fact anything over a 100 minutes could be considered negative (however only relative values are involved). On the other hand, if if takes less than 10 minutes to get home, one is too close to home and might have to run too many errands or go home for lunch. Thus being only 1 minute from home is less desirable. *Data Values.* The data values for the three candidate systems are shown.

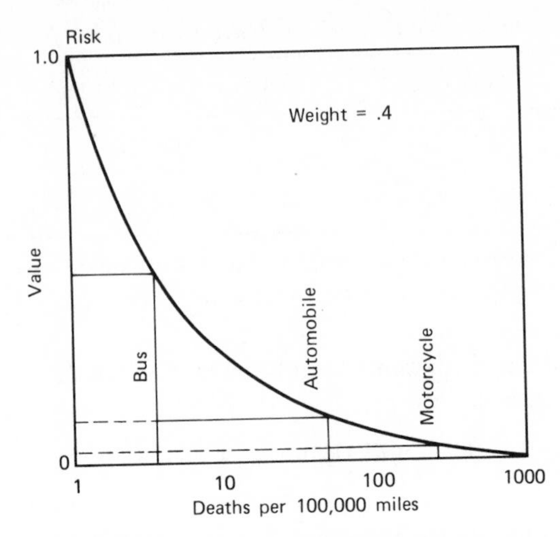

Figure 14.10. Risk value function. *Measurement Scale.* 1 to 1000 deaths per 100,000 miles on a logarithmic base (assumed scale). *Value Judgment.* Safety is rather important and bias toward high values for safety is shown by a concave curve. *Weights.* The time parameter has to do with the quality of life, the risk parameter with the length of life. Being biased toward quality of life, the author has assigned weights of .6 to the time scale and .4 to the risk scale, a 3 to 2 bias. *Data Values.* The data values are shown for the risk scale.

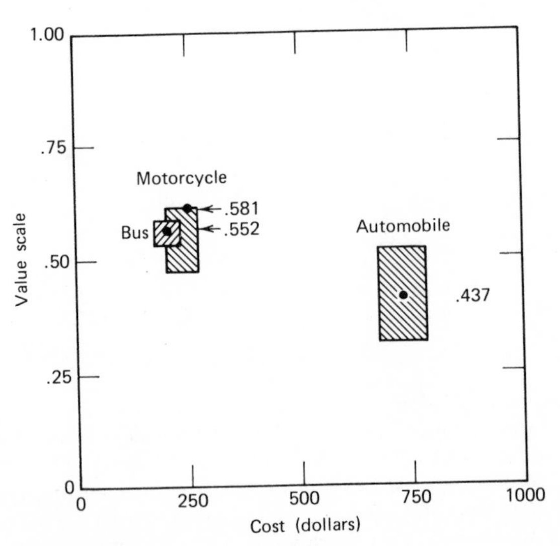

Figure 14.11. Results of problem evaluation with limits of uncertainty. NOTE. Ranges of uncertainty are not necessarily symmetrical because of the nonlinearity of value functions.

bus, since the areas of uncertainty overlap. The situation is indecisive for these two candidate systems, although either one appears to be decisively preferable to the automobile.

Structured value analysis is only an aid to a decision maker. It does not dictate the decision.

The author decides to use the car for a completely different reason: he can keep his golf clubs handy in the trunk. In other words, the governing parameter had not been entered into the system. The implication of the decision can be tested by going back and deriving a value set for time and risk where the auto ranks the highest. In this manner, the decision maker can understand the real impact of his decision. As it turns out, restructuring the time value function to peak at 20 minutes instead of 10 minutes is sufficient to accomplish this.

What has been learned is that even though he thought getting home in 10 minutes would be very desirable, 20 minutes is even more acceptable if golf can be worked in every once in awhile. The opportunity to play golf is a hidden value that has been brought out. A new value set, adding golf as a parameter, could be easily constructed but is not warranted, since the implications of the decision have already been understood and a meaningful decision has been made.

14.3.4 Sensitivity Analysis

The foregoing problem was quite simple, and adequate information was available for decision making. In more complex systems, however, it is important to understand the implications of the value set before using it for decision making. Sensitivity analysis is a technique that may be used to obtain this kind of information. Two types of sensitivity analysis have been used.

Parametric Analysis. Each parameter is varied by a fixed amount in turn while all others are held constant and the result is computed for each variation. The parameters that change the output computation the most are deemed most critical at the variation range used.

Statistical Analysis. Data sets are generated from a random number table and are computed for each set. As many as 1000 sets may be sampled to assure statistical convergence. The output values are tabulated and plotted to determine the nature of the set (unimodel or multimodel) and the range of "expected values." These "expected values" do not correspond to reality; they are a tool to determine the sensitive ranges of activity.

14.3.4.1 Parametric Analysis

Three types of parametric analysis have been implemented. They are as follows:

- Midrange as base value
- Selected data set as base value
- Variation of weights and curvature using 1 and 2

Midrange Base Value. The midrange of each parameter scale is used as a basic value. Each parameter is varied positively and negatively by a fixed percentage, and the structured value is computed for each variation. The parameters are then ordered by the magnitude of the change in value.

Any fixed percentage can be entered as a variation, and an "auto spread" is available that automatically steps through 10 evenly spaced percentage steps from 10 to 100%. This assures that sensitivity is examined across the whole range of parameters and that breakpoints and nodes are not missed.

Selected Data Set Base Value. This method is identical to the foregoing except that the values of a selected data set are used as the base value in place of midranges. In this manner the sensitivity of the data set is examined for its entry into the value set.

Variation of Weights and Curves. When a weight for a particular parameter is varied, weights for all other parameters are adjusted accordingly to assure that the weights still sum to unity. When a curve is varied, the desired curvature is entered.

14.3.4.2 Statistical Analysis

Up to 1000 random data sets may be generated and computed. The results are tabulated and plotted in histogram form with mean, standard deviation, minimum, and maximum calculated.

14.3.5 On-Line, Interactive Computer Implementation

All computational methods described earlier have been implemented in a conversational mode-interactive time-sharing set of programs. The programs are self-teaching and easy to run. In this mode, immediate feedback of changes, error ranges, parameter criticality, and impact of multiple value sets provides a tool for investigating the nature and form of intangible values. Critical parameters can then be the subject of focused, further evaluation.

14.4 FINAL COMMENTS

This chapter has demonstrated that subjective and intangible values seldom can be valued precisely. Methods for balancing gain and loss values, no matter how elegant, are no better than the precision of the value judgments always involved in such techniques. Nevertheless, considerable information can be garnered as an aid to decision makers as long as the limitations are not exceeded. The difficulty of assigning values to societal revealed preferences to be attempted in Part D stems from this basic condition. Accuracy and precision of such valuation will always be severely limited. Thus the gross balance method discussed in Chapter 12 becomes particularly suited for decision making at the societal level.

REFERENCES

1. J. Golden and R. Kuehnel, "A Preliminary Study of a Concept for Categorizing Benefits," MTR-6569. McLean, Va.: The MITRE Corporation, January 1974.
2. Richard Wilson, "Examples of Risk-Benefit Analysis," *CHEMTECH,* October 1975, pp. 604–607.
3. Stephen M. Barranger, Bernie R. Judd, and O. Warren North, "The Economic and Social Costs of Coal and Nuclear Electric Generation: A Framework for the Coal and Nuclear Fuel Cycles," STI Project MSU-4133, as discussed in *Proceedings of Quantitative, Environmental Comparison of Coal and Nuclear Generation, and Their Associated Fuel Cycles Workshop,* MTR-7010, Vol. I. McLean, Va.: The MITRE Corporation, August 1975, pp. 15–94.
4. W. D. Rowe, "Decision Making with Uncertain Utility Functions," Doctoral thesis, American University, Washington, D.C., 1973.
5. Robert Schlaifer, *Analysis of Decisions Under Uncertainty.* New York: McGraw-Hill, 1969, p. 137.
6. C. West Churchman, E. Leonard Arnoff, and Russell L. Ackoff, *Introduction to Operations Research.* New York: Wiley, 1957, Chapter 6.
7. R. Duncan Luce and Howard Raiffa, *Games and Decisions: Introduction and Critical Survey.* New York: Wiley, 1967.
8. W. D. Rowe, "Cost Effectiveness Determination by Structured Value Analysis," WP-1580-Rev 1. McLean, Va.: The MITRE Corporation, February 1969.
9. W. D. Rowe, "The Application of Structured Value Analysis to Models Using Value Judgments as a Data Source," M70-14. McLean, Va.: The MITRE Corporation, March 1970.
10. W. D. Rowe, "Proposed Methodology for TAS-70 Cost/Benefit Analysis," WP-8344-Rev 1. McLean, Va.: The MITRE Corporation, November 1969.
11. J. D. Dukowicz, W. Fraize, E. Keitz, S. Poh, and J. Stone, "Advanced Automotive

Power System Structured Value Analysis Model, MTR-6085. McLean, Va.: The MITRE Corporation, October 1971.

12. Edwin L. Keitz, "Application of Structured Value Analysis in Determining the Value vs. Performance of Air Quality Monitoring Networks," M70-27. McLean, Va.: The MITRE Corporation, April 1970.
13. J. T. Stone and S. H. Stryker, "Program for the Elderly—Structured Value Analysis Model," MTR-6144. McLean, Va.: The MITRE Corporation, February 1972.

Part D

EVALUATION OF REVEALED SOCIETAL PREFERENCES FOR RISK ASSESSMENT

Chapters 8 to 11 presented qualitatively a variety of risk factors. Such discussion promotes better insight, but it is important to estimate quantitatively the relative impact of these risk factors. Although one might propose a series of experiments in behavioral and psychological frameworks to address this problem, an available technique is to examine the collective behavior of society through evaluation of existing data. This part examines existing data bases relative to societal risk before using these data in an attempt to quantify the relative impact of risk factors as risk referents for use in the process of determining acceptable levels of risk.

The use of existing societal behavior as a risk reference has been called the method of revealed societal preferences. Chapter 15 examines problems involving data obtained in this manner, explains how risk rates are determined, and reviews existing data bases that are useful in developing revealed societal preferences. The data bases themselves are summarized in Appendix A.

Two aspects of risk valuation are addressed: relative risk and absolute risk. Relative risk, covered in Chapter 16, provides an initial look at the effect of risk factors on risk valuation through comparison of different kinds of risk. Absolute risk, evaluated in Chapter 17, represents an effort to evaluate quantitatively the risk acceptance levels for different kinds of risk based on revealed societal preference. Although accidental risks provide the most straightforward data base, levels of risk from disease are also dealt with.

15

Aggregated Societal Data for Risk Assessment

15.1 REVEALED SOCIETAL PREFERENCES

One means of obtaining insight into risk behavior is to investigate societal acceptance of existing risks, through historical review of the frequency of different consequences from a variety of causes. Since it appears that society has accepted certain risks in the past, there is some justification for assuming that these levels of risk are or have been socially acceptable. Otway calls the use of existing societal behavior of this nature the method of revealed societal preferences.[1]

15.1.1 Behavior and Risk Attitudes

Societal preferences as revealed are used to provide a risk referent, based on the idea that "society is as what society does." There is considerable merit in this approach, as opposed to methods that seek to determine risk preferences from observation of response to experimental situations, including the use of questionnaires and polling. These latter approaches, which involve the psychological and psychometric study of behavior in identified groups or strata of society, attempt to measure attitudes toward risk as opposed to risk behavior.[1] These methods tell us a great deal about understanding risk behavior, but they have limited value in assessing real risk situations.

15.1.2 Limitations in Assessing Risk Attitudes

Experiments set up to simulate risk behavior are artificial. Subjects know at the outset that their real risks are limited by the professionalism and liability of the experimenters. That is, the subjects themselves differentiate scenerios from real situations.

Second, the use of questionnaires and polls to gain information on risk behavior is inconsistent with the situational aspect of risk exposure. It is easy to say one thing about a risk situation if it is hypothetical, but quite another thing to react to a real situation. A person does not always behave under stress according to his own predictions. For example, if a disaster occurs frequently, the reaction of people will differ from what it would be when they are not well acquainted with it, or know about its occurrence from heresay only. Before they experience the disaster, they are more careless or more confident; when faced with it, they are less sure about the remedy procedures.[2] Values change accordingly.

Finally, in cases of people actually experiencing real risks, thus providing "targets of opportunity" for examining behavior under real risk, stresses are often clouded by subsequent rationalization and mental blocks, which are a basic human reaction to stress.*

Essentially, all three approaches for measurement of risk attitudes by experimental methods encounter difficulties in associating results obtained in artificial situations† to behavior under actual risk situations. However the method of revealed societal preferences has its own set of limitations.

15.1.3 Limitations in Measuring Revealed Societal Preferences

Two major areas that limit the usefulness of the revealed societal preference approach involve the aggregated nature of the data that are available and the perception of and action by society in relation to historical risk information.

15.1.3.1 Problems with Aggregated Data

Generally, data involving the historic frequency of different types of risk consequence are taken either for society as a whole or for broad classes and strata in society. For example, accident data are presented by age, sex, type of accident, month and year, death, fatalities, illness, and property damage. Work accidents and mishaps to the general population are segregated. These categorizations are easily obtained, since most of the information is collected when an event occurs, and there are institutional channels for collecting, verifying, and aggregating the data. The motivation for obtaining these data initially arose from actuarial needs to develop effective insurance systems for profitable operation. Unfortunately, these data are made available because they are obtainable; they do not represent the kinds of information that are important for risk assessment. For example, a major area of interest in risk assessment is the determination of the difference in risk behavior for voluntary and involuntary risks. Such data are not readily available, and since motivation in accepting risk is hard to measure, the

* See Section 9.1.3 for a discussion of this phenomenon.

† In this context a debriefing after experiencing a real risk has some artificial aspects.

source reports in accidents do not often include adequate information to make differential conclusions of this nature. Thus available data often are not useful for risk assessment, but one makes use of what one can get.*

Data that are available are collected by different methods for different technological systems. Mole[3] points out that in Great Britain no common definition of injury is used when collecting information of casualties in road, rail, and airplane passengers. For both rail and public transport the reported injury rate is one to two orders of magnitude greater than the death rate, but death is the only kind of injury for which unambiguous statistics are available. When immediate and delayed deaths due to injury are involved, however, even the validity of statistics on fatalities is uncertain.

Another problem is that the very process of aggregating data tends to obscure cause and effect. Statistical analyses can be made of aggregated data, including all varieties of regression and correlation studies; but, at best, these indicate the existence of statistical relationships, not causal ones.

A further problem relates to the effect of the process of quantification and communication of risk data on societal behavior, with its concomitant problems of risk identification and perception of risk in society.

15.1.3.2 Historical Perception of Risks in Society

Although major disasters, natural and man-made, have been reported and documented since biblical times, compilation of data on risk is mainly a feature of the twentieth century. Society as a whole is generally risk aversive,[4] but we have been able to measure the extent of various risks and compare their magnitudes only since such information has become available. This has led to a process of risk identification that, along with man's belief in his ability to control his environment, has made risk aversive action and the pursuit of increased safety a societal goal.

It is difficult to estimate the extent to which such statistics on risk have been an initiator of risk aversion or whether inherent risk aversive attitudes generate the need for these statistics. Undoubtedly reinforcement of risk aversion and safety consciousness occurs in both directions. This means that society reacts dynamically to risks and risk information, and that any historical perspective on risk behavior must show trends and changes in concern when appropriate. Mole indicates that:

> Degree of concern about particular kinds of risk is certainly greatly magnified by the immediacy of television pictures and by the reporting habits of the media, but special concern about a simultaneously occurring group of casualties may possibly

* The design of data systems to acquire meaningful data, along with the costs, has not been considered here. Such a study is essential but must await better definition of the types and specifications of data needed.

be determined not only by social conventions but also by a basic human characteristic with evolutionary advantages and therefore not eradicable by reason and information.[3]

What has happened in the past may no longer be acceptable practice for new technological systems, especially when major innovations may affect the whole world for extended periods.[5] This change is reflected for large consequence, low probability accidents such as those associated with liquefied natural gas tankers and nuclear power plants, and for systems that can affect the environment globally for long periods, such as ozone layer changes due to supersonic transport flights at high altitudes. The implementation of the National Environmental Policy Act of 1969 (NEPA) and the establishment of an Office of Technology Assessment in the Congress in 1975 are examples of institutional changes reflecting these new concerns. The uncertainties involving the magnitude of probabilities and consequences in new major technological systems are as much a part of the concern as are the actualities in existing systems. Until the uncertainties are reduced one way or another, the new technological system will remain suspect.

15.1.4 Differences in Estimation of Risk for New and Existing Technologies

It has been pointed out by Levine[6] that the use of historical data must result in the underestimation of risk, since many low probability events may not have occurred between the time of the inception of an existing technology and/or the period for which risk information is compiled. This is undoubtedly true, but the perception of risk for existing systems is what is real, and this perception is based on experience. Furthermore, when a low probability, high consequence accident does occur, risk aversive pressure is focused on reduction or elimination of such consequences through new control mechanisms. For example, as a result of the crash into a mountain during a landing approach of a commercial transport aircraft in Virginia in 1974, Federal Aviation Agency rules have been changed to require equipment showing altitude and aircraft controller procedures to monitor altitude, as well as to decrease ambiguities in communication.

Thus for existing systems, risk estimates for determining societal risk levels and preferences must use the societal perception of risk, not computed risk. History is what has happened, not what might have happened. Underestimated or not, historic values for existing systems reflect the best perception of risk for such cases.

New systems face a new ball game, justifiably or not. Modeled risk analysis is now a prerequisite, both institutionally and socially, for such systems. New systems will have to abide by new ground rules.

These new requirements may have considerable impact on innovative capacity. This is one of the hidden costs of saying no as revealed by Dyson,[7] but some of the risks are so large in our exponentially growing world society that reasonable concern is a necessity. The ability to respond to such concerns reasonably takes time, money, and resources that are not always available to individuals or small organizations, forcing the restriction of innovation and technological development to large institutions. This penalty may have to be paid, although there are many steps that might be taken to minimize it. It is too broad a topic to be dealt with here.

15.1.5 Objectives of Analysis of Historic Risk Data

There are two major objectives to be addressed in the analysis of risk data. The first involves the relative levels of historic risk acceptance for different types of risk and the impact of different risk factors as described in Part B. The second objective is to establish risk levels that society is experiencing for different types of risk.

15.1.5.1 Establishing Risk Comparison Factors

Risk comparison factors can be established for different types of consequence as well as for different types of risk.

Different Types of Consequence. Risk data are generally available for fatalities, illness, property damage, life shortening, and productive days lost, which can be measured with reasonable objectivity. The values associated with these different consequences and awareness of their interrelations provide one type of risk comparison. However aggregation into or conversion of these values into a single scale is in itself a value judgment and is not part of the examination of historic data. Each value of a type of consequence is meaningful by itself. If there is later found to be value to aggregation with all its limitations, such steps may take place; but are not to be confused with observations.

Another area involves data on consequences of types that are less amenable to objective measures, such as those involving aesthetic values and the quality of life. The EPA[8] lists at least 30 different quality of life factors under six main headings:

- Economic environment
- Political environment
- Physical environment
- Social environment
- Health environment
- Natural environment

Many of these are consistent with the hierarchical lists of consequences based on Maslovian ideas given previously. The same approach is taken by Molitor,[9] but considerably more effort in this area is necessary to relate quality of life factors to risk consequence valuation. The lack of suitable definitions of such consequences and the virtual absence of data in this area preclude any consideration in this treatment.

Different Types of Risk. It is evident that man accepts different levels of risk for different types of risk (e.g., voluntary vs. involuntary risk.)[2] Different types of risk thus can be compared, to ascertain the relative level of risk acceptance for each type of consequence associated with the risks. Thus the levels of fatalities, injuries, and property damage from a voluntary risk associated with a given event can be compared with the level of fatalities, injuries, and so on, for a similar involuntary risk.

The relationships are called risk conversion factors (RCF) and they can be explicitly expressed. Since they are often estimated from a variety of data sources, estimates of uncertainty in such comparisons should be made evident. A formal notation for such factors is presented in the next section.

15.1.5.2 Societal Risk Levels

Societal risk levels are obtained by direct observation of societal data and by extrapolation from risk comparison factors. In the first case, an adequate data base must be examined to establish the level of risk that society is experiencing for given types of risk and type of consequence. In the second case, a risk level for a type of risk for which adequate data are not available may be estimated through use of an RCF, relating the risk type in question to a risk for which an adequate data base exists. This extrapolation is subject to considerable error, but often it is the only approach possible. Both approaches are used in this treatment.

15.2 METHODS FOR PRESENTING RISK DATA

There are many different methods for presenting and comparing risk data. Two methods are of particular interest here: calculation of risk rates, and calculation of losses to life expectancy.

All methods suffer from limitations in data. For each type of risk event (storms, air crashes, etc.), the distribution of the number of events over a given period of reporting and the magnitude of the events can be found. Histograms of frequency and magnitude of events have a variety of functional forms, and the use of statistical descriptors of central values can be especially misleading because many events are not purely random or independent. For example, there are more airplane flights, increased passenger loads, and better safety procedures than there were 20 years ago. Conversely, railroad passenger traffic has decreased during the same two decades. Better reporting methods also exist, but often the information is presented in a biased manner. As a result, the use of central statistical measures can yield only gross estimates. When possible, the range of uncertainty should also be estimated.

15.2.1 Risk Rate Format for Societal Risk

For a particular class of events i, such as commercial passenger aircraft accidents or marine accidents, a number of such accidents or events N_i will occur in a given period of years t_i. The mean number of accidents per year N_i is computed by the formula

$$\bar{N}_i = \frac{N_i}{t_i} = \text{mean number of accidents or events per year} \quad (15.1)$$

For each event j of class i, there will be a number of consequence measures for consequences of differing nature:

F_{ij} = number of total fatalities for accident ij
F_{ij1} = number of fatalities under voluntary risk conditions
F_{ij2} = number of fatalities under involuntary risk conditions
I_{ij} = number of total injuries
I_{ij1} = number of injuries—voluntary risk
I_{ij2} = number of injuries—involuntary risk
D_{ij} = cost of event in dollars

Other dissimilar consequences (illness, quality of life factors, etc.) can all be addressed in the same manner as long as clear definitions exist.

$$F_{ij1} + F_{ij2} = F_{ij} \quad (15.2)$$

$$I_{ij1} + I_{ij2} = I_{ij} \quad (15.3)$$

The mean number of fatalities, injuries, or costs is derived for each factor by taking the sum of the magnitude of each event and dividing it by the number of events in question. Thus

$$\bar{F}_i = \frac{1}{N_i}\sum_j F_{ij} = \text{mean number of fatalities per accident of type } i \quad (15.4)$$

$$\bar{F}_{i1} = \frac{1}{N_i}\sum_j F_{ij1} = \text{mean number of fatalities, voluntary risk} \quad (15.5)$$

$$\bar{F}_{i2} = \frac{1}{N_i}\sum_j F_{ij2} = \text{mean number of fatalities, involuntary risk} \quad (15.6)$$

In the same manner, the mean number of injuries $\bar{I}_i$, $\bar{I}_{i1}$, $\bar{I}_{i2}$ and the mean cost D_i can be calculated. Whether the mean (as opposed to the mode) is a good central measure depends on particular distributions of frequency and magnitude of events.

If the frequency of events is high enough to provide some measure of statistical convergence, rates of fatalities, injuries, and costs can be computed for individuals and populations at risk. The populations at risk are denoted as follows:

P_i = total population at risk
P_{i1} = population subject to voluntary risks
P_{i2} = population subject to involuntary risks

Then the number of fatalities, injuries, and costs per year for each class of accident or event N_i is of the form

$$\bar{N}_i \times \bar{F}_i = \text{mean number of fatalities per year} \tag{15.8}$$

$$\bar{N}_i \times \bar{I}_i = \text{mean number of injuries per year} \tag{15.9}$$

$$\bar{N}_i \times \bar{D}_i = \text{mean year costs} \tag{15.10}$$

The risk to an individual is

$$\bar{f}_i = \frac{\bar{N}_i \times \bar{F}_i}{P_i} = \text{mean probability of death to an individual at risk per year} \tag{15.11}$$

$$\bar{k}_i = \frac{\bar{N}_i \times \bar{I}_i}{P_i} = \text{mean probability of injury to an individual at risk per year} \tag{15.12}$$

The death rate per 100,000 people at risk f_i is

$$f_i = \bar{f}_i \times 10^5 = \frac{\bar{N}_i \times \bar{F}_i \times 10^5}{P_i} \tag{15.13}$$

and the injury rate per 100,000 people at risk k_i is

$$k_i = \bar{k}_i \times 10^5 = \frac{\bar{N}_i \times \bar{I}_i \times 10^5}{P_i} \tag{15.14}$$

The voluntary and involuntary risk rates can be found accordingly.

The use of the population at risk as a divisor involves some danger of misrepresentation. The larger the population at risk, the smaller the individual risk for a given consequence and its probability. Thus lower risk estimates for individuals result if larger populations are stated. This problem arises because the degree of exposure to risk is of primary concern; that is, for a *measured* frequency of occurrence of a consequence from actual data, the smaller the group exposed to the risk, the higher the risk to an individual member of the exposed group. Overstatement of the population exposed can lead to understatement of the individual risk.

The proper populations of risk must be identified for each risk, and it is important to recognize that all members of the population do not necessarily experience the same risk. Thus the risk rates often must be broken into subsets for different population exposures. The total exposure is properly calculated by integrating the probability of risk for a given individual or group of individuals over the total population. Since the ability to accurately determine the probability for each individual is limited, the resultant calculation may be no more accurate or meaningful than average values taken over the whole population exposed or over subgroups. A large number of exposure levels can be used if reasonable determinations of the exposed population and probability of occurrence at each level is available. One convenient approach is to divide the population into three different exposure classes: average, protected, and exposed. Some individuals are more protected or more exposed than the average individual; for example, those who live in earthquake-prone regions or close to the sea in areas where high tides and floods occur, have higher exposure to risks than the average. On the other hand, those who live inland are not exposed to tidal risks, and those who live in areas of low seismic activity are more protected than the average from these risks.

When it is desirable to express the degree of containment of risk (i.e., the separation of exposed and protected populations), a degree of containment index may be convenient. This can be computed for any consequence; but to illustrate the concept, we use fatalities. Essentially, the ratio of risk between the exposed population and the unexposed population is desired.

$$\text{containment index} = \text{C.I.} = \frac{\bar{f}_i}{\bar{f}_i'} \tag{15.15}$$

where $\bar{f}_i$ is the risk to an individual in the exposed population and f_i' is the risk to an individual in the protected population. Since P_i refers to the exposed population, $T - P_i$ is the protected population, where T is the total population.* Then

$$\text{C.I.} = \frac{(T - P_i)}{P_i} \times \frac{g_i}{g_i'}, \tag{15.16}$$

where
$$\begin{aligned} N \times F_i &= g_i \\ N_i \times F_i' &= g_i' \\ P_i &< T \\ g_i' &\leq g_i \end{aligned}$$

* For purposes of the index, the total population T is usually taken to be the total U.S. or world population. Alternatively, P_i is the protected population if the populations considered are not collectively exhaustive.

However this index ranges from unity to infinity, with small changes in factors causing large excursions of the index over parts of the range. A more amenable index* is derived by smoothing the range in the form

$$\text{C.I.} = \frac{(T - p_i)}{P_i} \times \log \frac{g_i + 1}{g_i' + 1} \tag{15.17}$$

Alternatively, when multiple populations are involved

$$\text{C.I.} = \frac{P_i'}{P_i} \times \log \frac{g_i + 1}{g_i' + 1} \tag{15.18}$$

15.2.2 Life Expectancy Models

A model for assessing risk data based on loss in life expectancy has been developed by Baldewicz et al.[10] The basis for this approach is set forth here, with some attempt to simplify the original notation.

Assuming that all insults for a given risk system are linearly independent, the total rate of loss of life expectancy, based on 10^6 exposure hours to each stressor, is taken as

$$\dot{L} = 10^6 \sum_i \omega_i \dot{L}_i = 10^6 \sum_i \omega_i \frac{L_i}{T_i} \tag{15.19}$$

where $\dot{L} = L_i/T_i$ (loss of life expectancy in years per exposure hour for the ith insult)
L_i = lost years of life expectancy
T_i = time of exposure in hours
ω_i = coefficient of insult intensity

Taking ω_i as unity for fatal insults, it is anticipated that the preponderance of other ω_i's will lie between 0 and 1. (Possible exceptions are: totally immobilizing disabilities or illnesses, for which peripheral insults to family, friends, and society are involved; or perceived future risks entailing undue irrational fears.)

Procedurally, implementation of (15.19) requires much more information than typically has been available for other risk assessment schemes. For example, treatment of multiple fatalities incurred in a large-scale technological system (e.g., in coal mining) requires age distributions of the victims as well as life expectancy data for the calculation of loss of life expectancy due to fatalities L_f for some appropriate actuarial period (taken here to be one calendar year). In general,

$$L_f = \sum_j n_{f,j} (A_j - a_j) \tag{15.20}$$

* By definition, an index is dimensionless.

where $n_{f,j}$ = number of fatalities with age a_j
A_j = associated life expectancy

Alternatively (15.20) can be written

$$L_f = N_f\,(\overline{A_j - a_j}) = N_f l_f \tag{15.21}$$

where N_f = total number of fatalities
l_f = average lost years of life expectancy per fatality for group at insult during actuarial period

These calculations can be made for other types of insult such as property damage expressed in dollars lost:

$$L_{pd} = N_{pd} l_{pd} \tag{15.22}$$

where N_{pd} is the assessed property damage in, say, dollars, and l_{pd} the "lost years" of life expectancy per assessed dollar of property loss or damage. The latter quantity requires further discussion. To first order, it is assumed that restoration of property simply involves labor (i.e., the intrinsic value of raw materials can be neglected). Furthermore, it seems reasonable to assume that the distribution of labor required in terms of skills and services will not differ appreciably from that residing in the economy at large. Therefore the "lost years" of life expectancy per assessed dollar of property damage, assuming a one-year actuarial period, will be taken as

$$l_{pd} = \frac{40/168}{I} \tag{15.23}$$

where I is the mean income of the nation's labor force and $^{40}/_{168}$ is the fractional time employed, assuming a 40-hour work week.

Problems of interpretation become further complicated when insults assume the form of "deferred" risks. For the special case of a deferred risk that causes premature death, the loss of life expectancy incurred by the population at risk (again, for a one-year actuarial period) will be taken as

$$L_{fd} = \frac{\sum_{a_j}\sum_{t_k} p_r(a_j, t_k) l_{fd}(a_j, t_k)}{P_r} \tag{15.24}$$

where a_j and t_k denote admissible ages and cumulative exposure years, respectively, for the subpopulation p_r at risk, P_r denotes the total population at risk

$$\left(\sum_{a_j}\sum_{t_k} P_r(a_j, t_k)\right)$$

and

$$l_{fd}(a_j, t_k) = \{[A_{j+1}(O) - A_{j+1}(t_k + 1)] - [A_j(O) - A_j(t_k)]\} \quad (15.25)$$

Here $l_{fd}(a_j, t_k)$ is to be interpreted as the decrement in mean life expectancy for an individual of age a_j undergoing unit increment in cumulative exposure years from t_k to $t_k + 1$.

Similar equations can be constructed for other types of insult, such as morbidity.

Based on this model, Baldewicz et al.[10] have calculated values for

l_f = average lost years of life expectancy per fatality
L_f = loss of life expectancy due to fatalities
N_f = number of fatalities per 10^6 exposure hours

for 10 risk systems, making some examination of nonfatal injuries and property damage. Subsequent chapters discuss the data bases and the results of their calculations in connection with specific risk systems, along with the conclusions of the UCLA investigators.

Unfortunately, this study has not provided a means of distinguishing between voluntary and involuntary risks or between ordinary and catastrophic events. More particularly, as pointed out by the authors, the results are very sensitive to the mean age of users of the system. The authors note that when a younger age group uses the newer forms of transportation, it results in higher values for loss of life expectancy for these systems.[11] Thus the development of risk conversion factors is fairly difficult.

Bunger[12] has taken a somewhat similar approach but has used a life table "cohort" of 100,000 people exposed to various incremental risks at different times in their lives (from birth to death), including built-in latency periods for deferred risks for a variety of occupational risks assumed to be voluntary. These cohorts are compared with a base cohort representing the life expectancy of the basic population. The excess fatalities occurring in different times of life are then used to derive a loss of life expectancy.

Additional results show the expected life shortening of an individual at the time he makes a decision to enter a hazardous occupation, and the number of additional fatalities per 100,000 people expected per year and for lifetime for different occupations.

15.3 FORMAT FOR RISK CONVERSION FACTORS

Risk conversion factors are used to compare the values of different types of consequences. The general format used to make such information available in an unambiguous manner is as follows:

$$\text{RCF}\begin{bmatrix}\text{risk type 1}\\(XX, \mathbf{YY}, ZZ)\\\text{risk type 2}\end{bmatrix}$$

where XX = a lower bound estimate
ZZ = an upper bound estimate
YY = an estimate of the most likely value
lower risk type = the basic risk
upper risk type = the risk that is compared with the base case

The RCF is dimensionless, and a reversal of the upper and lower limits is equivalent to finding reciprocals of XX, **YY**, ZZ, and interchanging the XX and ZZ results.

$$\text{RCF}\begin{bmatrix}\text{risk type 2}\\(1/ZZ, 1/\mathbf{YY}, 1/XX)\\\text{risk type 1}\end{bmatrix}$$

For example, the RCF for comparing catastrophic risk rates for fatalities with similar risk rates for all accidents is

$$\text{RCF}\begin{bmatrix}\text{catastrophic fatalities}\\(1.9 \times 10^{-2}, \mathbf{2.0 \times 10^{-2}}, 2.1 \times 10^{-2})\\\text{all accident fatalities}\end{bmatrix}$$

This means that to obtain the fatality risk rate that occurs for catastrophic risks from the rate from ordinary accident risk rate for fatalities, the latter (base case) risk rate is multiplied by the RCF. Thus if the risk rate for all accidents is 2.7×10^{-4} fat/yr, we have

$$\text{catastrophic risk rate} = 2 \times 10^{-2} \times 2.7 \times 10^{-4} = 5.4 \times 10^{-6} \text{ fat/yr*} \quad (15.26)$$

Conversely

$$\text{RCF}\begin{bmatrix}\text{all accident fatalities}\\(42.6, \mathbf{50.0}, 52.6)\\\text{catastrophic fatalities}\end{bmatrix}$$

and

$$\text{ordinary accident risk rate} = 50 \times 5.4 \times 10^{-6} = 2.7 \times 10^{-4} \text{ fat/yr} \quad (15.27)$$

An RCF for one type of consequence, such as fatalities, is not necessarily the same for other types (injury, property damage, etc.).

* Note that as stated earlier, the notation, fat/yr, refers to fatalities per year per individual member of the population unless otherwise specified.

15.4 DATA BASES FOR HISTORICAL DATA ANALYSIS

A variety of data bases reports statistics on risks. The data are historical and often suffer from arbitrary definition of classes of risk. In general, three classes of statistics are available: (1) financial losses, primarily data from the insurance industry; (2) health information, data from a variety of sources indicating illness and death from disease; and (3) accident statistics, information from insurance companies, regulatory agencies, and so on.

A number of different data sources were used here, and the data must be processed before they are used in determining risk conversion factors for risks and consequences of different types and in establishing absolute risk rates. Thus data from different data sources are employed in a single analysis, which can result in errors arising from different assumptions made in compiling data. The methods of compilation are not always obtainable.

15.4.1 Basic Data Sources

Primary data sources were used wherever possible, but the main sources are listed below.

- General statistical information

  Statistical Abstracts of the United States[13]
  Insurance Facts[14]

- Accident data—United States

  CONSAD Report to the EPA[15]
  U.S. Nuclear Regulatory Commission Reactor Safety Study[16]
  U.S. Occupational Safety and Health Administration News Bulletin[17]
  Accident Facts[18]
  Baldewicz et al.[10]
  Chauncey Starr's publications[19, 20]
  Bunger et al. on occupational risk[12]
  Wiggins derived data[21]

- Accident data—Great Britain

  Mole[3]
  Pochin[22]

Other sources of data are used as appropriate and are referenced directly when used.

15.4.2 Synthesized Data

The author has taken data from many sources and combined these in various ways to present a variety of risk relationships. This process, termed "synthesized data" by the author, is presented as required, either in the text or in Appendix A.

Two particular examples are given here.

The first set of data is from a paper about catastrophic accidents by Chauncey Starr.[20] From this data base, a comparison of death rates for natural and man-made disasters has been made (Table 15.1). The data for the time period t_i, the magnitude of the catastrophe in terms of deaths per event F_i, and the frequency of the event per year N_i, are taken directly from Starr's paper. The time period indicates the period during which the number and magnitude of events were recorded; the three magnitude entries indicate the mode of the distribution of event magnitudes, the largest magnitude, and the average magnitude of the event F_i. By multiplying the average magnitude value by the frequency of events per year, and dividing by the population involved, the probability of death for a single individual in that population can be computed. This is, of course, an average based on linear assumptions (column 7 of Table 15.1 for each type of disaster). This probability of death for a single individual is directly related to the death rate per 100,000 persons per year by multiplying the probability of death by a factor of 10^5 (column 9).

Since some individuals are more protected or more exposed than the average person, however, it is important to know about the variability of risk. Columns 8 and 10 show a possible death rate per 100,000 per year for the most protected and most exposed individuals, respectively. The final column lists the factors that were used to multiply the average value to determine the range of the protected and exposed populations. Conclusions drawn from these data are addressed in subsequent sections.

Insurance Facts[14] reports higher numbers than the Starr data for the probability of death from hurricanes and tornadoes for the United States. For hurricanes, the U.S. population is exposed to 7.4×10^{-7} fat/yr as the mean value for 1923–1972. For tornadoes over the same period, the mean value is 1.25×10^{-6} fat/yr. The difference is explained by the realization that *Insurance Facts* reports all events with fatalities, whereas the Starr data are for catastrophes only.

A second set of data was developed for the Environmental Protection Agency through a contract arrangement with CONSAD Research Corporation. The resulting report, entitled "The Consequences and Frequency of Selected Man-Originated Accident Events,"[15] is an exhaustive investigation into man-made catastrophic accidents for the years 1953–1973, inclu-

Table 15.1 *Comparison of Death Rates from Natural and Man-Made Catastrophic Events*

Type of Event	Time Period[a]	Magnitude (deaths/event)[a] Mode	Maximum	Average	Frequency[a] (event/yr)	Probability Death for One Individual[b]	Death Rate/100,000/yr[c] Protected	Averaged	Exposed	Range Assumption[d]
Natural disaster										
Earthquakes	1920–1970	30,000	180,000	25,000	0.50	3.5×10^{-6}	0.00	0.35	35	$0, 10^2$
Flood; tidal wave	1887–1969	2,000	900,000	28,000	0.54	3.9×10^{-6}	0.0	0.39	39	$0, 10^2$
Hurricanes	1888–1969	400	11,000	1,205	0.41	1.5×10^{-7}	0.0	0.015	1.5	$0, 10^2$
Tornadoes (U.S.)	1900–1969	72	689	78	0.74	3.0×10^{-7}	0.003	0.030	0.3+	$10^1, 10^1$
Typhoons, cyclones, blizzards	1890–1970	10,000	300,000	37,240	0.18	2.0×10^{-6}	0.02	0.20	2.0+	$10^1, 10^1$
Totals						9.9×10^{-6}	0.023	0.99	79	
Man-made disasters										
Major air crashes	1965–1969	82	155	78	6.0	1.3×10^{-7}	0.0	0.013	1.3	$0, 10^2$
Major auto crashes	1966	10	40	19	11.0	6.0×10^{-8}	0.0006	0.006	0.06	$10^1, 10^1$
Explosions	1950–1968	10	100	26	2.0	1.5×10^{-8}	0.00015	0.0015	0.15	$10^1, 10^2$
Major fires	1960–1968	12	322	35	0.67	6.4×10^{-9}	0.00006	0.0006	0.006	$10^1, 10^1$
Major railroad crashes	1950–1966	19	79	30	1.0	8.4×10^{-9}	0.0	0.0008	0.08	$0, 10^2$
Major marine accidents	1965–1969	32	300	61	6.0	1.0×10^{-7}	0.0	0.010	1.0	$0, 10^2$
Totals						3.2×10^{-7}	0.00081	0.032	2.6	

[a] *Source:* Data from C. Starr, "Benefit-Cost Relationships in Socio-Technical Systems: Environmental Aspects of Nuclear Power Stations," IAEA-SM-146/47. Vienna: International Atomic Energy Agency 1971, p. 900.

[b] Based on a linear assumption using a 1970 U.S. population of 203×10^6 and a 1970 world population of 3.56×10^9 as appropriate.

[c] Averaged death rate per 100,000 for total population, a protected population in the sense a tidal wave is unlikely to occur inland, an exposed population in the sense that a coastal inhabitant has a higher exposure to a tidal wave event.

[d] Multipliers to get protected and exposed values, respectively, from the averaged value.

Table 15.2 *Summary of Risk Rates for Man-Made Nonmilitary Catastrophic Event Consequences for the United States (based on CONSAD data)*

	Voluntary Risks		Involuntary Risks	
Consequence	Mean Number/yr	Risk Rate[a]	Mean Number/yr	Risk Rate[a]
Fatalities	421.7	2.1×10^{-6}	21.2	1.1×10^{-7}
Injuries	453.4	2.3×10^{-6}	107.2	5.4×10^{-7}
Property damage (millions of dollars)	\$70.08	35¢[b]	\$3.82	0.02¢[b]
Number of events	26.76	1.1×10^{-7}	1.0	5×10^{-9}

[a] Per individual based on a U.S. population of 2×10^8.
[b] Per member of U.S. population.

sive. The selection criteria were the three conditions for a catastrophic accident used previously.* The original data consisted of 814 distinct entries. However many of these entries resulted in deaths involving populations outside the United States. To assure that consistent populations and risk data were used, 53 events involving foreign military air crashes, 14 foreign commercial aircraft accidents, 3 events in foreign buildings and structures, and 15 foreign marine events were deleted. These data, derived in detail in Appendix A (Tables A1.1 through A1.7), show fatalities, injuries, and property damage for all risks from catastrophic events, number of events, and involuntary risks for seven classes of man-made accidents.

Table 15.2 summarizes these data. Risk rates to individuals are based on a U.S. population of 2×10^8 people.

Deficiencies in data arise from a number of sources, but data on risks suffer particularly, from the condition that most data are derived for purposes other than risk analysis. As a result, the data cannot easily be segregated into the different categories of risk for analysis. As an example, most data on accidents are derived without regard to whether involuntary or voluntary risks were involved or whether the consequences were ordinary or catastrophic. The difficulty in determining property damage for catastrophic risks is a case in point. Appendix B.1 provides an explicit example of measurement difficulties of this type, using data from the coal mining industry.

Another difficulty lies in the manner in which data are presented. Some-

* Ten or more fatalities, 30 or more injuries, or $3 million in damages.

times probabilities of a consequence are given for different magnitudes of the same consequence. In other cases an average magnitude of the consequence is given with a single probability of occurrence, which means that data must be manipulated to assure that like figures are being compared. Appendix B.2 furnishes means for manipulation to compare the probability for an average consequence magnitude with an array of probabilities for different magnitudes of the same consequence.

REFERENCES

1. Harry J. Otway, "Risk Assessment and Societal Choices," IIASA RM-75-2. Laxenburg, Austria: International Institute for Applied Systems Analysis, 1975, p. 7.
2. Helga Velimirovic, "An Anthropological View of Risk," IIASA RM-75-55. Laxenburg, Austria: International Institute for Applied Systems Analysis, 1975, p. 17.
3. R. H. Mole, "Accepting Risks for Other People," *Proceedings of the Royal Society of Medicine,* Vol. 69, February 1976, p. 109.
4. Velimirovic, "An Anthropological View of Risk," p. 32.
5. *Ibid.,* p. 33.
6. Saul Levine, U.S. Nuclear Regulatory Commission, personal communication to the author, June 25, 1975.
7. Freeman J. Dyson, "The Hidden Cost of Saying NO," *Bulletin of the Atomic Scientists,* Vol. 31, No. 6, June 1975, pp. 23–27.
8. *The Quality of Life Concept, A Potential New Tool for Decision Makers.* Washington, D.C.: Environmental Protection Agency, Office of Research and Development, March 1973.
9. Graham T. Molitor, "A Hierarchy of Needs and Values," in *Quality of Life Concept,* Part II, p. 203.
10. W. Baldewicz, G. Haddock, Y. Lee, Prajoto, R. Whitley, and V. Denny, "Historical Perspectives on Risk for Large-Scale Technological Systems," UCLA-ENG-7485. Los Angeles: UCLA School of Engineering and Applied Science, November 1974.
11. *Ibid.,* p. 77.
12. Byron Bunger and John R. Cook, "Use of Life Table Methods to Compare Occupational Radiation Exposure Risks and Other Industrial Risks," Meeting of the Health Physics Society, June 30, 1976.
13. *Statistical Abstracts of United States,* Government Printing Office, 1971, 1972, 1973, 1974.
14. *Insurance Facts,* 1973 and 1974 editions. New York: Insurance Information Institute.
15. CONSAD Research Corporation, "Consequences and Frequency of Selected Man-Originated Accident Events." Environmental Protection Agency Contract Report No. 68-01-0492. Washington, D.C., 1975.
16. "Reactor Safety Study: An Assessment of Accident Risks in U.S. Commercial Nuclear Power Plants," WASH-1400 (NUREG-75/014). Washington, D.C.: Nuclear Regulatory Commission, October 1975.
17. *OSHA News Bulletin*, USDL-73-256, U.S. Department of Labor, June 15, 1973.

18. *Accident Facts,* 1973 and 1974. Chicago: National Safety Council. Ill.

19. Chauncey Starr, "Social Benefit Versus Technological Risk," *Science,* Vol. 165, No. 3899, September 19, 1969, pp. 1232–1238.

20. Chauncey Starr, "Benefit-Cost Relationships to Socio-Technical Systems," *Environmental Aspects of Nuclear Power Stations,* IAEA SM-146/47. Vienna: International Atomic Energy Agency, 1971, p. 900.

21. J. T. Wiggins, "Balanced Risk: An Approach to Reconciling Man's Need with His Environment." Paper presented at Engineering Foundation Workshop on Risk-Benefit Methodology and Application, Asilomar, Calif., September 25, 1976. Papers edited by David Okrent.

22. E. E. Pochin, "Occupational and Other Fatality Rates," *Community Health, London,* Vol. 6, No. 1, 1974.

16

Estimation of Relative Risk Factors

This chapter estimates a number of the risk factors discussed in Part B, using existing data bases, where possible, to calculate the magnitude of risk conversion factors. This preliminary analysis serves as a means of arriving at an initial characterization of the risk conversion factors and the methods employed for this purpose.

Several factors are not amenable to such analysis because they involve personal value judgments or unnecessary aggregation resulting in masking of uncertainty. These are the factors involving the nature of consequences, as discussed in Chapter 9. We begin with these, since their treatment affects the manner in which other factors are estimated.

16.1 FACTORS INVOLVING THE NATURE OF CONSEQUENCES

16.1.1 Classes of Consequences

The hierarchical classification of consequences presented in Table 4.2 contains values related to the "quality of life" for both individuals and groups. Thus determination of the relative weights among these classes is a much broader problem than determining risk comparison factors. On an individual basis, assignment of weights is the result of personal value judgments. Such was the case for Dr. Hoke's FLU scale in Chapter 6. Individual judgments of this nature are personal and subject to change as the individual's attitude and situation change. Although psychological and preference testing provide insight into individual tastes, aggregation to find weights for society as a whole is, indeed, more complex. The whole concept of the quality of life is involved and receiving considerable attention[1] although little progress has been made to date.

Such weights could be used here effectively if they were available and universally accepted. In their absence, however, it is not necessary to wait for the debate to end for the purpose of risk assessment. Chapter 5 dealt with the question of single and multiple risk referents. Retaining the different classes of consequence as risk measures results in multiple risk referents and multiple risk conversion factors. This approach may be more complex, but it prevents masking of risk relationships.

16.1.1.1 Multiple Risk Conversion Factors

For each risk conversion relation (i.e., between one type risk and another), a separate risk conversion factor can be shown for each class of consequences. For example, there will be separate RCFs between ordinary and catastrophic risk for both fatalities and injuries. Wherever possible, such multiple relationships will be shown.

16.1.1.2 Data for Classes of Consequences

There is sparse data on the different classes of consequences arrayed in Table 4.2. The data that are available fall into relatively few classes:

- Fatalities—accidental
- Fatalities—disease
- Injury—accidental
- Illness—disease and accident related
- Property damage
- Years of life-shortening
- Earnings foregone because of illness or injury
- Earnings foregone because of life-shortening
- Insurance settlements

Data toward the aesthetic value end of the hierarchal scale are scarce. Whenever data are available for risk conversion factor calculation for different classes of consequences, they are given. Generally, the analysis is restricted to fatalities, injuries, and property damage.

For example, here are the CONSAD data totals for the 21-year period covered:[2]

Number of incidents	810
Deaths	13,782
Injuries	12,472
Reported property damage	$2.075 billion

One is tempted to determine the reported property damage per death or per injury. This can be done, but it must be recognized that such a result represents a report of the ratio of deaths to property damage, not the amount spent to avoid a fatality, nor the indirect costs of the fatality, nor compensation paid.

Furthermore, as Mole states, "It seems impossible at the present time to assess social attitudes to serious injury as compared with death (to which all the data refer) since even the experts are uninformed about the lifetime sequelae of serious injury."[3]

16.1.2 Variation in Cultural Background and Values

The perception of risk in different cultures and strata of society is far from uniform. Variations in cultural background arise in different countries from dissimilar paths of historical development, from racial and religious differences, and from unequal states of economic development. There is also great variation in culture within a country because of the existence of different castes, social and economic levels, educational opportunities, religion, and so on. Some general conclusions about the effect of different cultures on risk valuation may be made, but with limited confidence.

16.1.2.1 Variation Across International Boundaries

Disaster Toll in Different States of Economic Development. There is little question that the value of life differs in modern and traditional cultures. As Kates points out, the economic level is a factor determining how man copes with risks from disaster.*

> To be poor is to be vulnerable—from the harmful acts of men and from the hazardous events of nature. It is only recently, however, that we could begin to specify the relative vulnerability of the poor nations to natural disaster: per capita death perhaps a hundredfold greater; relative per capita damage perhaps tenfold greater than the rich nations.[4]

Kates goes on to differentiate the toll of disaster from the ways in which different societies cope with disaster. He begins with fatalities.

> 1. *Fatalities.* Political rhetoric or journalistic impressions to the contrary, there is little social leveling in natural disaster. Death, the ultimate human loss, is relatively rare, but death comes unequally in the thirty or so major disasters (excluding drought) that occur each year. The annual probability of an individual dying in a natural disaster in the rich countries is in the range of 10^{-6} (one or more chances in a million, per individual, per year), 10^{-5} in the developing world, and 10^{-4} in Ban-

* Reprinted by permission from reference 4.

Table 16.1 *Comparative Death Rates from Natural Disasters by National Income*[a]

	Per Capita National Incomes			
Annual Death Rate (per million population)	Least Developed (≤ $200)	Lesser Developed ($200–$1000)	Industrial (≥ $1000)	Total
High > 200	2	14	1	17
Moderate and low ≤ 200	10	24	19	53
Totals	12	38	20	70

Source: Robert W. Kates, "Natural Disasters and Development." Wingspread Conference Background Paper, Racine, Wis., October 19–22, 1975.
[a] The figures are somewhat distorted by the exclusion of drought, which affects much more the least developed countries, and by the bias in the overall data compilation toward the developed industrial countries.

gladesh. But no one really knows precisely* the frequency of disaster occurrence and the annual toll in lives, material wealth, and productive activities.[4]

The basis for this information is contained in Appendix A.2, from which it can be safely inferred that the world's developing countries, which have 66% of total population, experience 95% of disaster-related deaths. This vulnerability of developing countries to death from disaster is not uniform. For example, Table 16.1 presents the relationship between death rates and national income for the 70 nations for which at least one disaster was recorded between 1947 and 1973.

2. *Economic Losses.* Postdisaster estimates are usually overestimates, often by a factor of two. Such early uncorrected damage estimates for catastrophic disasters tend to find their way into international records, while small losses from recurrent localized hazard events are undercounted. The absurdities of national accounting

* The Disaster Research Unit (DRU) of the University of Bradford in their recent overview of data sources concluded that:

"First, there is no common definition of disaster, no universal scale of disaster measurement and consequently little compatibility between the data. Secondly, there is no accounting for disaster losses except in individual appraisals of a specific disaster situation. Thirdly, most data contains a bias which results from its specific viewpoint on disaster. Fourthly, at the present moment the best data is available from UNDRO (United Nations Disaster Relief Office) and the Natural Hazard Research Group. The USAID (United States Agency for International Development) provides the most information, but the internal and external consistency of the data is much in question."[5]

practices often turn postdisaster aid or reconstruction costs into positive gains of gross national product and these may be further increased by disaster-induced inflation. Disaster prevention costs are buried in governmental budgets or are unknown for nongovernmental production units. And many long-term social costs are never accounted for. They are concealed or go unrecognized in the "average" statistics. All of the foregoing are unintentional distortions, and added to these must be the many intentional distortions. The politics of disaster can either lead to excessive claims or equally to the denial of losses. Nonetheless, using as guides the scattered estimates detailed, it is possible to make an "educated guess" as to global losses and costs of prevention.

Natural hazard taxes the global economy by perhaps $40 billion a year, $25 billion of which is in damage losses and the remainder in the costs of prevention and mitigation. Of the $25 billion, perhaps $10–15 billion in damages comes in the form of disasters, large catastrophic losses, bringing the "average" disaster loss to $350–500 million. Similarly, the annual death toll from hazard might be estimated at around 250,000, about half of which occurs during major disasters.

In contrast to the exceptional regional differences in death rates, losses from natural disaster are more proportional to the world income distribution. Therefore, perhaps three-fourths of global hazard losses occur in the wealthy countries, although the proportional burden of losses are much higher in developing countries. The Natural Hazard Research Group has compiled careful estimates of loss for three major natural hazards: drought, flood and tropical cyclone, comparing the costs and losses for each, in one industrial and one developing country. These are shown [in Table 16.2].[4]

3. *Risk Conversion Factors for Natural Disasters in Different Economies.* Risk conversion factors can be estimated for both fatalities and property damage on the basis of Kates' data. The range of uncertainty is estimated by the author. The following risk conversion factors were determined:

RCF [rich country—fatalities/natural disaster (.05, **.10**, .15)* developing country—fatalities/natural disaster]

RCF [developing country—fatalities/natural disaster (6, **10**, 20)† rich country—fatalities/natural disaster]

RCF [rich country—fatalities/natural disaster $0.5, \mathbf{1.0}, 1.5 \times 10^{-2})$ poor country—fatalities/natural disaster]

RCF [poor country—fataliities/natural disaster (60, **100**, 200) rich country—fatalities/natural disaster]

* It should be noted that these RCFs are given to two significant places in order to show the range meaningfully. This should not be construed to mean that the measurement of these values is that precise.

† Rounding errors may prevent direct computation of an RCF from its reciprocal RCF. The accuracy of these values is limited to the data presented for these calculations and does not preclude the existence of other values from other sources.

Table 16.2 *Selected Estimates of Natural Hazard Losses*

Hazard	Country	Total Population	Population Risk	Annual Death Rate/ Million at Risk	Losses and Costs per Capita at Risk			Total Costs (% of GNP)
					Damages Losses	Costs of Loss Reduction	Total Costs	
Drought	Tanzania	13	12	40	$ .70	$ .80	$ 1.50	1.84
	Australia	13	1	0	24.00	19.00	43.00	0.10
Floods	Sri Lanka	13	3	5	13.40	1.60	15.00	2.13
	United States	207	25	2	40.00	8.00	48.00	0.11
Tropical cyclones	Bangladesh	72	10	3000	3.00	.40	3.40	0.73
	United States	207	30	2	13.30	1.20	14.50	0.04

Source: Kates, "Natural Disasters and Development."

RCF [rich country—total cost/natural disaster (3, **12**, 30)* poor country—total cost/natural disaster]

RCF [poor country—total cost/natural disaster (.03, **.08**, .30)* rich country—total cost/natural disaster]

RCF [rich country—cost as percentage of GNP (.050, **.054**, .058) poor country—cost as percentage of GNP]

RCF [poor country—cost as percentage of GNP (18, **19**, 20) rich country—cost as percentage of GNP]

Variation in Modern Cultures. Otway et al.[6] have made some initial investigations across national boundaries of industrial societies. Otway[7] replicated in Austria a survey originally conducted in Canada by Golant and Burton,[8] who attempted to obtain ordinal rankings for various hazard conditions. As reported by Otway[7] for the Canadian group, the effect of experience with specific risks (i.e., experienced respondents versus inexperienced) was the most important factor in determining response. This was not found in the Austrian sample, where the most important determinant was the subjects' self-rated ability to imagine themselves in particular risk situations (i.e., "good" imaginability vs. "poor"). This beginning research effort is continuing; it provides early insight into some nonobvious cultural differences.

Similar Risks in Different Countries. Mole[3] has compared air, rail, and bus travel fatalities per 10^8 passenger miles traveled in the United States and Great Britain for a 40-year period. In this case, the trends and the rates for risks of similar types in different countries are essentially the same. Interesting enough, the 40-year trends in both countries show a very large reduction in fatalities over time for air travel in comparison to rail or bus. Mole suggests that pressure for safety in air travel may be due to worldwide compilation of statistical information and press coverage, as well as to more recent technological initiation of air travel as a new system. Problems that have worldwide attention may very well imply greater concern for safety and more risk aversive pressure precisely because of the wider exposure.

Preferences in Different Countries. Another technique is to compare revealed societal preferences in different countries. Data, however, are taken by different methods, and thoroughness of compilation and degree of analysis varies. Furthermore, when evaluating industrialized nations, dif-

* Based on Table 16.2

Table 16.3 *Selected Occupational Fatality Rates in Some Industrialized Countries*

Country[a]	Metal Working Industries (fat/10^6/yr)	Chemical Industries (fat/10^6/yr)
Federal Republic of Germany	114 ± 6	103 ± 6
France	118	169
Great Britain	136 ± 5	87 ± 5
United States	174	60
Mean	135.5	104.8
Standard deviation	27.4	46.0

[a] Foreign data from Pochin,[7] U.S. data from *OSHA News Bulletin.*[10]

ferences are seldom marked. For example, occupational risk rates for a few industrial nations for comparable occupations are reported by Pochin.[9] Taken together with similar data from the United States (Table 16.3), this provides a first approximation* for a very limited sample. The variation as indicated by the computed standard deviation for each case is not large and is most likely well within error caused by differences in measurement definition and methods.

Pursuit of cultural differences in risk valuation across international boundaries is of interest, but such data are neither necessary nor meaningful for use within the United States. Since this study is primarily aimed at the U.S. population, further effort here is left to the future and other investigators.

16.1.2.2 Variation within National Boundaries

Cultural differences within the United States stem from many factors—income, level of education, minority group membership, and so on. Virtually no risk data are compiled in a manner that incorporates variations, although some inferences are possible from generalized data.

Distribution of Losses Within a Disaster

The global inequality characteristic of natural disaster impact is mirrored by the loss distribution within a given disaster. Whether the death toll is great or small, there are still major differences in who suffers, who pays. Those who lose their lives,

* The Pochin data were provided primarily to show the different fatality rates in different occupations. The international information was used to substantiate these differences, not to show international differences.

their families who grieve, the injured who suffer disablement and pain, those who are permanently dislocated, and the damaged who are made poorer—all differ in number affected and in magnitude of suffering by an order of difference. More numerous but much less affected are those indirectly impacted, the disturbed adjacent to the disaster area, the voluntary donors of aid and relief, and the involuntarily taxed. The sequence of return after disaster is class related. . . . For those on the bottom of the social hierarchy, 85% never rebuilt their lives in San Francisco [1906 earthquake]. And, unfortunately, more recent studies still show that even with disaster aid and conscious welfare policies, those on the bottom of the social structure have the greatest difficulty in reconstructing their lives.[4]

Size of Community and Risk Preferences. Another avenue of investigation is cultural differences involving the sizes of cities. *Accidents Facts*[11] reports the following for 1973:

City Size Based on Population	Accidental Death Rate per 100,000 Population (All Causes)
> 500,000	32.1
350,000–500,000	42.5
200,000–350,000	41.0
100,000–200,000	35.7
50,000–100,000	29.2
25,000–50,000	36.3
10,000–25,000	31.4
All reporting cities	33.9

Except for cities between 200,000 and 500,000 populations, there is little to suggest that an aggregate difference exists between the cultural values of residents of small towns and large cities.

Economic Classes and Risk Preferences. Another possibility involves the capability of people at the higher end of the economic scale to minimize exposure to certain kinds of risk, that is, to become members of a protected population. Such inferences must take into account that certain types of voluntary risk are attractive to high income members of the population seeking means of using leisure time. Leisure activities are often much more risky than occupations.[9] Conversely, the same population may be particularly averse to risks imposed on them involuntarily. The extent to which populations may either expose or protect themselves in comparison with normal populations was estimated in Table 15.1. The desire and/or ability of a person to choose an exposed or protected position involves both knowledge of the situation and resources to implement protection when desired. Individual propensity for risk and one's position on the Maslovian hierarchy may be primary determinants. This question also involves the ability to avoid and control risks (see Section 16.2).

If any case, aggregate accident statistics in this area mask the ability of individuals to avoid certain risks (e.g., involuntary risks) while consciously exposing themselves to other risks for some direct benefit. The benefits may be associated with personal satisfaction from excitement, such as high risk exposure from leisure-time activities. The latter voluntary risks may more than offset any risk aversive action elsewhere. For example, many environmentalists oppose small risks imposed on them and society from new technologies, yet as active mountain climbers they voluntarily undertake great personal risk.

16.1.3 Common Versus Catastrophic Risks

For the purpose of investigating events in the past that can be considered catastrophes, the definition of catastrophe used is an event that results in one or more of the following conditions: 10 or more fatalities, 30 or more injuries, and \$3 million or more in damages. Difficulties arising from this definition were discussed in Chapter 9.

Data derived from Table 15.1 provide individual risk rates for deaths from worldwide natural catastrophes at a level of 9.9×10^{-6} per year, and for man-made disasters at a level of 3.2×10^{-7} per year. Data from the CONSAD report (Appendix A.1) indicate a man-made disaster rate of 2×10^{-6} per year for this country, nearly seven times greater than the data from Table 15.1.

The following data are available for catastrophes involving fatalities in terms of death rate, 10^5 people/yr; are

Natural catastrophes	0.99	fat/10^5/yr
Man-originated catastrophes	0.032 – 0.22	
Total catastrophes	1.022 – 1.21	

These may be compared with total death rate data in Table 16.4.[11] A review of other years shows that the values change very little from year to year.

On this basis, some risk conversion factors can be derived:

RCF [Catastrophic accidents—fatalities / ordinary accidents—fatalities] (.018, **.020**, .022)

RCF [ordinary accidents—fatalities / catastrophic accidents—fatalities] (47, **50**, 55)

RCF [catastrophic accidents—fatalities / total death rate] (1.1, **1.2**, 1.3) $\times 10^{-3}$

RCF [total death rate / catastrophic accidents—fatalities] (770, **830**, 910)

Table 16.4 Death Rate by Cause of Death

Class	Death Rate/100,000/yr	
Natural catastrophes		0.99
Man-made catastrophes		0.032–0.22
Common accidents (1972)		
Motor accidents	27.2	
Work related	6.8	
Home	13.0	
Public nonmotor vehicle	11.3	
Total		58.3
Disease (1969)		
Heart disease	367	
Cancer	160	
Stroke	103	
Pneumonia	31	
Diabetes mellitus	19	
Arteriosclerosis	16	
Total		696
Homicide, suicide, other		199
Total		954

Source: Accident Facts, 1973.

Wilson, however, suggests a slightly different approach:

We can correctly compare risks . . . or death per year, so long as no more than one person is involved (at risk) at a time. This would include black lung disease, radiation cancers, or asphyxiation by gas in a domestic kitchen. In each of these hazards, only one person dies at a time, and sometimes we add a pollutant and, by chance no one dies.

However, when an accident involves more than one person—say N persons—then such a comparison may no longer be valid, and I assume, as a guess, that a risk involving N people simultaneously is N^2 (not N) times as important as an accident involving one person. Thus a bus or airplane accident involving 100 people is as serious as 10,000, not merely 100, automobile accidents killing one person.[12]

On this basis, an RCF is a function of the number of fatalities N:

$$\text{RCF}\left[\frac{\text{catastrophic accident—fatalities}}{\text{ordinary accident—fatalities}}\right](N^{-2}) \qquad \text{RCF}\left[\frac{\text{ordinary accident—fatalities}}{\text{catastrophic accident—fatalities}}\right](N^{2})$$

where N is the difference between number killed in catastrophic and in ordinary accidents.

Mole furnishes some qualitative insights indicating that a rigorous formula such as Wilson's might be too formal.

> Degree of concern about particular kinds of risk is certainly greatly magnified by the immediacy of television pictures and by the reporting habits of the media, but special concern about a simultaneously occurring group of casualties may possibly be determined not only by social conventions but also by a basic human characteristic with evolutionary advantages and therefore not eradicable by reason and information.[3]

16.1.4 Military Versus Societal Risk Bases

One method of comparing military and societal risk bases is to estimate the death rates from military and commercial airline catastrophes, using the proper populations at risk (voluntary risks only). For the military, the population is the number in the armed forces. One could argue that military transport of dependents and civilian military employees should also be included, but such data are difficult to determine. For commercial aircraft, the total U.S. population was used, including military personnel flying commercially.

The results in Table 16.5 cover three years, 1970–1972. The ratio of military death rate to that of civilians for air travel ranges from 36 to 53. Thus a risk conversion factor between military and civilian risks for aircraft would be about

$$\text{RCF}\left[\begin{array}{l}\text{civilian accident—fatalities}\\ (2.0,\ \mathbf{2.5},\ 3.0 \times 10^{-2})\\ \text{military accident—fatalities}\end{array}\right. \qquad \text{RCF}\left[\begin{array}{l}\text{military accident—fatalities}\\ (35,\ \mathbf{40},\ 55)\\ \text{civilian accident—fatalities}\end{array}\right.$$

It must be noted that these RCFs are for one form of military accidental death and do not involve combat fatalities.

Another method of estimating combat fatality relationships is to follow the lead of Starr,[13] who used the war in Vietnam as a test case. During the war's heavy years, he estimated that for a U.S. military population of 500,000, about 10,000 deaths per year occurred. This is an individual death rate of 2×10^{-2} per year, or 2000 deaths per 100,000 per year. This is about twice the total U.S. death rate of 954 per 100,000 per year, given in Table 16.1, and is about 35 times the rate for accidents. When the figure for age distribution is adjusted, however, assuming participants' ages range from 18 to 24, the picture of war versus peacetime changes. *Accident Facts* for 1973 reveals that the total death rate from all causes for males in this age group

Table 16.5 *Comparison of Death Rates for Commercial and Military air Crashes, 1970–1972*

Year	Total Military Personnel[a]	Number of Military Airline Deaths			Death Rate per Individual	Commercial Airline[b]
		U.S.	Foreign	Total		
1970	3,065,508	76	129	205	6.7×10^{-5}	1.2×10^{-6}
1971	2,698,657	34	47	81	3.0×10^{-5}	1.0×10^{-6}
1972	2,504,845	24	0	24	9.6×10^{-5}	6.9×10^{-7}
Mean					3.65×10^{-5}	9.6×10^{-7} [c] 1.02×10^{-6} [d]
Standard deviation					2.89×10^{-5}	2.6×10^{-7} [c] 4.9×10^{-7} [d]
RCF	Commercial-military				.019–.028	

[a] U.S. Budget.
[b] U.S. crashes and populations of 2.03×10^8.
[c] Three years shown.
[d] Twenty-one years.

is 186.5 per 100,000, and for accidents, 109.4 per 100,000.[11] In this case, the ratio is about 11:1 for war deaths to all causes of death for this male age group, and about 18:1 for war to accidental death. On this basis:

RCF ⌈military combat—fatalities
(10, **20**, 40)
⌊civilian accident—fatalities

RCF ⌈civilian accident—fatalities
(.02, **.05**, .10)
⌊military combat—fatalities

RCF ⌈combat fatalities
(2, **10**, 20)
⌊all fatalities—combat age males

RCF ⌈all fatalities—combat age males
(.05, **.10**, .50)
⌊combat fatalities

16.1.5 Natural Versus Man-Originated Risks

Risks with consequences of any magnitude may be caused by natural phenomena or by man-originated events. However the distinction between naturally occurring and man-originated risks, generally, is of concern when related to catastrophic events. A variety of classifications of catastrophic risk affect the possible modes of comparison for natural and man-originated events.

16.1.5.1 Avoidability and Controllability of Catastrophes

Risks from natural causes can be classified as unavoidable or avoidable. Totally unavoidable risks are completely beyond the control of man and are treated as acts of God. These risks involve such major natural phenomena as the sun exploding, major solar flares, impact of very large meteorites, and volcanic eruptions. Other natural risks and man-originated risks involve different degrees of avoidability and controllability.

1. For risk to be avoidable, individuals or groups, by changing their exposure to potential risks, must be able to reduce the chances of being involved.
2. For risk to be controllable, society must be able to take action to prevent a catastrophe or at least mitigate it.

Avoidability means an ability or at least a choice of choosing one's site of activity and residence to reduce exposure to risks, that is, finding a location with lower probability of disaster. The option also exists to restrict participation in some activity to avoid risk exposure. Avoidability of risk exposure is the only means of risk reduction for some natural disasters. Relocating to seismically stable areas to avoid earthquakes and moving to high ground to avoid floods from tidal waves are examples.

The attempt to control natural disasters is a recent undertaking of man, as civilization has become increasingly prone to both natural and man-originated disasters.

As civilization has progressed, man has congregated in large cities; at the same time economies of scale have demanded that industrial facilities become ever larger. Therefore, there is now potential for man-made catastrophes of unprecedented magnitude. Moreover, as civilization progresses we live longer, and want to live longer still. We want to make our lives ever safer. Accidents, and even natural disasters which were once dismissed as acts of God, are now considered to be under control. It's not that we deny the existence of God; it's just that there is no need to blame Him for our incompetence.[12]

Kates argues strongly against the idea that control of natural disasters is effective in traditional or modern societies.

The global toll of natural disaster rises; at least as fast as the increase of population and material wealth, but probably faster. Disasters may be less frequent but more catastrophic, are more costly in lives and relative wealth in developing countries, increasingly costly in absolute wealth in industrial nations.[4]

The reasons for such increases are described by Burton et al.

In examining [hazard] adjustments, it is useful to make a distinction between those which seek to rearrange or manipulate nature and those which involve a rearrangement or alteration of human behavior. The former may be equated with the technological approach to hazard problems, the latter with the social or behavioral approach. The technological approach emphasizes the construction of dams and levees to control floods; sinking of new and deeper wells in periods of drought; cloud seeding to increase rainfall in subhumid areas; and the buttressing of potential slide areas in earthquake zones. All these actions are directed to affecting the cause of the problem or to a modification of the hazard itself.

The social or behavioral approach emphasizes the careful planning of flood-plain land use; more cautious use of water and/or curtailed water use in times of drought, and use of legislative guides to encourage better building design for earthquake resistant structures. The prevailing public approach has been to offer immediate relief and then to turn to the technological approach. Dams follow floods, irrigation projects follow droughts.

We find that when carried out in isolation without adequate reference to social considerations, the technological adjustments may lead to an aggravation of the problem rather than an amelioration, as when upstream reservoir construction encourages increased invasion of a Tennessee Valley flood plain of Chattanooga. Commonly, the benefits received are short run, and involve the elimination of numerous "small" losses at the cost of greater long-term losses often of a catastrophic nature.

Control of floods seems to induce more rapid development of flood plains, plus a relaxation of emergency preparations. Thus a consequence of adopting the technological fix is the relaxation of preparations for other, more extreme, action.[14]

16.1.5.2 Classification of Disasters

The idea of controllability of natural disasters requires a classification of disasters more extensive than simply natural and man-made. Each of these has two aspects of interest.

Natural Disasters.

Unavoidable Natural Disasters. Beyond the control of man to avoid or mitigate (e.g., impact of large meteorites).

Avoidable Natural Disasters. Man has some ability, often limited, to mitigate risk of disaster (e.g., floods) by avoidance or control means.

Man-Originated Disasters. Man-originated disasters occur because of an activity of man to set the scene for a disaster, followed by an action to trigger it. For example, a hazard exists to people living downstream from a man-made dam. A subsequent disaster may be precipitated by natural triggers such as a large earthquake, or by man-made triggers (sabotage, poor construction, inadequate maintenance, etc.). Flooding, for example, could be triggered either by man or by nature. The design of the dam may have made the risk of flooding more pronounced, or the flood may represent a water volume well beyond the design capacity of the dam.

Man-Triggered Disaster. Activities of man have not only created the exposure to risk, but also trigger the disaster.

Man-Caused Disaster. Activities of man originate the exposure to risk, but natural phenomena trigger the disaster.

There is a difference in the risk acceptance of the two types of natural risk, based primarily on the probability of occurrence. Unavoidable natural risks are by their existence basic thresholds of risk that man cannot avoid. These thresholds are examined separately. The difference in acceptance of the two types of man-originated disaster is less clear, since there is a paucity of data.

16.1.5.3 Disaster Data

Natural disaster data used here are derived from three sources: Kates, Starr, and the Reactor Safety Study. Kates' data, reproduced in Figure 16.1, demonstrate that:

> Over the 27-year period, the average number of disasters* has remained relatively constant, about 30 per year, or, if anything, declining somewhat. At the same time,

* Disasters are defined by Kates operationally as all geophysical natural events (with the exception of drought), causing at least $1 million damage or 100 dead or injured. This definition differs from that used for man-originated events and will cause the RCF for relating man-originated risks to natural risks to be understated.

death rates have been climbing significantly (albeit distorted by the occurrence of the largest disaster of the century, the Bangladesh cyclone of 1970). The areal extent of disaster, measured by the increase in the number of large-area disasters, seems to have increased significantly as well.[4]

The average death rate shown is about 10 fat/yr/10^6 people, or 1 fat/yr/10^5, with a range between 0.5 and 2 fat/yr/10^5. This compares very closely with the Starr data in Table 15.1, which give a value of 0.99 fat/yr/10^5. For man-originated disasters, the Starr data cite a value of 0.032 fat/yr/10^5, whereas the CONSAD data give 0.22 fat/yr/10^5.

The Reactor Safety Study[15] provides information to allow a comparison

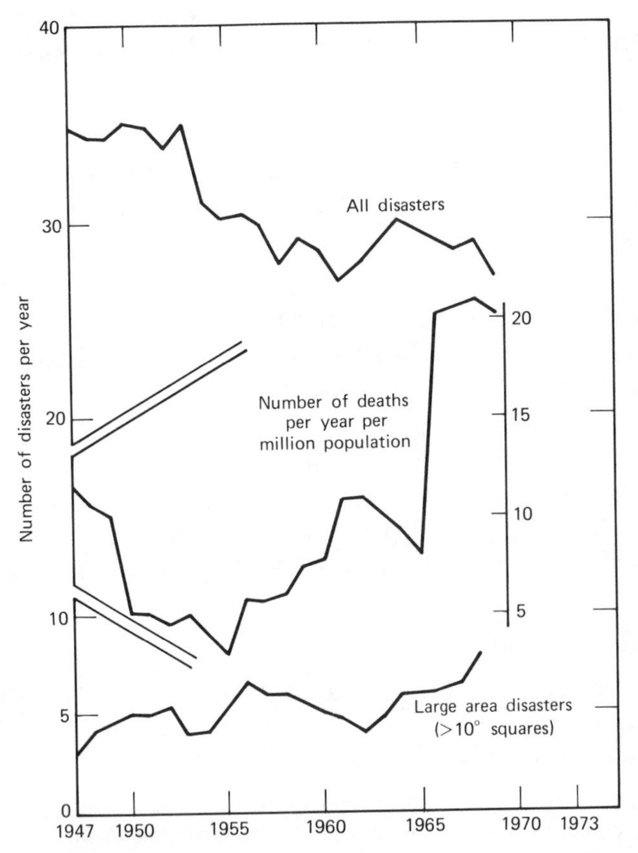

Figure 16.1. Global disasters, 1947–1973, 5-year moving average. *Source:* J. Dworkin, "Global Trends in Natural Disasters 1947–1973," Natural Hazards Working Paper No. 26. Boulder, Colo.: University of Colorado, Institute for Behavioral Science, 1974. Reprinted by permission.

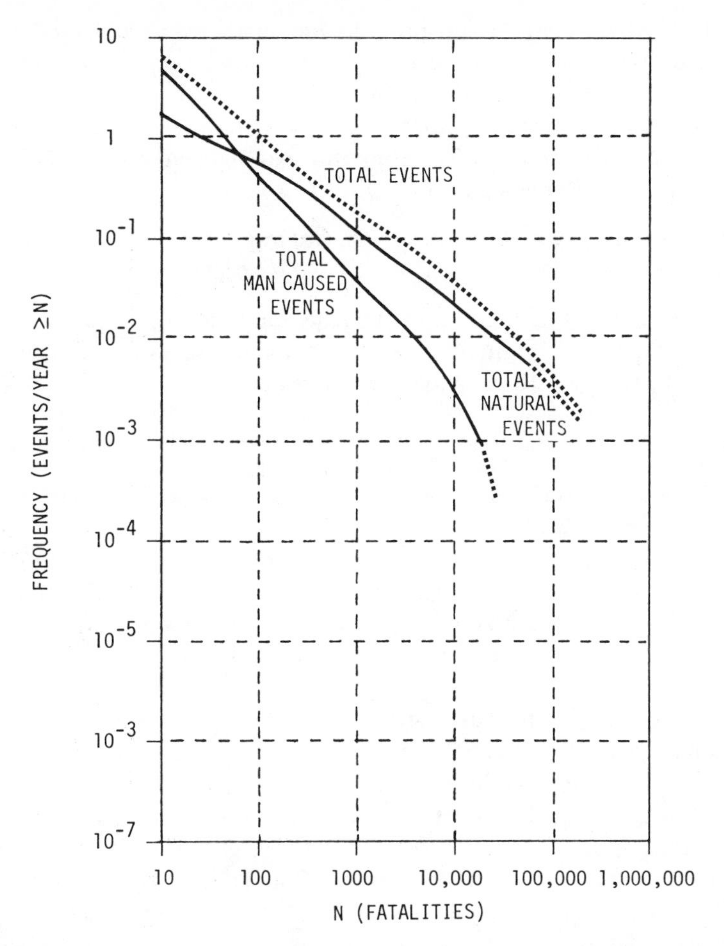

Figure 16.2. Frequency of total natural and man-made events with fatalities greater than N. *Source:* Ref. 15.

of natural and man-originated catastrophes as a function of the number of fatalities per event (Figure 16.2).* The data indicate that man-caused events have risk rates for events that exceed those for natural events where the events have smaller numbers of fatalities per event, but this condition is reversed at higher numbers of fatalities per event. A high dependence on the size of the consequence is indicated. However this difference at low levels of N may well be caused by the difference in definition of natural disasters and by the failure to report many naturally caused accidents. At the higher end

* Basic risk data from the Reactor Safety Study are reproduced in Appendix A.3.

it is affected by the limitation prior to the 1950s in the potential size of man-made disasters. Only since the advent of nuclear weapons, biological warfare, and newer forms of technology has the size of consequences of man-made disasters begun to approach those of natural disasters. Thus the data base for the Reactor Safety Study for man-made and natural disasters might better have been restricted to the last 25 years.

16.1.5.4 Risk Conversion Factors for Natural and Man-Originated Disasters

Risk conversion factors are calculated using the ratio of midrange of values for natural and man-originated disasters for the best estimate, and worst ratios for the ranges with subsequent rounding off.

RCF [natural catastrophe—fatalities / man-originated catastrophe—fatalities] (2, **10**, 50)

RCF [man-originated catastrophe—fatalities / natural catastrophe—fatalities] (.02, **.1**, .5)

The Reactor Safety Study also compares property damage from natural and man-originated disasters. This information (Appendix A.3.3) indicates little difference in the shape and ratios of the curves from those for fatalities. Apparently, then, the foregoing RCFs are also suitable for property damage.

Fires and dam failures are accidents that can be triggered naturally. Unfortunately, the data bases for accidents of these types do not easily lend themselves to understanding the triggering causes. Data included in Appendix A.3.1 from the Reactor Safety Study show that dam failures and fires are the major contributors to man-made large event disasters. On this basis an estimate of an RCF between man-caused and man-triggered natural disasters is questionable at best. However the range could be no less than unity; that is, no different at one extreme and no higher than the natural/man-made ratio at the other extreme.

RCF [man-caused catastrophe—fatalities / man-triggered catastrophe—fatalities] ((1, **2**, 5))*

RCF [man-triggered catastrophe—fatalities / man-caused catastrophe—fatalities] ((.2, **.5**, 1))

16.1.6 Knowledge as a Risk Factor

Here we want to determine not whether particular knowledge involves risks, as discussed in Chapter 9, but whether the process of risk identification

* The double parentheses indicate an estimate made by the author in the absence of hard data. Such estimates are to provide some feeling for possible ranges. If they are useful in this manner, they are justifiable.

makes society risk aversive. Certainly one must know of a risk to be aware of it, but our concern here is with (1) societal recognition of a new class of risks and its reaction to them, and (2) attempts to institute risk aversive action by making information of risk levels and avoidance techniques easily available to the public.

16.1.6.1 Recognition of New Risks

Risk identification changes the perception of risks. When new risks are discovered, either through better understanding of existing systems or through origination of systems involving new risks, anxieties are created for those exposed to such risks, actually or potentially. Those who are not exposed may also be concerned if the risk recipients are identifiable. A good example is the furor in the wake of the discovery of the toxicity of the pesticide Kepone, and contamination of plant workers and the environment. The production of Kepone ceased, cleanup and control procedures were instituted, and nationwide concern was expressed for the individuals who were incapacitated as well as for the continuing contamination of rivers and fish.

When new risks are identified, they are rarely quantified. This lack of quantification prevents a risk assessment in a manner whereby a risk agent can compare the new risk with other risks. The anxiety caused by the risk cannot be alleviated until the scope, magnitude, and ability to take risk aversive action are known. Thus a new risk may be magnified out of proportion to its real value by lack of knowledge of the scope of the risk. Lack of risk quantification can lead to overvaluing of a risk; that is, if quantified, the risk might have had a lower value. Furthermore, subsequent quantification does not always change the preconditioning that a qualitative risk may cause. The initial impression, erroneous or otherwise, is often hard to reverse. This leads to the dilemma of ascertaining the propriety of early reporting of identified risks versus waiting until adequate quantification occurs in a particular situation.

Early reporting of risks may be the only way to focus attention and resources to allow quantification. Failure to report new risks prior to quantification may lead to charges of "cover-ups" and loss of credibility. Early quantification of indentified risks may be the best compromise.

16.1.2.2 Information and Risk Aversion

There is an expectation that publishing data on risks will make people aware of the risks, create anxieties, and as a result stimulate risk aversive action in the form of increased use of safety measures, and the exercise of greater caution and restraint. However there is no evidence that the qualitative warning on a package of cigarettes has reversed the trend in smoking, a voluntary decision of the smoker. On the other hand, it is common practice by police and safety authorities to show movies of automobile carnage to new drivers, and to display wrecked automobiles. This seems to constitute a short-term reminder to drive safely and, hopefully, there is some carryover to the long term. Conversely, familiarity with a risk may result in increased caution from understanding the consequences in a measured, personal manner, or the reverse by building indifference to the risk.

Some efforts to quantity the effect of knowledge on risk have been undertaken. Kunreuther[16] has used insurance premiums for different accidental risks to examine the difference among preferences for purchasing insurance when the risks to the purchaser are unknown, then quantified to various degrees. Knowledge of risk and induced anxiety does indeed increase purchases of insurance, but the method of imparting information and the degree of effect are not quantified. Marshall[17] has attempted to model the value of information to a risk taker and to place value for him on announcing information concerning risks. However the mathematical formulation is based on unspecified utilities or risk agents, and no quantification of effect of the value of information on risk has been obtained. Risk conversion factors cannot be developed at this time.

16.2 FACTORS INVOLVING TYPES OF CONSEQUENCE

16.2.1 Voluntary Versus Involuntary Risks

Though we continue to use the complex definition of voluntary and involuntary risks derived in Chapter 8, to compare results provided here with other sources having different definitions, it is desirable to review some of Starr's conclusions on this subject.

16.2.1.1 Use of Starr's Definitions

Starr's definition of voluntary and involuntary risks is somewhat different from the author's:

> Societal activities fall into two general categories—those in which the individual participates on a "voluntary" basis and those in which the participation is "involuntary," imposed by the society in which the individual lives. The process of empirical optimization of benefits and costs is fundamentally similar in the two cases—namely, a reversible exploration of available options—but the time required for empirical adjustments (the time constants of the system) and the criteria for optimization are quite different in the two situations.[13]

The case for the difference in time constants is explored subsequently in terms of observing the effective discount rate for voluntary and involuntary societal risks.

Starr's definition of voluntary and involuntary activities is also more general:

> In the case of "voluntary" activities, the individual uses his own value system to evaluate his experiences. Although his eventual tradeoff may not be consciously or analytically determined, or based upon objective knowledge, it nevertheless is likely to represent for that individual a crude optimization appropriate to his value system.
>
> "Involuntary" activities differ in that the criteria and options are determined not by the individuals affected but by a controlling body. Such control may be in the hands of a government agency, a political entity, a leadership group, an assembly of authorities or "option makers," or a combination of such bodies. Because of the complexity of large societies, only the control group is likely to be fully aware of all the criteria and options involved in their decision process.[13]

With these general criteria, Starr derives a risk multiplier of three orders of magnitude difference between voluntary and involuntary exposure for equivalent benefits (Figure 16.3). Otway and Cohen[18] reject Starr's analysis on the basis that Starr's[19] methodology and results are "excessively sensitive to the assumptions made and the handling of data and that the existence of simple mathematical relationships, based upon the revealed preferences method, is unlikely."[18]

Otway and Cohen performed a regression analysis on the same data base, but with entirely different results, leading them to reject Starr's conclusions,

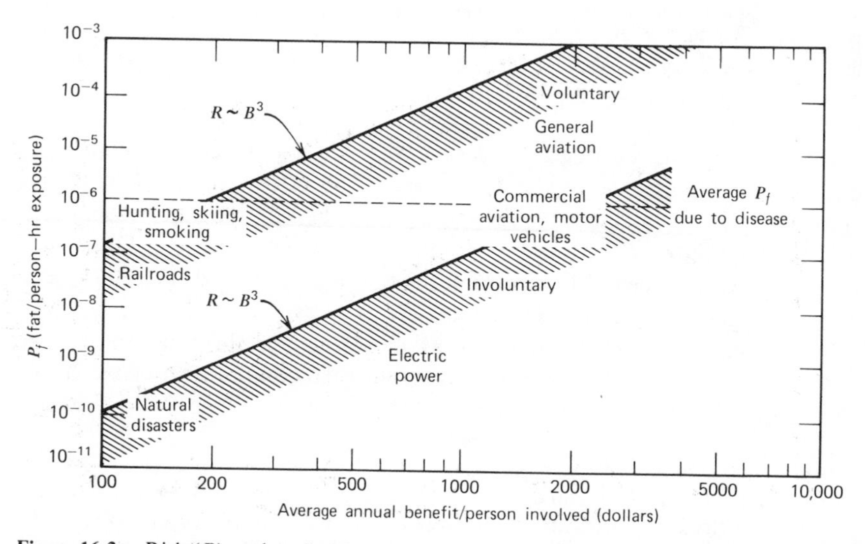

Figure 16.3. Risk (R) vs. benefit (B): voluntary and involuntary exposure. *Source:* Starr.[13, 19] Reprinted with permission of American Association for Advancement of Science.

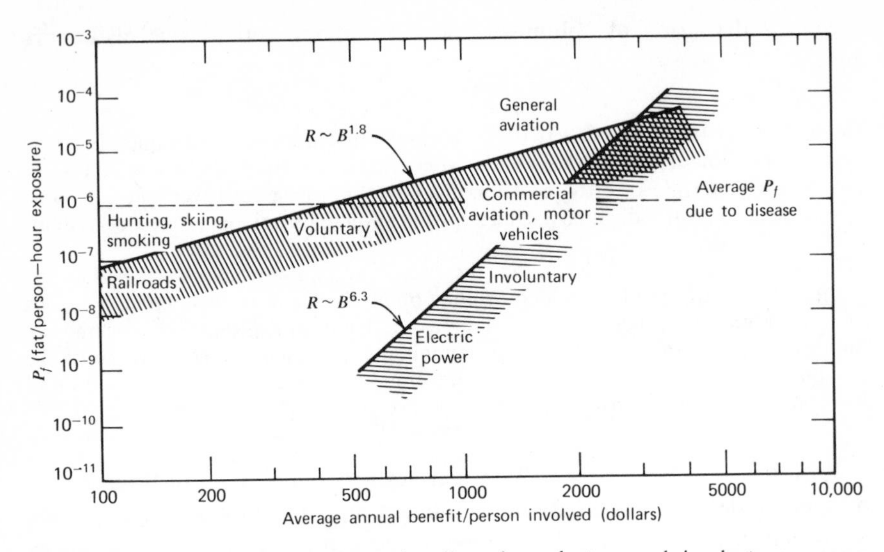

Figure 16.4. Risk vs. benefit: regression lines for voluntary and involuntary exposure (Original Starr data.) *Source:* Otway and Cohen.[18]

namely:

1. The indications are that the public is willing to accept "voluntary" risks roughly 1000 times greater than "involuntary" risks.
2. The statistical risk of death from disease appears to be a psychological yardstick for establishing the level of acceptability of other risks.
3. The acceptability of risk appears to be crudely proportional to the third power of the benefits (real or imagined). . . .[13]

For voluntary risks in society, Otway and Cohen found a relation indicating risk to be proportional to the 1.8 power of benefit as opposed to Starr's third power. For involuntary risks in society, the relationship between risk and benefit follows a sixth power proportionality. The results of Otway and Cohen are reproduced in Figure 16.4; the authors concluded that all three of Starr's assumptions are insupportable.

However, it may well be that Otway and Cohen are both premature and overly critical of the Starr approach. In a personal communication to Otway, Starr makes the following comments:

The principal objective of my studies was to establish a set of plausible hypotheses and quantitative criteria which could be used for national decision-making on the acceptable level of risk associated with large-scale technical systems. In order to approach this issue, I looked for historical patterns which might suggest some broad principles.

As indicated in all the papers I have written on this subject, I consider the hypotheses which I have suggested as both very preliminary and simplistic in what must eventually be a more complicated analytical process.

The detailed examples given in my papers served only as a basis for extensions to plausible generalizations. They were not intended at any time to prove the validity of these generalizations.

It appears to me that in undertaking your critique you may have reversed the approach to the issue as described above. In particular, the points you make . . . on the inadequacies of measuring all the parameters of revealed preferences were all mentioned as problems and discussed in my papers. Apparently you believe that because of these difficulties one cannot base a decision-making approach on a study of historical behavior. I believe that is a fundamental philosophical point, but your paper does not really provide either alternatives or a full discussion of these issues. . . .

I believe your criticism of the relationship between voluntary and involuntary risk is quantitatively not valid. First, there were four points on the involuntary curve—natural hazards, electric power, commercial aviation, and motor vehicles. Second, although benefit calculations would necessarily be very difficult, they were based on the relative value to the individual of having the technical system vs. not having it in our present societal structure. In the particular case of electricity, for example, one has to estimate the probable change in the GNP per capita if electricity were not available. I believe that if you examine that situation you will find that the benefit quantity which I used is probably correct. Certainly, the calculation that you present in your paper is not applicable.*

Notwithstanding the inaccuracies in this approach, the Otway-Cohen results (Figure 16.4) are that involuntary risks become more acceptable only at the top end of the benefit scale, and at the lower end a factor greater than 1000 is to be expected. The crossover point at about $3000 benefit per person is an interesting figure, but the very inaccuracies of the method specified by Otway and Cohen preclude any judgments except that it seems a very high value for individual indirect benefit.

16.2.1.2 More Restricted Definitions

Data derived on the basis of the more restrictive definition used here show a much smaller risk multiplier. CONSAD data (Tables A1.1 and A1.3) provide one basis for comparing total risks and involuntary risks, respectively. In making such a comparison, however, the size of the populations at risk must be taken into account.

Commercial Airlines. If it is assumed that all citizens have access to commercial airlines, and that risks to people on the ground are to the same

* Letter from Starr to Otway, March 29, 1975, supplied by C. Starr. Starr data reprinted by permission of the American Association for the Advancement of Science.

population,* then for an average value of 206.2 voluntary fatalities per year (total risk minus the involuntary risk) and an average value of 2.1 involuntary fatalities per year, a factor of nearly 100 results. Since a smaller higher risk population around airports was not used, one would expect this factor to be on the high side. However the Reactor Safety Study shows a factor of between 30 and 70 for the probability of total air crashes with fatalities between 1 and 100 over events with the same size consequences for persons on the ground.[15] The probable value of the factor most likely lies between 10 and 100.

Military Air Crashes. For a military population of 3 million for voluntary risks, a total of 88.6 fatalities per year on the average and a total U.S. population subject to the mean involuntary risk of 5.0 fatalities per year, the risk factor is

$$\frac{83.6 \text{ fat/yr}}{3 \times 10^6 \text{ people}} \div \frac{5.0 \text{ fat/yr}}{2 \times 10^8 \text{ people}} = 1115$$

This value is perhaps 20 times or more greater than that for commercial airlines.

Railroads. Assuming like populations for voluntary and involuntary risks on railroads, since all have access to trains and rights-of-way, a factor of greater than 20 results, based on annual averages. However since only two involuntary events are listed, the validity of the result is questionable.

Marine and Mines. The data show no involuntary fatalities for catastrophic marine and mining accidents. The number of involuntary deaths for marine accidents involves a very small population living near or on water. No conclusions can be drawn.

Although the 1972 *Statistical Abstracts of the United States* provides data on the occupational fatalities of all industries, including mineral recovery, the record of involuntary deaths from such events as subsidence are not as easily available. The difficulties in obtaining data on involuntary risks from mines are discussed in Appendix B.1, since this area is a good example of data deficiencies.

Bus, Auto, and Truck. Pedestrian and bystander fatalities for catastrophic events are relatively high, with 2.2 involuntary fatalities per year versus a total of 16.4 voluntary fatalities per year on the average; a ratio of about 7.5 : 1.

Total. Aggregating data in Tables A1.1 and A1.3 for total fatalities over the 21-year period results in a 21 : 1 ratio. The average fatalities per year

* Risk near commercial airports is higher than for the general population.

result are in the same ratio. The validity of such aggregation is questionable.

16.2.1.3 Avoidability of Risks and Alternatives to Risks

When it is possible to avoid risks by simply choosing not to accept them, the existence of reasonable alternatives makes such choices more attractive. Therefore the avoidability of risk must always be considered in conjunction with available alternatives. When the only alternative is to avoid the risk, the threshold of action is a function of the perceived degree of satisfaction by the risk taker of his status quo. When the threshold is exceeded, the risk taker will flee or avoid taking the risk. In Chapter 11 it was hypothesized that the threshold of avoiding the risk was a function of the degree of perceived satisfaction with the status quo, the type of risk, and the magnitude of the risk. At this time, the hypothesis remains untested, since data about people who have chosen to avoid risk by fleeing permanently are sparse. Even fewer data are available on the perceived degree of satisfaction with the status quo. Unfortunately, the design of experiments to gather such data is difficult because the concept is so subjective.

The investigation of the effect of different alternatives is another proposition. Hedging and insurance are means used to alter the risk consequences to the risk taker or to spread the risks among larger numbers of risk takers, respectively, particularly when the risks are convertible to financial terms.

The existence of alternatives does affect the valuation of involuntary risks if they are not originated by man. For example, people living in earthquake- or tornado-prone areas have the option to move elsewhere. Having such an option gives the risk taker and the risk evaluator an illusion of voluntary risk, resulting in a lower negative valuation of the consequence. A different contention is that unless attractive risk alternatives are available, a benefit-risk tradeoff cannot be made. Thus the risk is still involuntary by definition, but the valuation of the consequence is sometimes similar to voluntary risk. For example, a fisherman who knows no other trade could not move inland to avoid hurricanes, unless there was a new, feasible, attractive occupational alternative to fishing. Knowing he could move inland, even without an alternative, gives the risk taker some apparent aspect of control over his destiny in the sense that he is "voluntarily" taking the imposed risk. Again, this apparent controllability changes the risk evaluation, but by definition here the risk is still involuntary. It is a case of involuntary risk when the risk taker changes his valuation of consequences through a perceived degree of personal control. If attractive alternatives do exist, and personal benefit-risk tradeoffs can be made, the risk is indeed voluntary.

16.2.1.4 Regulated Voluntary Risks

Examination of regulatory actions to prevent exposure to hazardous substances provides another aspect of revealed preferences. Assuming that standards for occupational risks set by the OSHA as a voluntary risk regulation and standards set to protect populations from exposure by the EPA as an involuntary risk regulation, we can make some general conclusions about these risks. Direct comparison of regulatory action between OSHA and EPA is not always possible, but Table 16.6 provides a few cases.

The range of 5 to 1 to 500 to 1 spans the same range as other evidence for these risk conversion factors. Further effort in such comparisons is warranted before more significant conclusions can be drawn.

It must be recognized that regulated voluntary risks differ from unregulated voluntary risks. The case in point, occupational risks, is regulated because there is some danger of employers exploiting employees without equivalent compensation, and there are few alternates available to employees who are experienced in a single occupation. For example, coal miners are not specially compensated for engaging in a high risk occupation, nor is it easy for a coal miner to find other employment without considerable retraining.

The average of the eight values is about 132; this is below the three orders of magnitude indicated by Starr. An RCF for regulated voluntary risk to

Table 16.6 *Ambient and Occupational Standards for Various Chemicals*

	Regulating Agency		
Pollutant	EPA	OSHA	Ratio
Beryllium	Ambient guideline,	2 $\mu g/m^3$—8-hour avg	200[a]
	0.01 $\mu g/m^3$—30-day avg	5 $\mu g/m^3$—ceiling	500[a]
Mercury (nonorganic)	Ambient guideline,	100 $\mu g/m^3$—ceiling	100[a]
	1.0 $\mu g/m^3$—30-day avg		
Alkyl mercury	Ambient guideline,	10 $\mu g/m^3$—8-hour avg	10[a]
	1.0 $\mu g/m^3$—30-day avg	40 $\mu g/m^3$—ceiling	40[a]
Sulfur dioxide	0.03 ppm—annual avg	5 ppm—8-hour avg	167[a]
	0.14 ppm—24-hour avg		36[a]
Carbon monoxide	9 ppm—8-hour avg	50 ppm—8-hour avg	5
Particulates	0.075 $\mu g/m^3$	5 $\mu g/m^3$	67

Source: Data from a private communication from Donald Goodwin, Environmental Protection Agency, Office of Air Programs, Raleigh, N.C.

[a] Effects of different measurement periods have been ignored, since translation is not directly possible.

unregulated risks may be obtained by comparing the difference between unregulated voluntary and regulated voluntary risk. Some estimates of the difference are shown below.

RCF [regulated voluntary risk / (.1, **.2**, .5) / unregulated voluntary risk]

RCF [unregulated voluntary risk / (2, **5**, 10) / regulated voluntary risk]

16.2.1.5 Voluntary Risk Conversion Factors

It was not possible to use data on decreased life expectancy for a voluntary-involuntary RCF, since neither Baldewicz et al.[20] or Bunger[21] provides any man-originated involuntary risk information against which to compare occupational (voluntary) risks. Data on voluntary and involuntary injuries are less precise, but generally they follow those for fatalities.

The RCF for voluntary versus involuntary risks, at least for catastrophic events, ranges from about 10 to 1000 for different accident sources, with the larger value representing military operations and the Starr estimate. Airline data provide the largest data base and give a multiplier of about 100. The higher values expressed by Otway and Cohen at low benefit levels are speculative.

RCF [voluntary risk—fatalities and injuries / (10, **100**, 1000) / involuntary risk—fatalities and injuries]

RCF [involuntary risk—fatalities and injuries / (10^{-3}, $\mathbf{10^{-2}}$, 10^{-1}) / voluntary risk—fatalities and injuries]

16.2.2 Discounting in Time

There are many aspects to discounting in time and very little data to support quantitative assessments of risk multipliers in this area. Furthermore, some special problems arise when latent risks, risk to progeny, and irreversible world commitments are involved. However an attempt is made here to describe time discounting factors.

16.2.2.1 The Discount Function

The discount function is usually considered to be a negative exponential function of the form

$$\text{discount function} = e^{-at} \tag{16.1}$$

where t is the number of time periods from the initiating event, and a is a constant reflecting the discount rate. When one applies this concept and attempts to identify the value of a for different societal discount functions, either probability of occurrence or the magnitude of the risk valuation can be discounted.

The underlying concept is based on the implication that a choice must often be made in the present time frame among two or more alternatives having different effects on future risks. The risk taker is assumed to be indifferent between two risks at the present time if the discounted value of the future increased risk ratio of the two risk alternatives is reduced to unity (i.e., the two risks appear to have identical future values). The resultant discount rate is called the effective discount rate. From this model, one can infer minimum levels of effective discount rate from observation of societal behavior, although in practice, such experience results from ad hoc decisions as opposed to informed, analytical choices.

The future risk ratio can be based on the lowest future risk condition, permitting the risk of the higher proposed alternative to be normalized to it. If the probability of occurrence of the lowest consequence, t periods in the future, is denoted as p_0, and p_j is the probability of occurrence of the higher risk consequence, t periods in the future, the ratio of p_j/p_0 is the increased risk that must be discounted to unity in terms of present value. On this basis, the present discounted value of the increased risk p_j is made equal to the present risk value of the least risk alternative, p_0, such that

$$p_j = p_0 e^{-at} = p_0(1 + i)^{-t} \tag{16.2}$$

The right hand form of the equation is in discount rate form, where

$$i = e^a - 1 \tag{16.3}$$

$$a = \ln (1 + i) \tag{16.4}$$

and i is the discount rate in decimal form. The percentage discount rate is found by multiplying i by 100%. The form of some of these functions is illustrated in Figure 16.5.

This form is reasonable if the exact delay time of the consequence is known. Most often the delay factor is itself a statistical function. For example, in the exposure of specific populations to increased levels of radiation, the delay of onset distribution tends to be sigmoid* or approximately Gaussian. For example, the Woodward and Fondiller study[22] on lung cancer in uranium miners predicts the onset of cancers with a mode of 12 to 14 years for a sigmoidal distribution. In another study[23] on the latency period for 185 cases of thyroid carcinoma among children with early childhood exposure to X-rays, the sigmoidal distribution shows a modal latency period of 9-years with a mean about 12 years. There are other similar studies available.

The proper discounting function would have to be integrated over the latency period distribution. However a central value estimate of the dis-

* The sigmoidal curve is skewed toward higher values than a Gaussian curve.

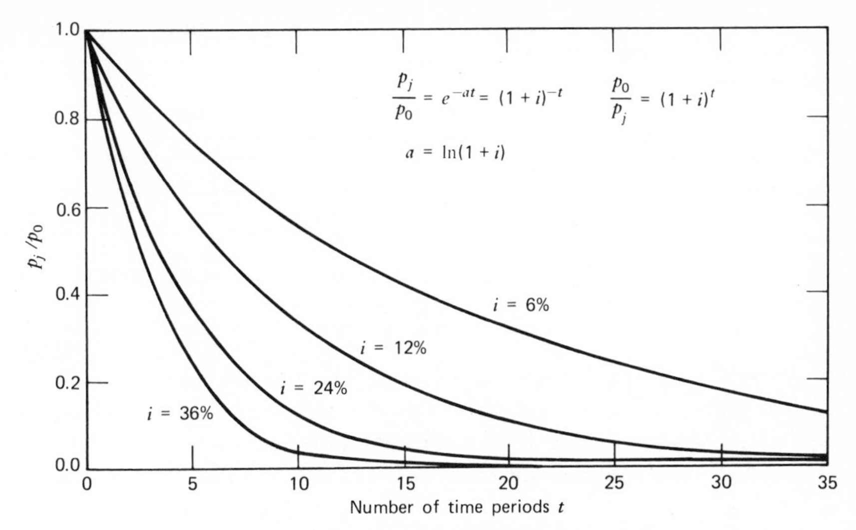

Figure 16.5. Form of various discount rate curves.

tribution, such as the mean or the mode, can be used to provide a first order estimate. Since only gross conclusions are drawn here, such use of central values may suffice.

16.2.2.2 Voluntary Risk Effective Discount Rates

A well-known voluntary risk with delayed consequences is cigarette smoking. Here, for the individual benefit of smoking, the individual increases his probability of contracting certain diseases if he continues the habit for a number of years. The 1974 report "The Health Consequences of Smoking"[24] provides the following ranges of increased risk of disease to smokers:

Disease	Range of Increased Risk of Smokers over Nonsmokers = p_j/p_0
Lung cancer	
All smokers	7.61–14.20
Heavy smokers	4.9–23.9
Chronic bronchitis	3.6–21.2
Emphysema	6.9–25.3
Coronary heart disease (male smokers)	2

The time of onset of the disease ranges from 5 to 20 years or more. The exact time distribution of disease onset and the distribution of increased risk

Table 16.7 *Discount Table for Effective Discount Rates (%) for Smokers Showing Possible Ranges:* $p_o/p_j = (1 + i)^t$

Mean Value of ratio of Increased Risk	Mean Years to Consequence Onset			
	5	10	20	25
5	38%	17%	8.4%	6.7%
10	50%	26%	12.2%	9.7%
20	82%	35%	16.2%	12.7%

are not well established. However one can speculate that the true distribution lies within certain ranges. With this in mind, a table of discount rates may be constructed (Table 16.7) for mean values of the ratio of increased risk and the mean value of the delay of onset distribution. For example, if the increased risk to smokers has a mean value of 10 and a mean value of onset of 20 years, the smoker who accepts this increased risk is using a 12.2% effective discount rate. The validity of this example may be questioned, but it indicates that the discount rate for this voluntary risk most likely exceeds 6%.

Another case involving both voluntary and involuntary risk regulations is that of vinyl chloride. Continuous exposure of one part per billion (ppb) of 1 million people to vinyl chloride results in 0.071 case of liver angiosarcoma and 0.15 case of all types of cancer with an approximate 20-year latency period.[27] Exposure to workers prior to regulation averaged about 350 parts per million (ppm) for a 7-hour per day, 5 days per week work schedule with a predicted incidence of 0.0052 per person-year exposure. The present OSHA standard for workers is 1.0 ppm for an 8-hour average. This in effect is a reduction of 350 to 1 and represents an effective discount rate of 34% for a 20-year latency period.

16.2.2.3 Involuntary Discount Rates

Some individuals potentially exposed to radiation from nuclear industry facilities are outside the fence of the plant and receive no direct benefit (e.g., salary) from the plant. Thus they are involuntary risk takers. If one assumes a linear dose-effect relationship as recommended by the BEIR report,[25] an estimated background dose rate of 150 millirems (mrem) per year and an exposure limit of 25 mrem per year as recommended by the EPA, the following discount rate calculation can be made:

$$\frac{p_j}{p_0} = \frac{175 \text{ mrem}}{150 \text{ mrem}} = (1 + i)^{-20} \qquad (16.5)$$

where 20 years is an estimate of the latency period from the BEIR report. The effective discount rate is

$$i = 0.77\% \tag{16.6}$$

This represents an estimate of the effective discount rate specified by the EPA* as an acceptable delayed involuntary statistical risk. This represents a discount rate of a factor of one to two orders of magnitude less than that for voluntary risks.†

As a check on this, the present standard of 5 rems per year for radiation workers who receive direct benefits for accepting risk up to these levels involves a discount rate of 22%, or comparing this value with the 0.77% just established, shows a ratio of 28:1 between voluntary and involuntary exposure discount rates established by government regulatory agencies. Note that our purpose is to estimate societal discount rates set by society or regulating bodies operating for society, not to determine acceptability.

Another regulatory example of prevention of exposure of involuntary populations to the delayed consequence of increases in cancer is the Delaney clause of the federal Food, Drug, and Cosmetic Act,[26] which states:

> That no [food] additive shall be deemed safe if it is found to induce cancer when ingested by man or animal, or if it is found, after tests which are appropriate for the evaluation of the subtlety of food additives, to induce cancer in man or animal, . . .

In this case, no discount rate is acceptable. Zero exposure (within an ability to define zero in terms of measurement) is the only recourse allowed. There is considerable controversy on this matter, but even if relaxation occurs, this author expects that prudent public health actions by regulatory agencies will keep the effective discount rate very low.

16.2.2.4 Progeny

The exposure to risk by an individual can result in a consequence affecting his progeny instead of himself. Any risk that involves genetic consequences directly affects future generations.

* In its uranium fuel cycle standard, EPA establishes a level of 25 mrem per year for exposure to individuals, whereas proposed Appendix I to Nuclear Regulatory Commission Regulation 10CFR50 recommends a design level of 5 mrem via air pathways and 3 mrem via water pathways to a maximum exposed individual. The NRC level represents an effective discount rate of 0.32%.

† It is important to note that for involuntary risks the ratio of risk reduction is always the *amount allowed* to the *least possible amount*. For voluntary risks this ratio may be the increased risk over not taking the risk, as in the case for smoking, or the amount of risk reduction implemented by a regulation (i.e., the ratio of the level of risk without and with the regulation).

Basically, although the consequence is delayed, the problem is also one of spatial distribution. Identification of progenic consequence recipients is different from identification of the individual exposed to a risk event. Furthermore, the effects may be to the specific offspring of the exposed individuals, or there may be a statistical increase of effects to future generations—for example, through changes in the genetic pool.

An example of increased risk to specific progeny is the exposure of young people, (most likely women in early pregnancy are more susceptible than others), to radiation exposure from occupational and medical situations, and accidents from related activities. Work and medical treatment bring specific benefits to the individuals at risk; an accident may not. However there is little evidence available to warrant conclusions. As larger populations are exposed, increases in genetic pool changes become more likely.

16.2.2.5 Irreversible Commitments

When events taking place in the present affect risk consequences in the future and cannot be altered once committed, an irreversible condition occurs. Contamination of the world environment from long-lived radionuclides (such as plutonium from nuclear weapons and nuclear power sources) and polychlorinated hydrocarbon contamination of the upper atmosphere ozone layer are two examples.

The problem is one of trading short-term benefits favoring an identified group against long-term involuntary risks for a large statistical population, in some cases yet unborn. Can such long-term involuntary risks be discounted? Although subject to change as society learns more about such commitments, the first conclusion would have to be answered negatively for five reasons. (1) Voluntary risks to individuals may be discounted by some risk takers for discount rates that range from 6 to 50% per year. (2) Involuntary risks to individuals are lower than these levels. (3) Involuntary risks to populations are regulated by government in this country in the fractional percentage discount range (e.g., 0.5%). (4) Irreversible risks would be expected to be discounted at some factor below involuntary risks to the population, the latter are already at fractional levels, and lower levels as required for irreversible commitments would tend to be meaningless in terms of discount. (5) Risks that are committed to the future with essentially an infinite time *will* occur (theoretically), since the "laws of probability" are perfect in eternity.

16.2.2.6 Latency Risk Conversion Factors

Risk conversion factors for latency effects involve an effective discount rate per year and the number of years needed for estimating future risks (i.e., the number of years of exposure for which risks in the future are committed).

$$\text{RCF}\left[\begin{array}{l}\text{immediate voluntary risk}\\ (10, \mathbf{20}, 30)\ \%\text{ per year}\\ \text{delayed voluntary risk}\end{array}\right.$$

$$\text{RCF}\left[\begin{array}{l}\text{immediate involuntary risk}\\ (.1, \mathbf{.5}, 1.0)\ \%\text{ per year}\\ \text{delayed voluntary risk}\end{array}\right.$$

The risk for T years of exposure is computed by:

$$\text{delayed risk} = \text{immediate risk} \times \left(\frac{1 + \text{RCF}}{100}\right)^T \tag{16.7}$$

The discount rates for irreversible risks cannot be discounted; that is, they have a 0.0% discount rate. This also may be true for progeny, but there is little evidence to support any conclusion because the problems affecting progeny are in a fluid state.

16.2.3 Spatial Distribution of Risks

There seems to be a great deal of qualitative information on spatial distribution of risks, especially identifiable versus statistical risk differences. For example, large sums of money are spent in rescuing children who have fallen into abandoned wells, but little is spent in eliminating the hazards. An ounce of prevention may be worth a pound of cure, but the investment in prevention is seldom made for low probability events. It is not until the cure to an identifiable individual or group is considered that sizable investments are made.

Qualitative data for voluntary risks are available, but they are more a measure of the propensity for risk taking than for consideration of spatial distribution of risks. Unfortunately, in the several cases involving involuntary risk and regulatory decisions that were investigated, decisions were not based on risk to risk recipients and/or statistical distribution of risk. The risks examined in these cases were legal risks: that is, what was the chance of being brought to court in a class action suit if no action was taken? Further investigation is in progress, and no quantitative data can be reported yet.

In chapter 8 some qualitative curves were introduced to aid in estimating the relationships. Since quantification is not possible at this time, the best that can be done is to speculate. In this case a speculation is made for the purpose of setting up a straw man. It represents the author's judgment only.

$$\text{RCF}\left[\begin{array}{l}\text{statistical risk agent}\\ ((1, \mathbf{10}, 1000))^*\\ \text{identifiable risk agent}\end{array}\right.$$

$$\text{RCF}\left[\begin{array}{l}\text{identifiable risk agent}\\ ((.001, \mathbf{.1}, 1.0))^*\\ \text{statistical risk agent}\end{array}\right.$$

* Double parentheses indicate that the estimate is a judgment, not based on data.

16.2.4 Controllability of Risks

Increased ability to control risks in terms of one's perception of controllability as an individual, and the degree of systemic control provided by technological and institutional processes generally are expected to make the value of consequences and risk acceptability somewhat higher (i.e., less conservative) than without controls. This is not true, however, for naturally occurring risks in comparison to man-originated risks as discussed in Section 16.1. Often uncontrolled acts of God are accepted at higher levels than controlled man-originated risks. Evidently this situation has not been accepted by technological societies, and man has attempted to extend his control over natural hazards. Kates[4] has indicated that this effort has led to fewer disasters, but it results in larger consequences when they do occur. Perhaps man's ability to control nature is only a perception of control and not a reality.

16.2.4.1 Perception of Controllability

For other than naturally occurring risks one's perception of controllability over his own fate is a personal, subjective condition. Examination of such conditions is most amenable to psychological testing and experiments, as opposed to searching for revealed preferences. Revealed preferences can only indicate broad behavior patterns, and to attempt to infer individual behavior from such data does not seem productive. Generalizations are all that are possible.

An example from the author's experience may illustrate this point. During the installation one of the first computer-controlled rolling mills in a steel plant in the early 1960s, considerable unrest and discomfort were expressed by the workers who were to operate the new automatic mills. Their total experience involved direct, manual control of rolling steel plates to specified dimensions, and they did not trust the automated equipment. They were put at ease, and full operation was achieved, when two dials were made available to the operators "for fine adjustments over the computer." The two dials and control knobs served no purpose other than to fill some holes that had been punched in error in the manufacturing process; they were unconnected to the system itself. Nevertheless, the operators could be seen carefully but contentedly adjusting the "fine tuning." Although not involving risk, the subjective perception of control was the dominant factor, not the objective situation.

Measurement of the extent of the subjective perception of control on consequence valuation at the psychological level through tests, experiments, polls, preference rankings, and so on, is at best difficult and imprecise. Investigation of behavioral patterns in connection with controllability research is even more difficult.

16.2.4.2 Degree of Controllability

In their analysis of average years of life expectancy lost per fatality, Baldewicz et al. investigated 10 risk systems, including their associated occupational hazards. The occupational aspect seems to involve voluntary risk and some perception of degree of control. Based on this analysis and the use of average years of life expectancy lost as a parameter, they found all 10 systems to have been essentially invariant over the past two decades. Several other conclusions—really, lack of conclusion in many cases—resulted.

From the historical trends [of] the rate of loss of life expectancy for the risk systems studies, it is concluded that (i) appreciable disparities exist in this factor for occupational hazards, despite nearly similar benefits for the populations at risk, (ii) federal legislation can have a significant impact on risk abatement, as has apparently been the case for coal mining, (iii) federal safety legislation efforts appear to be most responsive to highly publicized disastrous accidents rather than to chronic, low-level hazards (both accidents and disease) which actually contribute more significantly to rate of loss of life expectancy, (iv) the introduction of a new technology (as in commercial aviation, 1956–1960), as well as lack of maintenance of old technology, (as in rail transportation) can result in risk level increases, and (v) natural hazard risk levels are extremely low and can further be reduced in some cases, either by collective or individual precautionary measures.[20]

Conclusions ii and iii indicate that legislated systemic control as well as less stringent control schemes can be effective in risk abatement.

The Reactor Safety Study has made some qualitative judgments on the degree of controllability.

. . . although a careful person can probably take some action to reduce his risk to some types of accidents, he certainly cannot make his total accident risk zero. Thus, is seems reasonable to assume that even the most careful of individuals could not expect to reduce his risks from accidents by more than a factor of 100. Furthermore, to achieve this would require a significant departure from the typical U.S. life style.[15]

Such a reduction (i.e., 100) might be a limit for all risks; but since different types of risk are viewed differently, and a variety of controls are available, reduction in certain kinds of risk can easily be achieved by three or four orders of magnitude. An example of this is control of poliomyelitis. Less than 20 years ago, this disease was virtually uncontrollable. Today with vaccines developed by Salk and Sabin, polio is virtually unknown. This represents a reduction in risk of more than 10,000. Though some risks are uncontrollable, others may be controlled completely. Reducing all man's risk on the average by a factor of 100 may well be the maximum achievement possible, however.

Making up the cost to achieve this risk reduction is not only an economic problem; also involved are lifestyle choices, possible restrictions on freedom of action, and possible intrusion of regulatory controls into areas of voluntary risk. However, such consideration applies to existing systems, and the question of how much increased risk society will accept for introduction of new technological systems with increased involuntary risks and indirect benefits is another matter. There is evidence that through new mechanisms that make society aware of new risks, such as the National Environmental Policy Act of 1969 and the increased emphasis on technology assessment, society is more concerned about accepting and controlling new risks than about existing risks. This behavior may not appear rational to the technological community, but it may indicate the public's growing suspicion and animosity toward ever larger technological institutions. The existence of an antitechnology movement is well documented. Reversal of such movement is dependent on the ability of government and technological institutions to demonstrate that the short- and long-term welfare of the public will be protected and enhanced as new programs are implemented. This implies a requirement for some degree of systemic control on all new technological systems.

16.2.4.3 Controllability for New Technological Systems

Controllability of new technological systems implies the need for implementation of systemic control of risk as discussed in Chapter 8. Table 8.3 presented a series of control factors and ranked these approaches. Since no data have been identified to evaluate controllability, any quantification of the desirability of different levels of control is totally a value judgment—in this case, the author's.

There is no question that the desirability of controlling risk depends on the magnitude of the risk, that there is little concern for small risks and increased concern for control of risks with large consequences. For establishing risk comparison factors on the basis of judgment, however, the magnitude of risk is held constant and is assumed to be accounted for separately in developing acceptability criteria.

The four control factors developed in Table 8.3 are given in Table 16.8 along with a judgment on the cardinal value of desirability of control for each. The overall level of desirability of control is defined as the product of the factors

$$C_c = C_1 \times C_2 \times C_3 \times C_4 \tag{16.8}$$

where C_c represents the condition for a given risk with controls. If C_1 and C_2 are "no control" and "uncontrolled" at the same time, then C_3 and C_4 are ignored (set at unity), since double counting would otherwise take place.

Table 16.8 *Control Factors and Assignment of Cardinal Values of Desirability*

Control Approach[a]		Degree of Control		State of Implementation		Control Effectiveness	
Factor	C_1	Factor	C_2	Factor	C_3	Factor	C_4
Systemic control	1.0	Positive	1.0	Demonstrated	1.0	Absolute	1.0
Risk management system	0.8						
Special design features	0.5			Proposed	0.5	Relative	0.5
Inspection and regulation	0.3	Level	0.3				
		Negative	0.2				
No control scheme	0.1	Uncontrolled	0.1	No Action	—[b]	None	—[b]

[a] Each approach is inclusive of those below it, except for "no control."
[b] These levels are included in "no control" and "uncontrolled" by definition.

Table 16.9 *Degree of Control for Accident Categories*

Category	Degree of Control
All accidents	Positive control from 1910 to 1960, but level control at best since 1960
Motor vehicle	Fluctuates, both generally level or slightly negative control
Work	Positive degree of control since 1930
Home	Positive degree of control since 1940
Public, nonmotor vehicles	
Falls	Positive degree of control
Fires, burns	Positive degree of control
Drownings	Slightly positive but leveling off since 1960
Ingestion of food, objects	Negative control until last three years
Firearms	Fluctuates, but generally level
Poisons (liquid, solid)	Negative control
Poison (gas)	Level control

Source: Accident Facts, 1973, pp. 10–13.

Thus the minimum value of C_c is 0.01 and the maximum value 1.0 from Table 16.8. This provides the scale range of 100, as suggested previously. The descriptions of factors may be precise, the values are judgments and are imprecise. The control approach and degree of control factors each has a range of 10, whereas the two remaining factors only have a range of 2 to 1 as chosen by the author.

When the value of C_c is compared with the case for no control, C_0, a risk conversion factor results. The value of C_0 is equivalent to the value of C_c for "no control" and "uncontrollable" simultaneously, such that C_0 is 0.01 by definition.

$$\text{RCF}\left[\begin{matrix}\text{control} \\ (C_c \times 0.01) \\ \text{no control}\end{matrix}\right.$$

There are many different combinations for this risk conversion factor, and the particular combination is determined for each system. The degree of control is observable for existing systems on either a relative or an absolute basis. Table 16.9 provides an overview for different accident categories. It should also be noted that there is a great deal of variability in existing systems that may not be acceptable for new systems.

16.3 MAGNITUDE OF PROBABILITIES AND OTHER FACTORS

Factors that involve special situations and the individual propensity for risk taking are best addressed by direct experimentation and subsequent generalization. Factors that involve the magnitude of probability of occurrence are subject to some degree of historical data evaluation and are of primary concern in this section.

16.3.1 Probability and Uncertainty

Much has been published on probability and how it is related to uncertainty through objective experimentation and degree of belief. The assignment of probabilities to occurrences implies the existence of information, but imperfect information. Uncertainty is the absence of information, and probability values help reduce uncertainty. The reduction is not complete, however, since elimination of all uncertainty would imply certainty of future outcomes. Probabilities estimate the likelihood of an event, never its actual occurrence. Increasing the number of observations before a trial increases the estimate of likelihood. Increasing the number of trials involved in a decision increases certainty of outcome. In the first case, this is analogous knowing the odds on a roulette wheel, but not knowing the outcome of the next spin. In the second case, it is analogous to the "house" in that many spins are made before receipts are reckoned.

Low probability events are usually measured by examining the probability of occurrence over some time period. High probability events are usually measured by the frequency of occurrence in a given time period or the reciprocal of this value. These may be related to risk behavior when considered in conjunction with anxieties derived from the uncertainty in occurrence of unwanted consequences.

With high probability events the occurrence of unwanted consequences leads to anxiety and anxiety responses. Conversely, the occurrence of desirable consequences with high probabilities leads to hope and dependence. For low probability events the occurrence of unwanted consequences leads to the hope they will not occur, and for desired consequences it leads to anxieties that they might not occur (or hope that they will occur). These subtle differences help in understanding differences in behavior for low and high probability risks. People hope low probability risks do not occur and are not concerned with risks below particular thresholds. Conversely, those subjected to high probability risks learn to live with them and even become contemptuous of them through familiarity. Iron construction workers are examples of the latter case.

16.3.2 Low Probability Levels and Thresholds

Some risks in society are low enough to be considered negligible.

> One important concept is that of neglibible risk, a risk which in practice is ignored, even by those well informed about it. It is greater than zero but, because it is ignored, does not require statement in numerical terms. . . . It may not be acceptable, however, in all circumstances.[3]

This is especially true of uncontrollable risks that are considered acts of God. The risk of being hit by a meteorite is an example of this type of risk. Since man is not a troglodyte, this represents an unavoidable natural disaster. Avoidable natural disasters or man-caused disasters* cannot be used effectively for establishing such thresholds absolutely.

The Reactor Safety Study has estimated the risk from meteorites (Figure A3.2). This estimate shows a constant value of 10^{-3} as the slope of the curve relating frequency of fatalities from meteorites in the United States to the number of fatalities per event. On this basis one would expect 10^{-3} fatalities from meteorites in the United States per year, regardless of the size of the event. This represents a risk to an individual of 5×10^{-12} fat/yr, based on a population of 2×10^8.

It is expected that most other unavoidable natural hazards are less than this value, and the total risk from all unavoidable natural hazards would be in the range of 10^{-11} to 10^{-10} fat/yr. Since this represents a threshold for acts of God, avoidable natural hazards or man-caused or man-triggered disasters will have thresholds well below this number, at least by a factor of 10.

The threshold for man-originated risks might therefore be in the range of 10^{-11} to 10^{-13} fat/yr, a very small number.

16.3.3 High Probability Risks and Spatial Distribution

High probability risks are rationalized in a variety of ways by those who are exposed to them. Chapter 8 covered spatial distribution of risk on a qualitative basis as a means of coping with high probability risks. Quantitative study is made difficult by the existence of other psychological factors, such as benefits in the form of thrills or excitement that accompany certain hazardous undertakings and the ability of man to adapt under pressure to a variety of undesirable conditions. As a result, further quantification has not been attempted.

* See Section 16.1 for definitions.

Table 16.10 *Interrelationships Among Risk Factors Based on Differences in Voluntary and Involuntary Risks*

Benefits	Risk Factor: Voluntary: Risk-Benefit Tradeoffs Possible on Direct Basis	Risk Factor: Involuntary: Inequities Prevent Direct Risk-Benefit Tradeoffs
Avoidability and availability of alternatives	Avoidable by definition	Unavoidable; no alternatives but reduction in exposure possible Natural risk; act of God (minimize exposure) Man-made risk: risk/benefit inequity (minimize exposure and flight)
	Existence of alternatives makes game theory possible, although expected utility is only one criterion for decision	Unavoidable; only alternative to flee Natural risk; perceived degree of control Man-made risk; perceived degree of control and flight based on degree of satisfaction with status quo
	All alternatives undesirable; becomes more like an involuntary risk	Avoidable; other alternatives Spatial distribution; spread risk, insurance, etc. Controllability; reduce risk, implement systemic control for man-made risk
Time discount	5% per year or higher; risk to progeny discounted only if new children not sought	Less than 1% per year; No discount for irreversible risks or risk to progeny
Statistical distribution	Depends on individual propensity for risk and perceived degree of satisfaction with status quo	Avoidable; spread risk, insurance, etc. Unavoidable; may not happen to risk taker; not only one at risk (i.e., risk is shared by others—identified risk takers, statistical risk)
Controllability	Random vs. perceived degree of control Degree of systemic control Man-made vs. natural	Unavoidable; no alternatives except flee or fight, if man-made; perceived degree of personal control Unavoidable, but can reduce exposure (probability and magnitude of consequences); degree of systemic control and flight (if man-made) Avoidable; existence of alternatives; perceived degree of personal control; degree of systemic control

16.4 RISK FACTOR INTERRELATIONSHIPS

Up to this point, risk factors have been discussed for convenience, as if they were independent. The risk factors are not necessarily independent of one another however, and interrelationships are important.

16.4.1 Voluntary and Involuntary Risks

Otway and Cohen[18] have shown that involuntary and voluntary risks may well have quite different relationships between acceptance levels and benefit streams. They indicate a 1.8 power relationship for voluntary risks and a sixth power relationship for involuntary risk.

In either case, risk acceptability is proportional to benefit as well as to the components that make up the definition of voluntary and involuntary risk. Other risk factors are also affected by differences in voluntary and involuntary risks. Many of these relationships are summarized in Table 16.10.

16.4.2 Size of Risk Event

The definition of ordinary and catastrophic risk directly affects the numerical treatment of these risks. Man-made and natural risk levels are dependent on the size of the event.

16.4.3 Relationship Between Risk Structure and Risk Factors

Each of the risk factors discussed affects risk structure in some manner. Not all are identical, and some major factors are related to their risk structure effects in Table 16.11. These factors affect risk valuation by altering

Table 16.11 Relationship between Risk Structure and Risk Factors (How Risk Factors Can Alter Risk Valuation)

Risk Factor	Consequence Description	Probability of Occurrence of the Consequence	Value of the Consequence to the Risk Taker (Evaluator)
Voluntary–involuntary	×		×
Avoidability and		×	×
alternatives	×	×	×
Time discount			×
Spatial distribution		×	×
Controllability	×	×	×
Propensity to take risks			×
Natural–man-made		×	×
Ordinary–catastrophic			×

Table 16.12 *Some Major Risk Conversion Factors (Based only on Fatality Data)*

RCF	developing country	(6, **10,** 20)	RCF	rich country	(.05, **.10,** .15)
	rich country			developing country	
RCF	poor country	(60, **100,** 200)	RCF	rich country	(.005, **.01,** .015
	rich country			poor country	
RCF	ordinary accidents	(47, **50,** 55)	RCF	catastrophic accidents	(.018, **.20,** .022)
	catastrophic accidents			ordinary accidents	
RCF	military accidents	35, **40,** 55	RCF	civilian accidents	(.02, **.025,** .03)
	civilian accidents			military accidents	
RCF	natural catastrophes	(2, **10,** 50)	RCF	man-originated catastrophes	(.02, **.1,** .5)
	man-originated catastrophes			natural catastrophes	
RCF	man-originated catastrophes	(1, **2,** 5)	RCF	man-triggered catastrophes	(.2, **.5,** 1)
	man-triggered catastrophes			man-originated catastrophes	
RCF	voluntary risk	(10, **100,** 1000)	RCF	involuntary risk	(.001, **.01,** .1)
	involuntary risk			voluntary risk	
RCF[a]	immediate voluntary risk	(10, **20,** 30) % per year	RCF[a]	immediate involuntary risk	(0.1, 0.5, 1) % per year
	delayed voluntary risk			delayed involuntary risk	
RCF[b]	statistical risk agent	((1, **10,** 1000))	RCF	identifiable risk agent	((.001, **.1,** 1.0))
	identifiable risk agent			statistical risk agent	
RCF	controlled risk	((10, **100,** 1000))	RCF	uncontrolled risk	((.001, **.01,** .1))
	uncontrolled risk			controlled risk	

[a] Must be compounded by number of years of exposure.

[b] Double parentheses represent estimates.

the manner in which people view the description of consequences, the value of consequences, or the probability of occurrences.

16.5 SUMMARY OF RISK CONVERSION FACTORS

These tentative risk conversion factors represent an initial attempt to indicate how such factors might be determined, what difficulties arise in using existing data, where revealed preferences are inadequate measures, and some straw man values for initial use. It is hoped that subsequent work by others will begin to refine the numbers given here.

16.5.1 Limitations in Developing Risk Conversion Factors

Only a limited number of risk conversion factors have been developed. These generally involve fatalities, and data for other types of consequence are seldom at hand, or are less accurate. Thus there is a major gap in the manner in which societal data are routinely collected.

Revealed preferences were not always evident in data bases, as in the case for spatial distribution of risks and systemic control of risks. In other cases aggregated data are inadequate to provide measures, and direct experimentation is warranted. This is especially true when the individual propensity for risk is involved, as in risk identification when knowledge is a factor and for high probability risks involving spatial distribution.

Finally, factors are not mutually independent, and variations in one affect others.

16.5.2 Major Risk Conversion Factors

For purposes of summarization, a number of risk conversion factors that have major implications have been identified (Table 16.12). Although they are for fatality data only, they help form a basis for observing and estimating absolute levels of risk.

REFERENCES

1. *The Quality of Life Concept, A Potential New Tool for Decision Makers.* Washington, D.C.: Environmental Protection Agency, Office of Research and Development, March 1973.
2. CONSAD Report. See Appendix A.1.
3. R. H. Mole, "Accepting Risks for Other People," *Proceedings of the Royal Society of Medicine,* Vol. 69, February 1976, p. 15.

4. Robert W. Kates, "Natural Disasters and Development." Wingspread Conference Background Paper, Racine, Wis., October 19–22, 1975.
5. A. Baird et al., "Towards an Explanation of Disaster Proneness," Occasional Paper, No. 10, Disaster Research Unit, United Kingdom: University of Bradford, June 1975 (from Kates[4]).
6. Harry J. Otway, Philip D. Pahner, and Joanne Linnerooth, "Social Views in Risk Acceptance," IIASA RM-75-54. International Institute for Applied Systems Analysis, Laxenburg, Austria: November 1975, p. 16.
7. Harry J. Otway, R. Maderthaner, and G. Galtman, "Avoidance Response to the Risk Environment: A Cross-Cultural Comparison," IIASA RM-75-14. Laxenburg, Austria: International Institute for Applied Systems Analysis, 1975.
8. S. Golant and I. Burton, "Avoidance Response to the Risk Environment." Natural Hazard Research Working Paper No. 6, Department of Geography, University of Toronto, 1969.
9. E. E. Pochin, "Occupational and Other Fatality Rates," *Community Health* Vol. 6, No. 2, 1972.
10. *OSHA News Bulletin,* USDL-73-256. U.S. Department of Labor, June 15, 1973.
11. *Accident Facts.* Chicago: National Safety Council, 1973.
12. Richard Wilson, "Examples in Risk-Benefit Analysis," *CHEMTECH,* October 1975, pp. 604–607.
13. Chaucey Starr, "Social Benefit Versus Technological Risk," *Science,* Vol. 169, September 19, 1969, pp. 1232–1238.
14. I. Burton, R. W. Kates, and G. I. White, "The Human Ecology of Extreme Geophysical Events," Natural Hazard Research Working Paper No. 1, University of Toronto, 1968 (from Kates[4]).
15. *Reactor Safety Study*: An Assessment of Accident Risks in U.S. Commercial Nuclear Power Plants, WASH-1400 (NUREG-75/014). Washington, D.C.: Nuclear Regulatory Commission, October 1975.
16. Howard Kunreuther, "Limited Knowledge and Insurance Protection," in *Risk-Benefit Methodology and Application Conference Papers,* UCLA-ENG-7598. Los Angeles: UCLA School of Engineering and Applied Science, December 1975, pp. 173–232.
17. John M. Marshall, "Optimum Safety and Production of Information When Risks Are Insured," *Risk-Benefit Methodology and Application Conference Papers,* UCLA-ENG-7598. Los Angeles: UCLA School of Engineering and Applied Science, December 1975, pp. 243–272.
18. Harry J. Otway and J. J. Cohen, "Revealed Preferences: Comments on the Starr Benefit-Risk Relationships," IIASA RM 75-5. Laxenburg, Austria: International Institute for Applied Systems Analysis, March 1975.
19. Chauncey Starr, "Benefit-Cost Studies in Socio-Technical Systems." Report and Colloquium on Benefit-Risk Relationships for Decision Making, conducted by the Committee on Public Engineering Policy of the National Academy of Engineering, April 26–27, 1971. Washington, D.C.: The Academy, pp. 17–42.
20. W. Baldewicz, G. Haddock, Y. Lee, Prajoto, R. Whitley, and V. Denny, "Historical Perspectives on Risk for Large-Scale Technological Systems," UCLA-ENG-7485. Los Angeles: UCLA School of Engineering and Applied Science, November 1974.
21. Byron Bunger and John R. Cook, "Use of Life Table Methods to Compare Occupational Radiation Exposure Risks and Other Industrial Risks." Paper presented at the meeting of the Health Physics Society, Buffalo, N.Y., June 30, 1976.

22. "Probable Numbers and Cost Through 1985 of Lung Cancer Cases," Woodward and Fondiller, Inc., 1967, Appendix 7. Hearing before the Subcommittee on Research, Development, and Radiation of the Joint Committee on Atomic Energy, 90th Congress, 1st Session, Part 2, pp. 975, 1007.
23. G. W. Delphin and S. A. Beach, "The Relationship Between Dose Delivered to the Thyroids of Children and the Subsequent Development of Malignant Tumors," *Health Physics,* Pergamon Press, Vol. 9, No. 12, December 1963, pp. 1385–1390.
24. "The Health Consequences of Smoking," U.S. Department of Health, Education, and Welfare, Public Health Service, January 1974.
25. *The Effects on Populations of Exposure to Low Levels of Ionizing Radiation.* Washington, D.C.: Advisory Committee on the Biological Effects of Ionizing Radiation, Division of Medical Sciences, National Academy of Sciences, National Research Council, 1972.
26. Federal Food, Drug, and Cosmetic Act, Section 409c(3)A, 1938.
27. Arnold M. Kuzmack and Robert E. McGaughy, "Quantitative Risk Assessment for Community Exposure to Vinyl Chloride." Washington, D.C.: Environmental Protection Agency, December 5, 1975.

17

Estimation of Absolute Risk Levels

Absolute risk levels for different types of risk are estimated directly from historic, societal risk data as revealed preferences. Derived risk levels are obtained by applying risk conversion factors to absolute risk levels of one type of risk to derive an absolute level for another type of risk. Absolute and derived levels for the same risk type can be compared when both are available as a test of validity. It should be noted that the term "absolute" is used in contrast to "relative" for RCFs. These levels are not necessarily related to the actual risk, but the perception of risk on a behavioral basis.

17.1 PROBLEMS IN ESTIMATING ABSOLUTE RISK LEVELS

The inadequacy of data bases on revealed preferences notwithstanding, a number of generic problems in making such estimates are evident. The most significant of these are discussed below.

17.1.1 Dynamics of Society

Since society is in a constant state of change, we must ask, how long should data be valid?

Data more than a generation old probably are not valid in a rapidly changing technological society; thus pre-1945 data cannot be used to reflect society's present values and behavior. Considering the changes in society during the 1960s and early 1970s, perhaps data 10 years old ought to be discarded. However both the CONSAD study[1] and the Baldewicz[2] work show a reasonable continuous process for the last 20 years. Thus a 20- to 30-year period for analyzing data seems reasonable.

It should be noted that this period is being examined to estimate behavior of society toward risks, not to estimate risk, as was done in the Reactor Safety Study.[3] Thus all hurricanes recorded historically can be analyzed to develop risk estimates for hurricanes. It is another matter to estimate present risk aversion behavior; only data that apply to today's society are relevant.

Our contemporary society is generally risk aversive, and identification of new risks, even in existing systems, changes societal behavior. The very process of risk estimation and evaluation affects societal behavior.

17.1.2 Historic Risks Versus Modeled Risks

Historic data always underestimate risk, since low probability events that could occur may not have had time to occur. However estimates of *behavior* toward historic and experienced risks do not necessarily err on the low side. This is an important distinction that technologists making analyses of risk for technological systems fail to recognize.

> . . . CONSAD data is necessarily incomplete in its coverage of low probability events because it includes only 21 years of experience. Thus an event having a probability of occurrence of 10^{-2} yr would not likely be contained in this 21 years of data, let alone accidents as unlikely at 10^{-4}–10^{-6}/yr. If these low probability events could have high consequences then the risks could be significant. For example, if a 10^{-2} yr event could kill 10,000 people, then the associated risk could be as high as 5×10^{-6} (i.e., $10^{-2} \times 10^4/2 \times 10^{-7}$), which is much higher than the risks contained in the CONSAD data.
>
> Actuarial data will always be necessarily incomplete because very low probability–high consequence events will not be included. Thus, the data collected needs to be complimented by risk analysis using predictive techniques.[4]

17.1.2.1 Risk Measurement and Risk Perception

The view just expressed is proper for estimating risk, not for evaluation. If society has perceived an institutional or technological change, effective since older events have taken place, the older data are not valid for the purpose of revealing preferences. This change may or may not have changed the actual risk, but the perception of risk is no longer the same.

Dam and levee failures in the United States (Table 17.1) provide an example. Dam failures before 1900 are the major contributor to the total fatalities, with one event, the Johnstown Flood, representing more than two-thirds of the total fatalities. Even if we neglect the new construction techniques that have been implemented since the turn of the century, better communication and transportation methods, which have enabled emergency procedures to reduce fatalities from impending dam failures, cannot be ignored. For example, it is evident from history that there was no warning before the Johnstown Flood, even though evidence of impending dam failure was available. Nearly all dams and levees are monitored today, and advance warning and evacuation procedures are generally available,[5] as

Table 17.1 *Dam and Levee Failures in the United States (1889–1972)*

Year	Location of Disaster	Type of Structure	Lives Lost
1889	Johnston, Pa.	Dam	2000
1890	Walnut Grove (Prescott, Ariz.)	Dam	150
1894	Mill River, Mass.	Dam	143
1900	Austin, Pa.	Dam	8
1928	St. Francis Dam, Calif.	Dam	450
1955	Yuba City, Calif.	Levee	38
1963	Baldwin Hills (Los Angeles)	Reservoir	5
1972	Buffalo Creek, W. Va.	Dam	125
Total			2919

Source: Reactor Safety Study.[3]

demonstrated after the collapse of the Teton Dam in southeast Idaho on June 5, 1976. Although nearly $1 billion damage to property and crops took place, only six deaths were attributed to this major disaster because of early warning, resulting in the evacuation of 30,000 people. Thus going back 50 years to 1926, we see that there were only 618 fatalities over a half-century (about 12 fatalities per year), as opposed to 33 per year over a longer period.

The expectation of society that controllable risks will be controlled is basic to the nature of risk aversive conditioning by safety-conscious institutions (government and private) and through coverage of adverse events by the media.

17.1.2.2 Modeled Risk Estimates

As suggested by Levine, it is now possible to use predictive techniques to determine the effects of low probability events and to model future risks based on event and fault tree analysis. Societal concern regarding adverse effects of new technology has made such analysis mandatory for new technological systems. The environmental impact statements required by the National Environmental Policy Act, as well as new institutions for assessment such as the Office of Technology Assessment in Congress, and the development of techniques for technology assessment as a professional discipline, are the visible evidence of this condition.

Existing systems are exempted under a "grandfather clause" from requirements to file impact statements unless new risks are identified in these existing systems—for example, when a significant adverse event occurs or a new analysis identifies unsuspected risks inherent in existing systems, arousing public awareness.

It may not be wise or desirable to treat new systems differently from existing ones, but society demands it. As a result, revealed preference data must rely on analysis of perceptions of existing systems.

17.1.2.3 Selection of Data Bases

The main source of man-originated and man-triggered catastrophic data is the CONSAD Report.[1] Levine has pointed out that data used in the original draft of *An Anatomy of Risk* are incomplete.

> The basic evaluation of data relating to accident consequences that was used in determining the risk standards presented do not appear to have a valid statistical basis. The assignment of a value for the fatalities per year per individual from catastrophic accidents is derived principally from CONSAD data. This data source is incomplete because (1) it does not include all significant contributors to the selection of values, and (2) it is actuarial in nature and cannot adequately cover contributions from low probability events. Thus the values appear to be significantly underestimated in *An Anatomy of Risk,* resulting in setting risk acceptability limits that are too stringent.
>
> To illustrate the incompleteness of CONSAD data, it is noted that it includes only accident experience with buildings and structures, airlines, railroads, autos, buses, trucks, marine transport, and mines. Other available sources of data are, for instance, included in WASH-1400 [the Reactor Safety Study]. As an example, the risk of involuntary catastrophic fatalities due to fires and dam failures can be derived from curves of these types of accident data as compiled in WASH-1400. Integrating the area under these curves (for > 10 fatalities) and dividing by the exposed population gives contributions of 2×10^{-7} and 2×10^{-6}, respectively. Inclusion of just these two additional historical risks in the value would raise it by more than a factor of 20. The existence of this type of discrepancy suggests that a diligent search is needed to identify all the principal contributions to the various factors since these values are supposed to be a measure of existing risk.[6]

One must agree about the necessity for a diligent search to identify all principal contributors to the various risk factors. In fact, this was the purpose of the CONSAD report, which covered *all* man-originated catastrophic events over the 21-year period 1953–1973. Contrary to Levine's statement, it incorporates all dam failures and fires. The difference between WASH-1400 and CONSAD data is twofold. First, the CONSAD data cover a recent 21-year period and WASH-1400 covers all periods for which data are available. Second, the CONSAD data are historic and the WASH-1400 curves are modeled from historic data, without taking into account changes in degree of control and the perception of control.

The choice of a 21-year interval is arbitrary. However these data, which exhaustively examined all pertinent events in the period, are available and provide the best source for estimating absolute risk levels. Modeled risk estimates are not and should not be included. Indeed, results from these data may underestimate risk and set stringent levels for absolute risk. However present societal preferences are based on these data, and most likely they have more statistical validity for this purpose than the WASH-1400 data.

17.2 CLASSIFICATION OF ABSOLUTE RISK LEVELS

Absolute risk levels are classified simultaneously by two different schemes. The first involves the type of risk, and the second the class of consequence. The symbol $A_{i,j}$ represents absolute risk levels, where the i identifies the type of risk and j the class of consequence. More specific aspects of this structure are covered below.

17.2.1 Classification by Type of Risk

Table 17.2 classifies absolute risk levels for sudden-event risks (accidents, war, homicide, suicide, etc.). Table 17.3 extends this classification to risks from disease, including malnutrition. The risks in each case are classified by two criteria—whether the consequence is immediate or delayed, and whether the risk agent is identifiable or is a statistical member of the population at risk. These two criteria form four conditions, designated A, B, C, and D. The breakdown under each condition is the same as for condition A.

17.2.1.1 Use of Conversion Factors

Conversion from an immediate-statistical risk (class A) to the other classes is accomplished by use of risk conversion factors or by direct observance of data. The risk conversion factor for delayed risk depends on the number of years of exposure and whether the risk is voluntary or involuntary. To convert a statistical risk to an identifiable risk, an RCF of .1 is used. A third index k, denotes the number of years a risk is delayed. A zero, or omission of the third index, indicates no delay.

The classification also indicates whether the risk is voluntary, regulated voluntary, or involuntary (V, RV, or I). On this basis, the absolute risk level can be identified by a format:

$$A(V)_{i,j,k}$$

where the symbol in parentheses tells whether the risk is voluntary or involuntary, and so on. The use of double parentheses

$$A((RV))_{i,j,k}$$

Table 17.2 *Classification of Absolute Risk Levels (Accidental and Other Risks)*

Risks	Voluntary (V)	Regulated Voluntary (RV)	Involuntary (I)
A. Immediate—statistical			
1. Natural			
a. Thresholds			×
b. Catastrophic			×
c. Ordinary			×
2. Man-triggered			
a. Catastrophic	×		×
b. Ordinary	×	×	×
3. Man-originated			
a. Thresholds			×
b. Catastrophic	×		×
c. Ordinary	×	×	×
B. Immediate—identifiable[a]			
C. Delayed—statistical[a]			
D. Delayed—identifiable[a]			
Other risks			
A. Homicide			×
B. Suicide	×		
C. War	×		×

[a] Same as A.

Table 17.3 *Classification of Absolute Risk Levels (Disease Risks, Including Malnutrition)*

A. Immediate—statistical			
1. Natural			
a. Thresholds			
1. Hereditary susceptibility			×
2. Uncontrollable vectors and conditions			×
3. Aging			×
b. Catastrophic			
1. Epidemics		×	×
2. Famine, drought, etc.			×
Ordinary			
1. Hereditary susceptibility	×		×
2. Uncontrolled vectors and conditions	×	×	×
2. Man-triggered			
a. Catastrophic	×		×
1. Epidemic	×	×	×
2. Famine, drought, etc.			×
b. Ordinary	×	×	×
3. Man-originated			
a. Catastrophic	×		×
1. Epidemic		×	×
2. Famine, drought, etc.	×	×	×
b. Ordinary	×	×	×
B. Immediate—identifiable[a]			
C. Delayed—statistical[a] (including genetic pool)			
D. Delayed—identifiable[a] (including progeny)			

[a] Same as A.

indicates an identified risk agent, as opposed to a statistical one in the first case.

In the classification of Tables 17.2 and 17.3, areas that are meaningful are represented by an $\times$. The remaining cases are of little interest. Not all the interesting cases are amenable to quantification because of lack of data.

17.2.1.2 Accidental and Disease Risks

The index i is sequentially used to indicate the types of risk under heading A for both accidental and disease risks. It should be noted that accidental and disease risks differ in many respects. Accidental risks usually involve relation to a specific event, although the effects may be delayed. Often in accidents involving toxic chemicals acute effects are observed immediately, but chronic effects from exposure may not appear for several years. Thus sudden death, injury, shock, and trauma are associated with accidents, although many chronic conditions, including cancer, may result. On the other hand, the onset of disease is relatively slow. Disease may be caused by a single event (e.g., exposure of one person to a virus by the sneeze of another), but the perception of disease development takes place over a period of hours to years. One difference between accidental risk and disease risk is the perception of time interval of onset.

Some diseases are hereditary, or at least the susceptibility to some diseases is enhanced by congenital factors. On the other hand, "accident-prone" conditions are evidently psychologically based.

Society tries very hard to reduce controllable accidents, and this is also true for most contagious diseases (polio, measles, malaria, cholera, etc.). However considerably less attention is paid to the *prevention* of chronic disease than to *therapy* to cure the disease. This tendency is especially noticeable for heart disease, lung cancer from smoking, and some other forms of cancer that can be prevented by a change in lifestyle. For example, lack of exercise, continuing psychological stress, obesity, and rich diets are all factors that increase the risk of heart disease. In this case, the increased risk is voluntary (as in smoking); the quality of life is traded off against the life-shortening implied. On the other hand, some disease risks are involuntary, such as exposure to carcinogens in the environment or regulated voluntary occupational exposure to disease-causing agents.

Risks from both accidents and disease have unique aspects that must be considered when establishing absolute risk levels.

17.2.2 Classes of Consequences

The index j is used to indicate classes of consequences. The lack of data has restricted these classes to mortality, morbidity, property damage and casualty losses, and life-shortening as a result of premature death.

17.2.2.1 Fatalities ($j = 1$)

Fatalities are indicated by $j = 1$ and are in units of fatalities per year per individual (fat/yr). However as previously established, the exposure to an individual is understood, and the units fat/yr refer to an individual. Exposure to larger numbers of risk agents are always noted specifically. Fatalities represent premature death from accidents, disease, or other risks. Morbidity prior to death is not considered.

17.2.2.2 Injuries and Morbidity ($j = 2$)

Nonfatal injuries and dehabilitation from disease involve morbidity. The units for injuries per year per individual (he/yr) refer interchangeably to injuries or health effects. The particular type of risk determines whether injuries or health effects predominate.

17.2.2.3 Property Damage ($j = 3$) and Insured Losses ($j = 4$)

Estimates of property damage are based on the number of events per year for a type of risk, the average dollar loss per event for a year, and the number of people who share the burden of payment. The units are dollars per year per individual ($/yr).

Two entries are shown. The first is an estimate of overall property damage ($j = 3$). The second entry is an estimate of insured losses ($j = 4$). Insured loss data are rather well documented, since they represent actual cash transactions, but they do not provide an estimate of total damage. Wide variations in such estimates are possible, depending on who is estimating total damage.

17.2.2.4 Reduction in Years of Life Expectancy ($j = 5, 6$)

Based on 1969 mortality tables, Bunger et al.[7] have made estimates of the reduction in years of life expectancy for average populations involving a cohort of 100,000 members. The approach is similar to that used by Baldewicz et al.[2] and Sather[8] except that Bunger adds the risk to the cohort over the time of exposure and latency period (if any), whereas Baldewicz and Sather subtract existing risks from the mortality tables. They use different data bases, of course, but report essentially the same order of results.

The mortality data given in $j = 1$ serve here as a basis for computing the reduction in years of life expectancy over an exposure period appropriate to the risk in question and involving latency effects if warranted. This computation is made for the number of years of life expectancy lost per individual, and the index $j = 5$ represents this case. The exposure period and latency effects are indicated for each type of risk.

To simplify risk comparison, a second computation was made, based on mortality rate for one year taking place for the average age individual.* This computation of years lost of life expectancy per year of exposure is represented by $j = 6$.

The average years lost uses units of years (yr), whereas the average years lost per year of exposure uses years per year (yr/yr).

Data are not always available for all classes of consequence for different risks.

17.3 ABSOLUTE LEVELS OF RISK

Accidental risks are examined first, generally, in the order indicated by Table 17.2. Risks from disease are considered next, in the order indicated in Table 17.3.

17.3.1 Accidental Risks: Immediate, Statistical

17.3.1.1 Naturally Occurring Risks

Thresholds of Concern ($i = 1$). Section 16.3.2 examined the concept of threshold of risk. Certain uncontrollable risks at very low probabilities are ignored by man and society as a whole. The estimate for natural risks was 10^{-11} to 10^{-10} fat/yr.

$$A(I)_{1,1} = 10^{-10} \text{ fat/yr}$$

Only fatality information is shown, since injury and property damage from thresholds of concern have little meaning. Life-shortening involves exposure at the foregoing rate over the total lifetime of an individual and represents

$$A(I)_{1,5} = 3 \times 10^{-7} \text{ yr} \qquad \cong 1 \text{ second}$$

$$A(I)_{1,6} = 4 \times 10^{-9} \text{ yr/yr}$$

The $A(I)_{1,5}$ value is particularly interesting because it represents a life-shortening of about 1 second. This could, indeed, seem to be a reasonable base threshold. The $A(I)_{1,6}$ value has no equivalent meaning but serves as a base value for comparison.

Catastrophic Risk ($i = 2$). *Accident Facts*[9] reports natural cataclysms for three years (1968–1970) which resulted in an average of 239 fatalities per

* The average age of an individual is 32 years, with an average life expectancy of 70.76 years for the data used.

year (standard deviation, 166). This represents a death rate of 1.2×10^{-6} fat/yr and compares quite closely with a value derived from man-made catastrophic risk of 10^{-7} fat/yr (see below) and an RCF of 10 for conversion from man-made to natural catastrophic risk (i.e., a value of 1×10^{-6} fat/yr). Since only one significant figure seems meaningful, the latter value is proposed. Injury and property damage are derived by use of risk conversion factors and rounded off.

$$A(I)_{2,1} = 1 \times 10^{-6} \text{ fat/yr}$$

$$A(I)_{2,2} = 5 \times 10^{-6} \text{ he/yr}$$

$$A(I)_{2,3} = \$0.2/\text{yr}$$

$$A(I)_{2,5} = 3 \times 10^{-3} \text{ yr}$$

$$A(I)_{2,6} = 4 \times 10^{-5} \text{ yr/yr}$$

Ordinary Risk (i = 3). An RCF of 50 may be used initially to convert from catastrophic to ordinary risk resulting in values of

$$A(I)_{3,1} = 5 \times 10^{-5} \text{ fat/yr}$$

$$A(I)_{3,2} = 3 \times 10^{-4} \text{ he/yr}$$

$$A(I)_{3,3} = \$10/\text{yr}$$

The property damage number seems to be somewhat high. Table 17.4 shows an average value of \$1.02 for insured, fixed property losses. Assuming that only 70% of insured losses are fixed property, and only one-half of all losses are insured, $A(I)_{3,3}$ would be closer to \$3 than \$10 per year per individual. Therefore:

$$A(I)_{3,3} = \$3/\text{yr}$$

$$A(I)_{3,4} = \$1/\text{yr}$$

where $A(I)_{3,4}$ refers only to fixed property losses.

The fatality number may be low. *Accident Facts*[9] data show an average value of 1344 fatalities per year (standard deviation, 40.5) for ordinary (cataclysms subtracted out) natural fatalities. This is a rate of

$$A(I)_{3,1} = 6.7 \times 10^{-5} \text{ fat/yr}$$

and using a fatality-injury factor of 5, we have

$$A(I)_{3,2} = 3.6 \times 10^{-4} \text{ he/yr}$$

Table 17.4 *Property Damage (Insured, Fixed Property Losses Only) from Natural Disasters*[a]

Year	Damage (millions) Total	Dollars per Individual in United States[b]
1963	$ 11	$ 0.06
1964	148	0.74
1965	652	3.26
1966	57	0.29
1967	160	0.79
1968	90	0.45
1969	185	0.92
1970	360	1.80
1971	160	0.79
1972	212	1.06
Totals	$2035	$10.16
Mean	204	1.02
Standard deviation	184	0.92

[a] Hurricanes, tornadoes, floods, earthquakes, windstorms, hail.
[b] Based on population of 2×10^8.
Source: Insurance Facts, 1973.

These are in reasonable agreement with rates estimated from risk comparison factors.

$$A(I)_{3,1} = 7 \times 10^{-5} \text{ fat/yr}$$

$$A(I)_{3,2} = 4 \times 10^{-4} \text{ he/yr}$$

$$A(I)_{3,3} = \$3/\text{yr}$$

$$A(I)_{3,4} = \$1/\text{yr}$$

$$A(I)_{3,5} = 0.2 \text{ yr}$$

$$A(I)_{3,6} = 2 \times 10^{-3} \text{ yr/yr}$$

*17.3.1.2 Man-Originated Risks**

Thresholds of Concern ($i = 4$). A risk conversion factor of .1 is used with natural risk threshold values to develop a threshold of concern for man-

* Taken out of order for better sequence of presentation of data.

originated risks. The latter thresholds are lower than acts of God.

$$A(I)_{4,1} = 1 \times 10^{-11} \text{ fat/yr}$$

$$A(I)_{4,5} = 3 \times 10^{-8} \text{ yr} = 0.1 \text{ second}$$

$$A(I)_{4,6} = 4 \times 10^{-10} \text{ yr/yr}$$

One may well argue the significance of a tenth of a second, but surely risks originated by man below this level are most likely ignored.

Catastrophic Risk ($i = 5$).

1. *Involuntary Risk.* The basic data are derived from CONSAD data on involuntary risk, given in Appendix A.

$$A(I)_{5,1} = 1 \times 10^{-7} \text{ fat/yr}$$

$$A(I)_{5,2} = 5 \times 10^{-7} \text{ he/yr}$$

$$A(I)_{5,3} = \$0.02/\text{yr}$$

$$A(I)_{5,5} = 3 \times 10^{-4} \text{ yr}$$

$$A(I)_{5,6} = 4 \times 10^{-6} \text{ yr/yr}$$

2. *Voluntary Risk.* Data are obtained directly from the CONSAD data base in Appendix A using a U.S. population of 2.1×10^8 people. However data for $A(V)_{5,5}$ uses an exposure of 2×10^{-6} fat/yr over the ages of 18 to 65.

$$A(V)_{5,1} = 2.1 \times 10^{-6} \text{ fat/yr} = 2 \times 10^{-6} \text{ fat/yr}$$

$$A(V)_{5,2} = 2.3 \times 10^{-6} \text{ he/yr} = 2 \times 10^{-6} \text{ he/yr}$$

$$A(V)_{5,3} = \$0.4/\text{yr}$$

$$A(V)_{5,5} = 6 \times 10^{-3} \text{ yr}$$

$$A(V)_{5,6} = 8 \times 10^{-5} \text{ yr/yr}$$

Note that the factor of 5 between fatalities and health effects (injuries) found for involuntary risks does not hold in this case. The ratio is 2.1:2.3 before rounding.

3. *Regulated-Voluntary Risk.* Occupational risk, especially when regulated by government agencies, is one aspect of regulated-voluntary risk. Most occupational situations are not regulated for catastrophic risk. Coal miners, persons employed in commercial marine vessel transport, and crews of aircraft are particularly subject to catastrophic accidents, however. The

aircraft industry is highly regulated, but coal mining came under intensive regulation only in May 1973, with the establishment of the Mine Enforcement and Safety Administration to administer the federal Coal Mine Health and Safety Act of 1969. Marine transportation is basically unregulated.

Catastrophic accidents in each of these areas were examined individually using the basic CONSAD data.[1] Occupational fatalities and injuries were totaled and averaged over the 21-year period (1953–1973). The number of workers in each industry was found by averaging the total industry employment over the same period. Data for air crews were derived from other sources, presented later (Table 18.3). The results are as follows:

	Air Transport	Marine Transport	Coal Mining
Average fat/yr	30	34.29	19.5
Average he/yr	1.5	13.71	1.7
Average workers/yr	1.03×10^{6}	4.37×10^{4}	1.6×10^{5}
Risk rate per worker—fat/yr	2.9×10^{-5}	7.8×10^{-4}	1.3×10^{-4}
Risk rate per worker—he/yr	1.9×10^{-6}	3.1×10^{-4}	1.1×10^{-5}

Marine transport, which is relatively unregulated, has a very high catastrophic occupational risk followed by the partially regulated and now fully regulated coal mine industry. Air travel is highly regulated and has the lowest value. The question then is how to derive an absolute risk level for catastrophic regulated-voluntary risk from the limited data. All three occupations have risk levels greater than voluntary risk levels, and all are characterized by lower injury levels than for fatalities. A first approximation might well use air travel data, which are about an order of magnitude higher than catastrophic voluntary risk levels for fatalities and about the same level for injuries. That is, injuries are valued about 10 times less than fatalities. Therefore

$$A(RV)_{5,1} = 3 \times 10^{-5} \text{ fat/yr}$$

$$A(RV)_{5,2} = 3 \times 10^{-6} \text{ he/yr}$$

It is difficult to develop a meaningful definition of regulated-voluntary property damage associated with occupational risk. Thus we use the same value as that for voluntary risk.

$$A(RV)_{5,3} = A(V)_{5,3} = \$0.4/\text{yr}$$

Ordinary Risk ($i = 6$)

1. *Involuntary.* Involuntary ordinary risk is derived from the catastrophic risk levels and a risk conversion factor of 50 because we do not have statistical data for this type of risk.

$$A(I)_{6,1} = 5 \times 10^{-6} \text{ fat/yr}$$

$$A(I)_{6,2} = 3 \times 10^{-5} \text{ he/yr}$$

$$A(I)_{6,3} = \$1/\text{yr}$$

$$A(I)_{6,5} = 1 \times 10^{-2} \text{ yr}$$

$$A(I)_{6,6} = 2 \times 10^{-4} \text{ yr/yr}$$

2. *Voluntary Risk.* Overall accident death rates are derived from *Accident Facts*[9] after natural causes are removed. This results in 113,713 fatalities (standard deviation, 770) per year over three years. Thus

$$A(V)_{6,1} = 5.7 \times 10^{-4} \text{ fat/yr} = 6 \times 10^{-4} \text{ fat/yr}$$

Accident Facts[9] reports 14,028,000 bed disability injuries in 1973, as well as 20,703,000 injuries that were not disabling but necessitated restricted activity, and 26,189,000 injuries that did not restrict the victims' activity, for a total of 60,921,000 persons injured.

Bed disabling	$= 7 \times 10^{-2}$ he/yr
Activity restriction	$= 1.0 \times 10^{-1}$ he/yr
No restriction	$= 1.3 \times 10^{-1}$ he/yr
$A(V)_{6,2}$ (total)	$= 3.0 \times 10^{-1}$ he/yr

Insurance Facts indicates that total insurance premiums for property damage of about $38 billion were written in 1972. Property damage losses of about $32 billion were experienced as follows:

Fire	$ 2.3 billion
Auto	19.1 billion
Work	10.4 billion
Other	0.2 billion

This results in a property damage figure per individual of about $160.

$$A(V)_{6,3} = 2 \times 10^{2} \text{ \$/yr}$$

Exposure to these risks are limited to the years 18 to 65 for life expectancy computations.

$$A(V)_{6,1} = 6 \times 10^{-4} \text{ fat/yr}$$

$$A(V)_{6,2} = 3 \times 10^{-1} \text{ he/yr}$$

$$A(V)_{6,3} = 2 \times 10^{2} \text{ \$/yr}$$

$$A(V)_{6,5} = 1 \text{ yr}$$

$$A(V)_{6,6} = 1 \times 10^{-1} \text{ yr/yr}$$

3. *Regulated Voluntary Risk.* The question of regulated voluntary risk versus unregulated voluntary risk was briefly considered in Section 16.2.1.4. It is not clear whether such risks are truly voluntary. However a risk conversion factor of .2 was estimated for such conversion.

$$A(RV)_{6,1} = 1 \times 10^{-4} \text{ fat/yr}$$

$$A(RV)_{6,2} = 6 \times 10^{-2} \text{ he/yr}$$

$$A(RV)_{6,3} = 3 \times 10^{1} \text{ \$/yr}$$

$$A(RV)_{6,5} = 1 \times 10^{-1} \text{ yr}$$

$$A(RV)_{6,6} = 1 \times 10^{-2} \text{ yr}$$

Including in the consideration occupational hazards regulated by OSHA, the average risk for all industries is 17×10^{-5} fatality per employee/year.[10] This represents 1.7×10^{-4} fat/yr and compares favorably with $A(RV)_{6,1}$.

17.3.1.3 Man-Triggered Risks

Man-triggered risk generally represents a voluntary exposure to naturally occurring risks. "Voluntary" is used in the sense of avoidance, but exposure is involuntary if suitable alternatives to avoidance are not available. Computations shown are by risk conversion factor derivation.

Catastrophic Risk ($i = 7$)

1. *Involuntary Risk.* Data here are derived from $A(I)_5$ with an RCF of 2.

$$A(I)_{7,1} = 2 \times 10^{-7} \text{ fat/yr}$$

$$A(I)_{7,2} = 1 \times 10^{-6} \text{ fat/yr}$$

$$A(I)_{7,3} = 4 \times 10^{-2} \text{ \$/yr}$$

$$A(I)_{7,5} = 6 \times 10^{-4} \text{ yr}$$

$$A(I)_{7,6} = 8 \times 10^{-6} \text{ yr/yr}$$

The value of 2×10^{-7} fat/yr for $A(I)_{7,1}$ is close to the estimate made by Levine[4] for dam failure—perhaps the major source of man-triggered involuntary risk.

2. *Voluntary Risk.* A man-triggered voluntary risk involves exposure to natural catastrophes when the option to avoid the risk is given up for a direct benefit to the risk agent (e.g., by moving into a fertile flood plain, fully realizing the risks and benefits). The difficulty of determining motivation and measuring the degree of risk assessment involved makes direct measurement of these risks and involuntary risks very difficult. Numbers can be derived but hardly substantiated, through RCFs.

Ordinary Risks ($i = 8$). Data are sparse in the area of ordinary risks for the same reasons discussed for voluntary and involuntary risks. Many man-triggered voluntary risks are occupational, and these (e.g., for commercial fishermen or sailors) are often regulated. Zoning and building regulations that prevent use of land below dams, in flood plains, and over earthquake faults, are examples of regulations restricting voluntary use.

Risk conversion factors can be used as as a first estimate for these risks. Only fatality date are shown, although other consequences may be computed.

$$A(I)_{8,1} = 1 \times 10^{-5} \text{ fat/yr}$$

$$A(V)_{8,1} = 1 \times 10^{-3} \text{ fat/yr}$$

$$A(RV)_{8,1} = 2 \times 10^{-4} \text{ fat/yr}$$

17.3.2 Accidental Risks: Immediate-Identifiable

Immediate-identifiable accidental risks differ from the preceding cases in that specific risk agents are involved and named. Data are obtained by using the estimated RCF of .1 given in Chapter 16.

$$\text{RCF}\left[\begin{array}{l}\text{identifiable risk agent} \\ ((.001, \mathbf{.1}, 1)) \\ \text{statistical risk agent}\end{array}\right.$$

It is used across the board for all types of consequence, since it is at best a gross estimate. Since any of the levels in the last section may be used to derive an immediate-identifiable absolute risk level by use of the RCF, only one illustrative case is given. The case is for a man-originated, involuntary, catastrophic risk, $A(I)_5$, but the specific recipients are identifiable.

$$A((I))_{5,1} = 1 \times 10^{-8} \text{ fat/yr}$$

$$A((I))_{5,2} = 5 \times 10^{-8} \text{ fat/yr}$$

$$A((I))_{5,3} = 2 \times 10^{-3} \text{ \$/yr}$$

$$A((I))_{5,5} = 3 \times 10^{-5} \quad \text{yr}$$

$$A((I))_{5,6} = 4 \times 10^{-7} \text{ yr/yr}$$

17.3.3 Accidental Risk: Delayed-Statistical

Delayed risks for fatalities and injuries are evaluated by taking the value for the risk experienced immediately and discounting by the number of years of exposure. The effective discount rates for fatalities and injuries are derived from risk conversion factors. Property damage, a straightforward economic calculation involving insurance for protection, is not considered here.

Different time scales of exposure are possible, and a third index indicates number of years of exposure. Delayed risks for life-shortening computations are found by calculating the risk with the consequence delayed by the latency period as opposed to using a discount rate.

Two cases are shown, one for involuntary risks with a discount rate of 0.5% per year and one for voluntary risks using an annual rate of 20%. Table 17.5 shows the effect of years of exposure on compounded effective discount rates.

Table 17.5 *Present Value of a Future Risk Set at Unit Value*

Years	Involuntary, 0.5%	Voluntary, 20%
1	.995	.833
2	.990	.694
3	.985	.579
4	.980	.482
5	.975	.402
10	.951	.162
15	.927	.065
20	.905	.026
25	.882	.0105
30	.861	.0042
35	.840	.0017
40	.819	.0007
45	.799	.0003

17.3.3.1 Man-Originated, Ordinary Involuntary Risk

Values for fatality information are derived from $A(I)_{6,1}$ by using the compounded effective discount rate of 0.5% per year, from Table 17.5.

$$A(I)_{6,1} = 5 \times 10^{-6} \text{ fat/yr}$$

$$A(I)_{6,1,5} = 6 \times 10^{-6} \text{ fat/yr}$$

$$A(I)_{6,1,10} = 8 \times 10^{-6} \text{ fat/yr}$$

$$A(I)_{6,1,20} = 1 \times 10^{-5} \text{ fat/yr}$$

$$A(I)_{6,1,30} = 2 \times 10^{-5} \text{ fat/yr}$$

Values for $A(I)_{6,5}$ are computed from life tables in which the risk of 5×10^{-6} fat/yr is delayed by the number of years specified.

$$A(I)_{6,5} = 1 \times 10^{-2} \text{ yr}$$

$$A(I)_{6,5,5} = 1 \times 10^{-2} \text{ yr}$$

$$A(I)_{6,5,10} = 1 \times 10^{-2} \text{ yr}$$

$$A(I)_{6,5,20} = 7 \times 10^{-3} \text{ yr}$$

$$A(I)_{6,5,30} = 5 \times 10^{-3} \text{ yr}$$

In this case the number of years lost is the delayed risk estimate, not the perception of delayed impact.

17.3.3.2 Man-Originated, Ordinary Voluntary Risk

Values for fatality information are derived from $A(V)_{6,1}$ using a 20% per year effective discount rate.

$$A(V)_{6,1} = 6 \times 10^{-4} \text{ fat/yr}$$

$$A(V)_{6,1,5} = 1 \times 10^{-3} \text{ fat/yr}$$

$$A(V)_{6,1,10} = 4 \times 10^{-3} \text{ fat/yr}$$

$$A(V)_{6,1,20} = 2 \times 10^{-2} \text{ fat/yr}$$

$$A(V)_{6,1,30} = 1 \times 10^{-1} \text{ fat/yr}$$

Values for $A(V)_{6,5}$ are computed in two ways: a risk of 6×10^{-4} fat/yr over the delay period specified for a person's whole life span and the same risk level exposure and delay from the ages 18 to 65. The values are as follows.

Exposure	Total Life Span (yr)	Age 18–65 (yr)
$A(V)_{6,5}$	1	1
$A(V)_{6,5,5}$	1	0.8
$A(V)_{6,5,10}$	1	0.6
$A(V)_{6,5,20}$	0.9	0.4
$A(V)_{6,5,30}$	0.6	0.2

Again, these values represent measured risk, not the perception of delay of consequences. The decrease in years results from computing causes of death as opposed to discounting the delayed fatality.

17.3.4 Risks from Disease

Risks from disease involve probabilities and consequence values quite different from those associated with accidents. Moreover, most data on disease report the results of disease inception and therapy and are disease specific. Primary emphasis on the health care system is directed at cures rather than prevention (especially in the case of cancer), and the source of a disease is often undetermined or, in specific cases, indeterminable.

Different diseases have different consequence values for morbidity and mortality. As a rule, terminal cancer is viewed as a worse situation than terminal heart disease because cancer patients usually experience more pain and suffering before death. People die every year from influenza, yet the mere mention of certain seldom-encountered killer diseases (e.g., plague, smallpox) can incite panic in a population.

It is easy to report the morbidity* and mortality statistics each year from disease, but this compilation does not reflect revealed preferences, since many of the numbers are determined by uncontrollable or partially controllable factors. In the past 50 years major success in preventing and controlling contagious disease has led to increased emphasis on chronic ailments. A population no longer faced with many contagious diseases finds chronic disease to be the major cause of death, other than aging. A large amount of research effort goes into the development of cures and treatments for chronic disease, while preventive health measures are less effective. Preventive measures for cancer involve decisions affecting individual behavior (e.g., to abstain from smoking). Preventive measures for heart disease involve programs of exercise, diet and dietary control, and control of exposure to stress. All these necessitate personal, voluntary decisions that affect the way and quality of life of the individuals at risk.

* In many cases (e.g., cancer) even morbidity data are sparse.

Implementation of preventive measures also requires identification of the causes of a disease and the ability to control exposure to the causes. Control in turn affects the probability of occurrence. The symptoms and character of a disease, the ability to cure it, the pain and suffering, and the manner of death, all affect the value of the consequence. Both aspects must be considered in evaluating risk from disease.

17.3.4.1 Probability of Occurrence and Controllability of Disease Inception

Table 17.3 deals primarily with the causative aspects of a disease, the manner of exposure, and the degree to which exposure can be controlled. Malnutrition is included as a disease because two major causes of malnutrition—drought and famine—must be addressed. Table 17.3 contains entries for immediate-statistical risks that are most appropriate for contagious disease. Delayed risks to both statistical and identifiable recipients are more appropriate for chronic disease.

To determine probability of occurrence, the causative event or events must be identified. Yet many diseases are very difficult to identify; cancer, for example, has many causes, and synergistic effects among multiple stresses are possible. Thus in a two-hit hypothesis for cancer induction featuring two stresses to a single cell, there might be a chemical carcinogen and a radiation event. Under this theory, any combination of two hits from a variety of causes (including hereditary susceptibility) induces a cancer cell to develop. At best, statistical evaluation of causes may be possible.

Such difficulties make quantification of causes impractical at this time, and the ensuing discussion is qualitative.

Naturally Occurring Disease Risks. Aging is a basic uncontrollable natural process (at least with present knowledge) that is the ultimate cause of death when disease, accidents, or other consequences do not intervene. Whether aging is caused by a limited number of cell division-renewal cycles, by accumulation of metabolites at the cellular level, or by some other process does not affect the condition of individual mortality at this time. Furthermore, the onset of old age often contributes to susceptibility to other risks, both accidental (e.g., fires in nursing homes) and disease (e.g., influenza).

Some members of the population are naturally more susceptible to certain kinds of disease, such as breast cancer and certain types of heart disease. In many cases, these traits are hereditary and are passed from generation to generation. This results in a population in which the distribution of susceptibility to different diseases ranges from susceptible to resistant.

There are also a number of uncontrollable pathogenic factors, such as the viruses that cause the common cold; Burkitts lymphoma, a cancer thought to be induced by a natural human virus, is another example. Uncontrollable conditions also exist, such as natural background radiation as a cause of cancer and genetic problems.

At present these processes represent thresholds of control. However as society's ability to control the related phenomena increases, the proportionate risk from them may decrease slightly. Unless physical immortality becomes possible, however, aging remains the limiting process. Aging is a threshold of concern and governs society's risk acceptance process. In fact, aging is the major underlying condition that people, individually and collectively, must face.

1. *Catastrophic Risks.* Natural catastrophic risk from disease stems from exposure to naturally occurring epidemics and changes in weather conditions that bring on drought and famine. These risks are basically involuntary, but exposure to epidemics of certain types is regulated by governmental restrictions on travel and the imposition of quarantine. Individual risk agents may expose themselves to disease by traveling to areas where diseases are prevalent. Returning travelers who may or may not contract the disease may be carriers responsible for epidemics. Medical personnal are exposed on a regulated voluntary basis.

2. *Ordinary Risks.* Ordinary risks occur naturally from uncontrolled vectors and conditions, and some parts of the populations are more sensitive to certain diseases. Ordinary risks differ from thresholds in that voluntary steps may be taken to minimize exposure. This is true for hereditary conditions, if they are known in advance, as well as for control of exposure to disease vectors and conditions. A person with a known hereditary condition may well opt for special precautions. Medical and public health personnel take on both voluntary and regulated voluntary risks. People who volunteer to participate in experiments constitute another aspect of voluntary risk to naturally occurring diseases.

Man-Triggered Disease Risks. Man-triggered disease risks are those that exist naturally but are exacerbated by man's activities.

1. *Catastrophic Risks.* Overcrowding due to enlarging populations and urbanization, increased travel with short transportation time, and inadequate attention to preventive public health practice, can lead to epidemics such as plague. Overfarming, overgrazing, lowering of water tables from overuse of wells, of failure to return groundwater to the water table, can result in drought, famine, and even permanent loss of land, as is occurring in parts of Africa today.

2. *Ordinary Risks.* Man's activities produce increased exposure to natural risks, both voluntarily and involuntarily. In the first case, ordinary jet travel involves slight increased risk of radiation exposure because of the higher levels of cosmic ray activity present at high altitudes. Involuntary risks result from activities that man undertakes without knowing the adverse consequences. For example, plowing up soil contaminated by anthrax causes renewed exposures.

Man-Originated Disease Risks. Disease risks that are caused directly by man's activities are referred to as "man originated."

1. *Catastrophic Risks.* Epidemics may be caused deliberately through the use of biological warfare weapons. Widespread exposure to vinyl chloride, which resulted in many cases of liver cancer before the carcinogenic toxicity of the substance became known, is an example of a man-originated disease risk that was imposed involuntarily.

Drought may result from man changing the ecological balance with either planned or unplanned catastrophic consequences. The development of the Aswan Dam in Egypt has changed the ecology of the lower Nile. Some consequences were expected, but others, such as lack of refertilization through periodic flooding, may not have been anticipated.

2. *Ordinary Risks.* There is growing evidence that the preponderant causes of cancer are environmentally caused stresses, many of them due to man's activities. It is estimated that 60 to 90% of all cancers are environmentally caused (i.e., not due to hereditary factors).[12] Societal stresses and anxieties are thought to contribute to hypertension and heart disease. Exposure to these man-made causes is often voluntary (as in smoking), regulated voluntary (as in occupational exposure of workers), and involuntary (when by-products of an activity affect individuals not directly involved). Pollution of air and water typifies the latter situation.

The initiation of a given disease may result from many sources characterized by differing degrees of control of exposure or prevention. Some sources are naturally occurring and affect all or parts of the populations; others are triggered or originated by man. Man-originated causes may also be essentially uncontrollable because of such irreversible effects as increased radiation exposure from fallout accompanying the atmospheric testing of nuclear weapons. Statistically identifying the relative contribution of different causes and having the ability to control the causes are prerequisities to further understanding of disease risks and wisely determining how society spends money for prevention and development of cures and therapeutic processes for disease. Table 17.6 outlines the factors affecting disease risk valuation when the probability of occurrence is affected. The format follows that used in Chapter 8.

Table 17.6 Factors Affecting Disease Risk Valuation (Factors Affecting Probability of Occurrence)

A. Magnitude of consequence
   1. Catastrophic
   2. Ordinary
B. Class of consequence
   1. Death
      a. Immediate
      b. Lingering, with and without hope
      c. After degenerate illness
   2. Illness
      a. Permanent-temporary
      b. Degree of debilitation
      c. Degree of pain and discomfort
   3. Loss of quality of life
   4. Loss of quantity of life; life-shortening
   5. Loss of income
      a. Individual
      b. Societal
   6. Cost of health care
      a. Individual
      b. Societal
C. Controllability of consequence
   1. Curable
   2. Uncurable
   3. Arrest only
   4. Partial cure
   5. Risks of cure
   6. Pain and discomfort as a function of control
D. Progress of disease
   1. Immediate trauma and full or partial recovery
   2. Delayed trauma and degeneration
   3. Controlled and spontaneous remission
E. Dynamics of disease control
   1. Increasing or decreasing probability of occurrence
   2. Increasing ability to cure

17.3.4.2 Valuation of Disease Consequences and Controllability of Disease Progress

Two questions must be addressed in the valuation of disease consequences. First, how does death or illness from disease differ from death and injury from accidents? Second, how are death and illness among different diseases valued? Intertwined with these questions is the ability to control the progress of the disease through diagnosis and therapy.

The answer to the first question, disease versus accidental consequences, involves a number of factors (the degree and duration of pain and suffering, sudden death vs. a lingering, known disease progression, decreased quality of life as a result of illness or injury, etc. With the preponderance of accidents, there is an initial shock and trauma (including death), then a period of recuperation (assuming the victim is still alive) that often ends in complete recovery. The same process can occur with heart attacks, but with most curable diseases there is a time lag between onset, diagnosis, and therapy not usually associated with injuries (although sequentially applied therapy may be used for injuries). Many chronic diseases for which there are limited cures have a reverse process: initiation and recognition of the disease when the patient is in good health, then a progressive decrease in health until death, arrest of the disease, or implementation of a full or partial cure. Such degradation and its predictability have an enormous effect on psychological and physiological factors alike. On this basis, cancer is one of the most dreaded diseases. Of course, accidents can result in maiming and permanent disability.

It should also be recognized that many diseases are not fatal, although they are dehabilitating for short or long periods, involve pain and discomfort, and affect the quality of life and the pursuit of one's livelihood. The perceived need as well as legal requirements to grant sick pay, and loss of productive capacity of workers, makes such illness an industrial problem as well.

Diseases such as cancer, which bear a long-term negative prognosis with little hope of cure or remission, involve high negative consequence valuation. Conversely, since some of history's most dreaded diseases can be cured or arrested, these afflictions (e.g., leprosy) are no longer feared as much as they were in the past. As a result, answers to the second question depend on the ability to cure a disease as well as the interval between onset and death. A heart attack may be a sudden and unexpected event that is shocking to the victim's family and friends, but the same kind of deterioration that accompanies cancer may be absent.

Finally, the ability to control diseases of different types is constantly changing as medical technology and practice improve. Yet the population's susceptibility to various diseases seems to be changing. For example, cancer has had a constantly increasing rate of occurrence since the mid-1950s, with a major (unexplained) increase of 5% in 1975. Whether this jump is attributable to elimination of competing risks, a more susceptible population, increased environmental factors, or other causes has not been ascertained. It does represent a "lack of control."

Table 17.7 Factors Affecting Disease Risk Valuation (Factors Affecting Consequence Valuation)

A. Cause of disease—voluntary vs. involuntary
  1. Naturally occurring causes
    a.Congenital (involuntary)
    b.Hereditary (involuntary)
    c.Hereditary susceptibility (involuntary—voluntary)
    d.Age level susceptibility (involuntary—voluntary)
    e.Uncontrollable vectors (involuntary)
    f.Uncontrollable conditions (involuntary)
  2. Man-triggered causes
    a.Synergistic causes (involuntary—voluntary)
    b.Exposure to natural causes (involuntary—voluntary)
  3. Man-originated causes
    a.Environmental (involuntary)
    b.Occupational (voluntary)
    c.Voluntary risk exposure (voluntary)

B. Discounting in time
  1. Immediate onset
  2. Delayed onset (1–30 years)
  3. Progeny
    a.Identified (1–30 years)
    b.Statistical—genetic pool (30–50 years)

C. Spatial distribution
  1. Identifiable risk agents
  1. Statistical risk agents

D. Controllability of exposure
  1. Uncontrollable
  2. Avoidable
  3. Controllable

Table 17.7 summarizes factors that affect the valuation of consequences for disease, based on factors identified in Part B.

17.3.4.3 Controllability of Cancer

The probability of occurrence of cancer from environmental stresses is higher than that from hereditary or viral causes. Some environmental causes are directly controllable; others are not.

The worker is exposed daily to dangerous chemicals in the dye factory. Bladder cancer?

The insurance salesman finds himself drinking and smoking heavily as his years on the road mount. Oral and lung cancer? His all-American high-fat diet may also be taking its toll. Colon cancer?

The lonely librarian who has never married and had children as she had hoped—breast cancer?—secretly envies the promiscuous divorcée whose sexual activity began at a young age. Cervical cancer?

The fair-skinned farmer has spent his life in the sun. Skin cancer?

And the woman who grew up in the shadow of an arsenic-emitting smelter never really escaped its noxious fumes. Lung cancer?[12]

It is estimated that about one-third of all cancers are potentially preventable, although the origins of cancer are not well defined. Four voluntary steps often suggested involve abstaining from basic societal practices, as follows:

1. Cutting out smoking.
2. Cutting down drinking.
3. Getting less sun.
4. Eating a more moderate low-fat diet (the evidence is suggestive, but difficult to prove).[12]

These voluntary steps are being widely publicized, but relatively few members of the population seem to be practicing such abstinence and risk aversion. Perhaps one must also consider the latency period and the ability to cure various types of cancer, such as skin cancer resulting from exposure to ultraviolet light. Early diagnosis is a key parameter in effecting cures. Table 17.8 lists some estimates of expected cancer deaths and the degree to which they may be avoidable.

We do not have enough information to develop absolute risk levels because the valuation of the consequences is not taken into account. It may not be possible to make inroads in this area by revealed preference methods. Studies of the views of different populations by polling and other experimental methods may be more appropriate.

17.3.4.4 Verbal Preferences for Risk Avoidance

A number of studies[13, 14] have been made using verbally expressed preferences for risk avoidance of hazards, including disease consequences as indicated in Chapter 9. However the degree of aversion for illness has been partly measured by a "seriousness-of-illness" scale. A list of 126 diseases has been scaled by patients and doctors with respect to the seriousness of the disease, with a rank correlation of .95 between both groups. Kates[16]

Table 17.8 *Preventable Deaths*

Cancer Type and Associated Factors	Expected Deaths	Preventable Deaths
Cigarette smoking: lung and some larynx	80,000	70,000
Alcohol		
Head and neck	7,500	5,000
Esophagus	6,000	
Industrial exposure		
Bladder	9,000	5,000
Liver	9,800	
High fat, low bulk diet		
Breast	30,000	5,000[a]
Colon and rectum	30,000	10,000
Other factors		
Uterine, cervix	7,500	3,000[a]
Melanoma, other skin cancer	5,000	1,500
Totals	184,800	99,500

[a] Achievable mostly through earlier diagnosis by application of known techniques. Figures are for 1974, as reported by Dr. Marvin A. Schneiderman, National Cancer Institute.

points out that if compared with the actual risk of death, both patients' and doctors' perceptions of seriousness are poor. Leukemia, cancer, and multiple sclerosis are all perceived as more serious than the more frequent causes of death—strokes and heart attacks. However the perception of the consequences of disease from this study is consistent with the factors affecting disease consequence discussed in Section 17.3.4.2.

Because of the difficulties of relating causes and consequences and in determining the value of consequence assignment for different diseases, the statistics that are measurable—gross levels of occurrence and mortality for diseases—have limited meaning.

17.3.4.5 Gross Levels of Disease Occurrence

Since source and degrees of voluntary and involuntary exposure for most levels of disease occurrence are not available, only gross levels of disease occurrence can generally be used. A table of death rates from all causes for recent 20-year periods is updated and reported annually in the U.S. Statistical Abstracts. All major disease categories are covered but are of little value for risk analysis.

17.3.4.6 Life Extension by Elimination of Disease

Using life tables, several investigators have determined the years that might be added to life expectancy if certain disease causes were removed. Schoen and Collins[16] have used California populations for 1950, 1960, and 1970, and the National Center for Health Statistics[17] has made similar calculations for 1960 data for the total U.S. population. Schoen and Collins investigated differences for males, females, blacks, and whites, as well as the total population. Life extension calculations from both sources are summarized in Table 17.9. Only the total population data from both sources are shown. These provide some means of quantitatively comparing disease risk with accidental risk, but the subjectivity of the consequence values for disease and accidents makes such comparisons relatively meaningless. Thus very little can be said at this time about the valuation of consequences from diseases.

Table 17.9 *Gain in Expectation of Life at Birth from Elimination of a Specified Disease*
(expressed in years added to life expectancy)

Disease	Total California Population			U.S. Population
	1950	1960	1970	1960
Respiratory tuberculosis	0.37	0.08	0.03	
Other infections and parasitic diseases	0.25	0.12	0.10	
Neoplasms	2.29	2.39	2.85	2.27
Cardiovascular diseases	17.75	14.72	17.50	10.97
Respiratory diseases—influenza and pneumonia	0.39	0.62	0.49	0.53
Digestive diseases	0.10	0.06	0.06	
Diabetes mellitus				0.22
Certain degenerative diseases	0.70	0.68	0.78	
Maternal mortality	0.04	0.02	0.01	
Certain diseases of early infancy	1.15	0.99	0.73	
Congenital malfunctions				0.36
Expectation of life at birth	69.24	70.75	71.97	

Sources: California data from Schoen and Collins;[16] United States data from National Center for Health Statistics.[17]

Table 17.10 *Summary of Absolute Risks Levels for Different Types of Risk*

Type of Risk		Class of Consequence					
Index	Description	fat/yr (1)	he/yr (2)	\$/yr (3)	\$/yr (4)	yr (5)	yr/yr (6)
	Naturally occurring						
1	Thresholds of concern (*I*)	1×10^{-10}				3×10^{-7}	4×10^{-9}
2	Catastrophic risk (*I*)	1×10^{-6}	5×10^{-6}	0.2		3×10^{-3}	4×10^{-5}
3	Ordinary risk (*I*)	7×10^{-5}	4×10^{-4}	3	1	0.2	2×10^{-3}
	Man-Originated						
4	Thresholds of concern (*I*)	1×10^{-11}				3×10^{-8}	4×10^{-10}
5	Catastrophic						
	Involuntary (*I*)	1×10^{-7}	5×10^{-7}	2×10^{-2}		3×10^{-4}	4×10^{-6}
	Voluntary (*V*)	2×10^{-6}	2×10^{-6}	0.4		6×10^{-3}	8×10^{-5}
	Regulated voluntary (*RV*)	3×10^{-5}	3×10^{-6}	0.4		6×10^{-2}	8×10^{-4}
6	Ordinary						
	Involuntary (*I*)	5×10^{-6}	3×10^{-5}	1		1×10^{-2}	2×10^{-4}
	Voluntary (*V*)	6×10^{-4}	3×10^{-1}	200		1	0.1
	Regulated voluntary (*RV*)	1×10^{-4}	6×10^{-2}	30		0.1	1×10^{-2}
	Man-triggered						
7	Catastrophic						
	Involuntary (*I*)	2×10^{-7}	1×10^{-6}	4×10^{-2}		6×10^{-4}	8×10^{-6}
	Voluntary (*V*)	4×10^{-6}	4×10^{-6}	0.8		6×10^{-3}	2×10^{-4}
8	Ordinary						
	Involuntary (*I*)	1×10^{-5}				3×10^{-2}	4×10^{-4}
	Voluntary (*V*)	1×10^{-3}				2	0.2
	Regulated voluntary (*RV*)	2×10^{-4}				0.2	2×10^{-2}

17.4 SUMMARY OF ABSOLUTE RISK LEVELS

Absolute risk levels for accidental risks are summarized in Table 17.10. This exercise and table demonstrate that different types of risk and classes of consequence cannot be compared as though they were identical risks. Each type of risk must be treated separately, and risk levels must be estimated on the basis of that risk alone or at least for similar risks in society.

No claim is made to the validity of the numbers given. Their primary purposes here are to illustrate the methodology needed to develop valid absolute risk levels, to show limitations in data acquisition and analysis, and to make a first cut at identification of meaningful classifications of risk. The existence of the numbers and their use in Part E do provide a first overview of the process. The values are shown to one significant figure and may well be in error by one or more orders of magnitude.

REFERENCES

1. CONSAD Research Corporation, "Consequences and Frequency of Selected Man-Originated Accident Events." Environmental Protection Agency Contract Report No. 68-01-0492. Washington, D.C., 1975.
2. W. Baldewicz, G. Haddock, Y. Lee, Prajoto, R. Whitley, and V. Denny, "Historical Perspectives on Risk for Large-Scale Technological Systems," UCLA-ENG-7485. Los Angeles: UCLA School of Engineering and Applied Science, November 1974.
3. Reactor Safety Study: An Assessment of Accident Risks in U.S. Commercial Power Plants," WASH-1400 (NUREG-75/014). Washington, D.C.: Nuclear Regulatory Commission, October 1975.
4. Saul Levine, personal communication, June 25, 1975.
5. Joseph M. Hans, Jr., and Thomas C. Sell, "Evacuation Risks—An Evaluation" (EPA-520/6-74-002). Washington, D.C.: Environmental Protection Agency, June 1974.
6. Saul Levine, *op cit.*, edited to reduce confusion stemming from references to a previous draft of *An Anatomy of Risk*.
7. Byron Bunger and John R. Cook, "Use of Life Table Methods to Compare Occupational Radiation Exposure Risks and Other industrial Risks." Presented at the meeting of the Health Physics Society, Buffalo, N.Y., June 30, 1976.
8. H. N. Sather, "Biostatistical Aspects of Risk-Benefit, The Use of Computing Risk Analysis," UCLA-ENG-7477. Los Angeles: UCLA School of Engineering and Applied Science, September 1974.
9. *Accident Facts*. Chicago: National Safety Council, 1973.
10. *Occupational Injuries and Illnesses by Industry, 1972*: Bulletin 1830, U.S. Department of Labor, Bureau of Labor Statistics, 1974, and unpublished data provided by the Bureau of Labor Statistics.
11. *Insurance Facts*. New York: Insurance Information Institute, 1973.

12. Christine Russell, "We're Caught in a Grim Game of Chemical Roulette," *Washington Star,* May 23, 1976.
13. Steven Golant and Ian Burton. *"Avoidance Response to the Risk Environment."* Department of Geography; University of Toronto, Department of Geography; Working Paper No. 6, 1969.
14. Harry J. Otway and J. J. Cohen, "Revealed Preferences: Comments on the Starr Benefit-Risk Relationships," IIASA RM 75-5. Laxenburg: Austria, International Institute for Applied Systems Analysis, March 1975.
15. Allen R. Wyler, Minoru Masuda, and Thomas H. Holmes, "Seriousness of Illness Rating Scale," *Journal of Psychosomatic Research,* Vol. 11, 1968.
16. R. Schoen and M. Collins, "Mortality of Cause, Life Tables for California 1950–1970." Demographic Analysis Section, Bureau of Health Intelligence and Research Review, California State Department of Public Health, 1973.
17. "United States Life Tables by Cause of Death: 1959–1961." Washington, D.C.: National Center for Health Statistics.

Part E

METHODOLOGICAL APPROACHES TO RISK ASSESSMENT

Chapter 18 develops a methodology for formally assessing risk and risk inequities. The methodology involves four steps: balancing direct gains and losses, balancing indirect gains and losses, ensuring that risks have been cost-effectively reduced, and ensuring that risk inequities are examined. In the final step risk referents are derived against which measured levels of risk from a specific undertaking can be compared. The methodology is then tested against two existing systems, air and rail transportation, to evaluate its validity.

Chapter 19 applies the methodology to four fuel systems for supplying electrical energy; namely, oil, natural gas (including liquefied natural gas), coal, and uranium-burning nuclear energy. Each system is evaluated independently and the results are interpreted. These applications are aimed at demonstrating the methodology: they do not provide a "final" assessment of risks of these systems.

Finally, Chapter 20 presents further thoughts on formal methods of risk assessment, along with some conclusions regarding future efforts.

18

Determination of Acceptable Levels of Societal Risk

Absolute levels of risk, as initially developed in Chapter 17, provide a basis for establishing risk references for determining acceptable levels of risk for technological projects that impose new risks on society. To use these risk references, there must be a means to relate them to the decision process for evaluating specific projects and undertakings. New federal projects that are determined to have *environmental* impact are subject to the filing of an environmental impact statement under the National Environmental Policy Act of 1969. A part of the impact statement requires a cost-benefit analysis of the impact on society. In general, such analyses have been primarily economic exercises augmented by an identification and qualitative evaluation of the impact of noneconomic health, safety, environmental contamination, ecological, and aesthetic factors. Risk has not been treated specifically in most cases, especially those involving accidental events.

The need to address risk specifically in environmental impact statements or other evaluations of new or existing technological programs, as well as providing a framework for such analysis, has been the motivation for this book. The development of a methodological approach for determining acceptable levels of risk is a natural progression from a basic understanding of risk factors.

A methodology is an open ended procedure, and the validity of any given methodology can be proved only by the pragmatic acceptance of its utility in solving actual problems, and the degree to which the results of its application are adapted. *No* methodology has yet been developed that is widely accepted or has been particularly useful in assessing risk. Thus an attempt is made to establish the present methodology because such a methodology ought to exist, although different approaches will have varying levels of utility. Methodological development is aimed at "proving an existence theorem," not at gaining acceptance of specific methods.

However, the development of a specific method will provide a straw man for others to knock down or to use as a point of departure.

18.1 PREREQUISITES FOR A METHODOLOGY FOR RISK ACCEPTANCE

Before undertaking the development of a methodology for risk acceptance, a number of questions must be asked. Is there a need for a methodology for risk acceptance? How and where shall it be used? What methods are already available? What alternative approaches can be employed?

18.1.1 Need for a Methodology

Man is naturally risk aversive, but he is willing to take risks to achieve specific benefits when the choice is under his direct control. When the risk is imposed by man or nature without direct benefit, however, risk aversive action dominates. The content of the news media, a reflection of society's news preferences, make it obvious that society is more concerned with adverse consequences than with benefits. Disaster reports and disagreeable news far outweigh the achievements and beneficial events.

The risk aversiveness of society, coupled with increasing awareness of new risks resulting from side effects of new technology, has focused increased attention on technological risk. Such awareness and concern are probably irreversible, since the knowledge base for technology assessment and risk identification is available to everyone. Consideration of societal risk in all technological decisions is rapidly becoming accepted, and increased regulatory attention is devoted to risk assessment. A methodological approach to assure reasonable perspective in assessing risk is necessary if the regulatory apparatus is to work in a visible way. Kates[1] has postulated two different theoretical models for regulatory approaches: the rational and bureaucratic models.

> In the rational model, the common assumptions are as follows: There exist known and shared objectives or values—for example, a goal to protect the health of the people. Given such a goal, decision-makers have sets of alternative policies—for example, ambient air pollution standards, requirements for seat belts in autos, or coliform-count thresholds for permissible swimming. For each alternative, there is a set of known or inferred consequences. The choice, then, of which alternative to codify is a choice of which acts to maximize the objective, sometimes subject to some specified constraints. This model, of course, is rarely found in practice, but an elaborate normative theory exists for its application (Marschak[2]).

In contrast to the rational model, the bureaucratic model assumes for a bureaucracy a life of its own, one not necessarily related to its function. It is the organizational requirements for survival or growth rather than its function that dictates its policy. There are many variants of the model, but in most of them, decisions arise from conflict or power situations with bureaucratic sovereigns, rivals and allies (Downs[3]). In such context, studies of bureaucratic standards-setting have either focused on how the regulatees control the regulated (Kohlmeier[4]); how the interaction takes place between varied interests (Chicken[5]); or on emphasizing particular processes, modifications of the rational model (Simon[6]); incremental decision-making (muddling through) (Lindblom[7]); or crisis response (Hart[8]).

Some methodological structure for bridging these extremes is required if the regulatory process for risk assessment is to work.

18.1.2 Use of a Methodology in the Regulatory Risk Assessment Process

Purely voluntary risks (i.e., those entailing only direct gains and direct losses to the risk agent) need not and probably should not be regulated by government. Unfortunately, there are relatively few conditions for which only voluntary risks occur. There is usually some indirect risk imposed on others who neither directly nor indirectly share in the gains. For example, the act of suicide has a consequence not only to the individual who dies but to his survivors, his insurance company, his creditors, and so on. Furthermore, the benefit of the act to the individual (if death may be thought of as a benefit insofar as it relieves one of the problems of living) is situational and, in some cases, irrational. Both the state and the church attempt to make this act as unattractive as possible. Thus government is involved in regulating voluntary as well as involuntary risks to some extent.

This occurs when indirect losses associated with a voluntary risk condition affect significant numbers of the population or identifiable recipients indirectly; regulatory action to ameliorate the risk inequity becomes necessary. Thus the purpose of a risk assessment methodology is not to balance direct gains and losses, but to ameliorate inequities in balancing indirect gains and losses. Risk assessment is used only after a favorable balance of direct gains and direct losses has been made,* and involuntary and regulated voluntary risks are considered. Thus a risk assessment methodology is neither a cost-benefit analysis nor a substitute for such analysis. Its purpose, recognizing that some levels of risk always exist, is to determine when imposed risks on segments of a society are low enough to be acceptable. There is little question that the balancing of indirect gains to

* Or in the case of a subsidy, when indirect gains exceed indirect losses.

society against imposed risks is a requirement in risk acceptance. Higher levels of risks may be acceptable under these conditions; moreover, total societal equity is rarely achieved in practice.

18.1.3 Existing Approaches to Risk Assessment

Kates[1] has identified four basic approaches for risk evaluation: risk aversion, balanced risk, cost-effectiveness of risk reduction, and risk-benefit balancing.

18.1.3.1 Risk Aversion

> Aversive risk methods are those of avoidance or minimization, with little consideration to comparison with other risks and benefits. These methods are deeply embedded in culture as taboos, in society as standards and regulations, and in a person as avoidance and aversion. . . .
>
> Societies as well as persons make aversive risk judgments with little or no reference to other risks or benefits. Much regulatory activity for safety or public health is intended to encourage maximum aversion. All zero tolerance standards for pollutants or impurities, as well as standards at the supposed dose-consequence threshold, are examples of aversive risk evaluation. . . .
>
> Whatever the disparate rules of aversion, its employment is ubiquitous. Most people, everywhere, practice aversions—making both absolute and relative rankings of risks to be avoided. Yet such a calculus may be only remotely related to the frequency of events or the magnitude of consequences. [An] important form of aversion may be independent of specific hazards; namely, an aversion to risk itself. The aversion to risk is defined generally in terms of the willingness to accept less gain or to pay more for a reduction of risk. Risk aversion practices are widespread and provide the behavioral basis for insurance and other risk sharing activities.[1]

Systematic safety methods[9] involving hazard identification and evaluation and subsequent danger reduction are bases of safety practices in the United States and are further examples of risk aversion.

18.1.3.2 Balancing Risks

Balanced risk methods seek to compare and equalize risk consequences. A generalized form for comparison is necessary and the consequence scales used in Table 9.1 provide a first approach for generalizing these comparisons.

> The most common form of risk balancing is to compare frequencies of mortality, morbidity, or damage in order to encourage some desired action or reveal some inconsistency. Risk exposures may be quite uneven between societies, within a society, within a single lifetime, and even within a day.[1]

Medical practitioners often balance risks. The insertion of a catheter into the heart bears a risk of about 1 in 2000 of causing death. The knowledge derived of blood flow in the heart provides therapeutic strategies that can save the life of a larger proportion of cardiac patients.

Two aspects of balancing risks are important. First, risks from a specific undertaking may be compared with similar risks experienced by society in the form of risk references or referents. Second, a fundamental notion of a balance is that risk thresholds exist—some level of risk that is acceptable which is greater than zero.

18.1.3.3 Cost Effectiveness of Risk Reduction

As described in Section 5.3, the cost effectiveness of risk reduction extends the question of how much risk is acceptable to how much society is willing to pay to avoid a risk. This is *not* the value of a life, but the cost of avoiding a possible statistical death in society as described in Chapter 13. Linerooth[10] has made an extensive survey of the methods for assigning values to loss of life. A survey by Rausa[11] indicates the following:

1. Public expenditure for risk aversion varies from $100 for automobile seat belts to $10 million for removal of strontium-90 from milk during fallout. These values are for averting a death.
2. Court and insurance compensation ranges from $10,000 to $3 million for loss of life.
3. Willingness to pay to avert a death ranges from about $20,000 for kidney failure support to $2 million for radiation risk.*

As discussed in Section 14.1, cost effectiveness of risk reduction is one step of a cost-benefit analysis. An overall balance of costs and benefits to society is made initially, if only on a gross value judgment basis. Then further benefit (reduction in risk) is sought. In other words, an implied overall cost-benefit balance precedes a cost effectiveness of risk reduction analysis. If the risks already exist, the only benefit is risk reduction.[12] In all such models, the key question is, how much spending is enough?

18.1.3.4 Benefit-Risk Analysis

When a formal balancing of benefits and risks is made before cost effectiveness of risk reduction criteria are applied, benefit-risk analysis results. In this variant of benefit-cost analysis, risk is a surrogate for social cost. The shortcomings of formal cost-benefit analysis have already been discussed

* Based on the U.S. Nuclear Regulatory Commission cited criteria of $1000 per man-rem of exposure.

(Section 14.1), but the object is to balance a direct benefit of the activity against the risk, to determine how much risk reduction is adequate.

Such a methodology is desirable because it gives equal consideration of gains and losses, but difficulties impeding such analyses[1] probably prevent anything more precise than a gross balance* approach to implementation. Any attempt closer to a complete rational-comprehensive approach introduces new biases:

> When comparisons are made between benefits and risks or benefits and costs, further biases are introduced. Whole classes of benefits or costs of risks are, by definition and practice, more uncertain and more difficult to estimate, measure, or even describe. This lack of symmetry tends to bias analysis to the immediate, the definite, the economic, and the developmental. Altouney,[13] for example, showed how in 100 conventional U.S. benefit-cost analyses of water supply projects, the benefits were consistently overestimated and the costs underestimated. The benefits of many economic development projects appear definite and immediate, the risks appear indefinite, fuzzy, and amorphous. On the other hand, between amenity and disamenity, there is a bias to the disamenity. Consensus can be more easily marshalled as to the ugly, dangerous or objectionable—rather than the beautiful, the safe or desired.[1]

The implementation of detailed benefit-risk analysis can only serve for identification of critical parameters, and as pointed out by Crocker,[14] a benefit-cost analysis *of benefit-cost analysis* quickly leads to an understanding of the limitations of such methods.

18.2 FORMULATING A METHODOLOGY FOR RISK EVALUATION

Risk assessment involves both risk determination and risk evaluation. A methodology for risk evaluation assumes that risks have been previously quantified or may be quantified by other efforts as a prerequisite for risk evaluation.

The initial step in developing a methodology based on a gross balance approach is to formulate structural model.† The model envisioned here attempts to use aspects of all four risk assessment methods described earlier in a number of sequential steps.

1. Direct gain-loss analysis.
2. Indirect gain-loss analysis.
3. Cost effectiveness of risk reduction.
4. Reconciliation of risk inequities.

* See Chapter 12.
† See Chapter 12.

18.2.1 Direct Gain-Loss Analysis

In direct gain-loss analysis we compare direct gains against direct losses; this classical cost-benefit analysis is usually made by an individual or institution undertaking a project or program. The individual or institution receives the benefits and accepts the costs, and makes an analysis that is primarily economic. Voluntary risks are taken to achieve specific benefits.

If this balance is negative, the motivation disappears for going ahead with the project or program. It will probably be dropped unless the balance is changed or new factors such as subsidization are introduced. A favorable balance will provide incentive for the program. Institutional barriers involving legal constraints, taxes and related incentives, and public opinion are factors that are not always quantifiable.

The responsibility for carrying out this analysis lies with the individual or institution, private or governmental, undertaking the project for direct gain. The analysis may be formal or informal and is primarily a tool for the sponsor. It is an open-ended analysis, since additional direct costs from subsequent steps will affect the balance, and new factors must be accounted for as they occur. Such a process is dynamic; and the sponsor will continuously review his position through its completion, perhaps only for economic reasons.

18.2.2 Indirect Gain-Loss Analysis

The indirect societal gains of a proposed activity must be balanced against the indirect societal losses of the activity. Risks are one aspect of the societal losses. This balance is the type of overall cost-benefit analysis sought in environmental impact statements under the National Environmental Policy Act of 1969 and is a goal of technology assessment activities. The need to accommodate local balance, national balance, and world balance often results in qualitative value judgments as opposed to numerical balances. This is chiefly due to the difficulties of measuring intangible values and obtaining adequate data. However the qualitative balancing often is precise enough to allow most neutral parties to agree on a ranking of four different levels: (1) favorable—the balance overwhelmingly favors benefits over costs at the societal level; (2) marginal—the balance is slightly positive or even in considering benefits over costs; (3) unfavorable—costs generally outweigh the benefits; and (4) unacceptable—costs far outweigh societal benefits.

At the government level, a sponsoring agency is usually responsible for the preparation of such analyses, which become part of the public domain.

18.2.3 Cost Effectiveness of Risk Reduction

Since society is generally risk aversive, the risk in obtaining a given indirect gain (benefit) must be minimized to the extent feasible, even for favorable indirect gain-loss balances. The costs of risk reduction are in part direct costs and must be factored back into the direct and indirect gain-loss analyses.

The crucial matter to be determined in this risk aversion process is when the risk reduction is low enough. In considering the direct gain-loss analysis, the concept of "as low as practicable" is one limiting consideration. One point at which an ALAP limit can be obtained is described as follows: when the incremental cost per risk averted is such that a very large expenditure must be made for a relatively small decrease in risk as compared to previous risk reduction steps. This implies a relative risk for the activity in question. Another alternative involves defining the average practice of the best industry processes. Regardless of which definition is selected, quantification of this level is affected by the level of balance in the first two steps. That is, when the benefits are overwhelming, there may be a tendency to spend more to reduce risk than otherwise. In a marginal situation, costs to reduce risk may begin to tip the decision balance for going ahead with the activity one way or the other, whereas favorable cases may call for spending on risk reduction to be as safe as possible* to buy public acceptance. Certain aspects of control of planned releases of iodine from nuclear power plants are considered by some to be examples of the latter situation. The marginal case is illustrated by particular coal mines whose profits have been lowered below the break-even point as a result of costs incurred in pursuit of conformity to new safety laws.

Thus the "as low as practicable" concept limit is arbitrary, as is that for risk acceptance. Some other reference is required to determine when cost effectiveness of risk reduction has reached an acceptable level. This methodology of risk evaluation relies on the development of such a reference, based on acceptable levels of inequitably imposed risk. Up to this step, the three preceding steps are well described by present practice.

18.2.4 Reconciling Identified Risk Inequities

When the overall indirect gain-loss analysis is made and is favorable, various inequities may still exist for specific value groups. Those who assume the risks may not always receive the benefits, or the risk may not be evenly distributed among the benefit recipients. If this condition occurs, the risk must be identified and the nature and type of risk must be ascertained.

* "Safe as possible" implies implementation of a level of safety well beyond the ALAP level.

These risks can then be compared against the level of risk that society is experiencing for similar types of risk.* These risk references are of the type of absolute risk level developed in Chapter 17, but they must be converted to risk referents before making comparisons. The latter step involves a societal value judgment of how much additional risk society should assume to obtain indirect benefits, and it is a function of the indirect gain-loss balance. Since risk references are static and generally historic, dynamic risk aversive behavior must also be considered in establishing the risk referent. This involves the systemic degree of control of future risks that is built into the new project.

Thus the two steps in establishing a risk referent involve a benefit-risk balance in the first case and risk aversion in the second case. Finally, the risks of the new program must be balanced against similar indirect benefits to derive a net risk. For example, the life-extending aspects of radiation therapy may be balanced against the increased somatic risk of cancer induction by the therapy. The net risk that results from such balancing is the risk to be evaluated, and it may be negative (a probable gain). If all net risks of each type are negative or zero, there is no risk inequity, and a risk acceptance comparison is not necessary. Risks can be balanced only if measured on the same measurement scale. This is not possible very often, and risk balancing is of limited application. It should be used when feasible.

The methodology prescribed for risk evaluation permits reconciliation of inequities based on acceptable levels of risk in the form of risk referents, after the initial three steps are implemented.

18.2.5 Use of Value Judgments

Assuming that all parameters in the four steps will be relatively imprecise, value judgments must be used in the absence of hard information. Several classifications of value judgments must be identified, beginning with private and public value judgments. The first is the type made by the sponsor and involves voluntary risk considerations. The second is related to the public interest and consists of two classes: technical and social value judgments. Technical value judgments are made by technical experts and cover economic analyses, risk determination for a project, and balancing of risks. Social value judgments are made by society or, if in the public domain, by due process of law. Such value judgments involve qualitative balancing of indirect cost and benefits and the degree of risk aversion that is acceptable.

Technical value judgments are made by experts; societal value judgments

* Activities causing risk are not compared. The risks of activities are compared with similar risks in society, independent of source.

are made by all participants. Separation of these types of judgment is thus necessary, and *all* value judgments should be made in a visible manner.

In this methodology, values assigned to the indirect gain-loss balance may involve both types of value judgment. Judgments in developing a risk referent, the amount of new risk one will assume for achieving an indirect benefit to all society, and the degree of control that is acceptable, are societal. Determination of the level of risk for a given technical project and the degree of controllability implemented are technical value judgments.

18.3 ESTABLISHING RISK REFERENTS FOR THE METHODOLOGY

18.3.1 Multiple Risk Referents

The absolute risk levels for involuntary risk $A(I)_{i,j,k}$ and for regulated voluntary risk $A(RV)_{i,j,k}$ provide a basis for analyzing each type of risk on its own merit. Furthermore, risks can be balanced against similar indirect gains at this level to develop net risk values for the activity under consideration. Thus each type of measured risk for the activity is compared with each similar type of risk as a referent. When all measured risks are lower than their risk referent counterparts, the residual inequitable risks are acceptable. If any type or types of risk exceed the referent(s), the inequities have not been reconciled. This does not mean that the project is permanently unacceptable, only that further risk reduction is necessary to make it acceptable. The project becomes unacceptable when the sponsors are unwilling to go ahead with the project because to incur additional costs would change the direct gain-loss balance, yielding an unfavorable situation. In any case, the project would have to meet acceptable levels of risk to proceed.

The absolute risk levels given in Chapter 17 represent one method of classification of risks based on data available to the author for the risk factors identified. Other classifications will not invalidate the methodology as long as consistency in application occurs.

18.3.2 Risk Proportionality Factor Derivation from Risk References

The absolute risk levels represent the author's best estimate of the level of risk of each type to which society, historically, has been subjected. Further research and analysis will be needed to verify or modify these values, which are the summation of all risks of a type and are exclusive of new activities.

Thus these levels (or their updated values) are risk references, representing the summation of involuntary and regulated voluntary risks to society. Voluntary risk is not considered.

18.3.2.1 Risk Proportionality Factor

In using risk references we assume that some additional proportion of the total societal risk would be acceptable to society for some very desirable indirect benefit. The establishment of that proportion is a societal value judgment, and the result is called a risk proportionality factor.

The risk proportionality factor is based on the concept that any activity undertaken by man produces some inequity in the balancing of benefits and costs to different groups in society. For example, an extremely beneficial program to society, such as elimination of cancer as a cause of death, might very well decrease the life span of those not susceptible to cancer, since the resultant lower death rate might increase the age of the population and competition for scarce resources, including food. Thus we must ask: how much increase in risk will society accept for a new beneficial activity?

First, the new risk must be compared to the totality of similar risks to which the involuntary risk takers are subject, since absolute risk by itself has little meaning until one reaches the threshold of uncontrollable risk, such as the probability of the sun exploding in one's lifetime or the risk of being killed by a meteor. There seems to be little question that if a single activity doubled man's total net involuntary risk, that activity would most likely be unacceptable. Strictly as a value judgment, however, an extremely beneficial activity to society as a whole might be acceptable if the increase of net involuntary societal risks were less than 10% of the total involuntary risk level. This number may then be used as a top level for the risk proportionality factor for involuntary risk.

Regulated voluntary risks are a somewhat different case. If the risks are occupational and/or there are no reasonable alternatives available, one may expect to assume a greater proportion of risk; perhaps, equal to all other risks.

> *Societal Value Judgment 1.* Determine the risk proportionality factor upper limit for a very favorable indirect gain-loss balance for involuntary risk.
>
> *Societal Value Judgment 2.* Determine the risk proportionality factor upper limit for a very favorable indirect gain-loss balance for regulated voluntary risk.

For the purpose of a straw man, the author has made a personal value judgment of .1 for the risk proportionality factor for involuntary risk $P(I)$

and a value of 1.0 for the risk proportionality factor for regulated voluntary risk $P(RV)$.*

There are two variations in voluntary risk. One set of these risks involves the operator of a technological system (e.g., a commercial airline). The second set involves the population that is voluntarily exposed to risk on some informed basis, with reasonable alternatives available (e.g., airline passengers). The first set of voluntary risks is of concern to the operator, the second to society as a whole, since people want to be *aware* of the relative safety of their choice. Thus evaluation of this type of voluntary risk is of interest.

The two types of risk are not unrelated, as illustrated by the airline case. Should passengers decide not to fly for safety reasons, the profit and loss statements of the airlines will be affected. Thus it is of interest to airlines to be risk aversive for their clientele.

Although voluntary absolute risk levels are used for examing the second type of voluntary risk, a risk proportionality factor of .1 is proposed.† If the population is risk aversive, it will not accept large risks if alternatives are available.

18.3.2.2 Risk Proportionality Derating Factors

A second set of social value judgments is now in order to determine how the risk proportionality factors would be derated for less favorable indirect gain-loss balances. Based on the evaluation of precision of cost-benefit analysis made in Section 14.1, it is evident that resolution of one part in three or perhaps one part in five is the best that can be expected from an indirect gain-loss balance. This involves a technical value judgment based on the specific methods used and the availability and precision of data. For purposes of illustration the author assumes that a precision of one part in five is possible such that the following conditions of indirect gain-loss balance are discernible:

1. Favorable balance.
2. Marginally favorable balance.
3. Indecisive balance.
4. Marginally unfavorable balance.
5. Unacceptable balance.

Next we need to know how the risk proportionality factor should be

* A factor of 1.0 represents a doubling of existing risk for the new activity. A factor of 0.1 is 10% of the present risk.

† The first type of voluntary risk is not of concern here.

derated for each of the foregoing balances for involuntary and regulated voluntary risk. These are societal value judgments.

Societal Value Judgment 3. Determine the form of a derating function for the risk proportionality factor for involuntary risks as indirect gain-loss balances become less favorable.

Societal Value Judgment 4. Determine the form of a derating function for the risk proportionality factor for regulated voluntary risks if different from that for involuntary risk.

The author has selected the following derating functions as straw men.

Balance	Involuntary Risk Derating Value, $D(I)$	Regulated Voluntary* Risk Derating Value, $D(RV)$
Favorable	1.0	1.0
Marginally favorable	0.1	0.2
Indecisive	0.01	0.1
Marginally unfavorable	0.001	0.02
Unacceptable	0.0001	0.01

The form of these functions is shown in Figure 18.1 for the left-hand ordinate scale. The slope of the involuntary risk derating function is twice that of the regulated voluntary function on the logarithmic ordinate scale. The use of different functions is based on Otway and Cohen's[15] analysis of Starr's data for risk-benefit functions that exhibit different slopes for voluntary and involuntary risks. The rationale is that there is less inequity in regulated voluntary risk than for involuntary risk, hence a lower derating function. Figure 18.1, using the right-hand ordinate scale compares risk proportionality derating factors with the slope of risk-benefit functions of Starr[16] and Otway and Cohen.[15] To make this comparison, the highest value shown by Otway and Cohen (a benefit of $3000) is made equivalent to a favorable balance, and each order of magnitude below this is made equivalent to a value on the rank scale. The values chosen by the author are much more optimistic than the analysis of either Starr or Otway and Cohen would warrant.† This implies a bias toward higher levels of acceptable risk in the form of risk referents, and these functions may not be adequately risk aversive.

The final risk proportionality function is the product of the risk proportionality factor P and the risk derating value D. However another risk aver-

* This regulated voluntary risk derating function is also used for voluntary risk.

† Assuming the validity of scales for the abscissa.

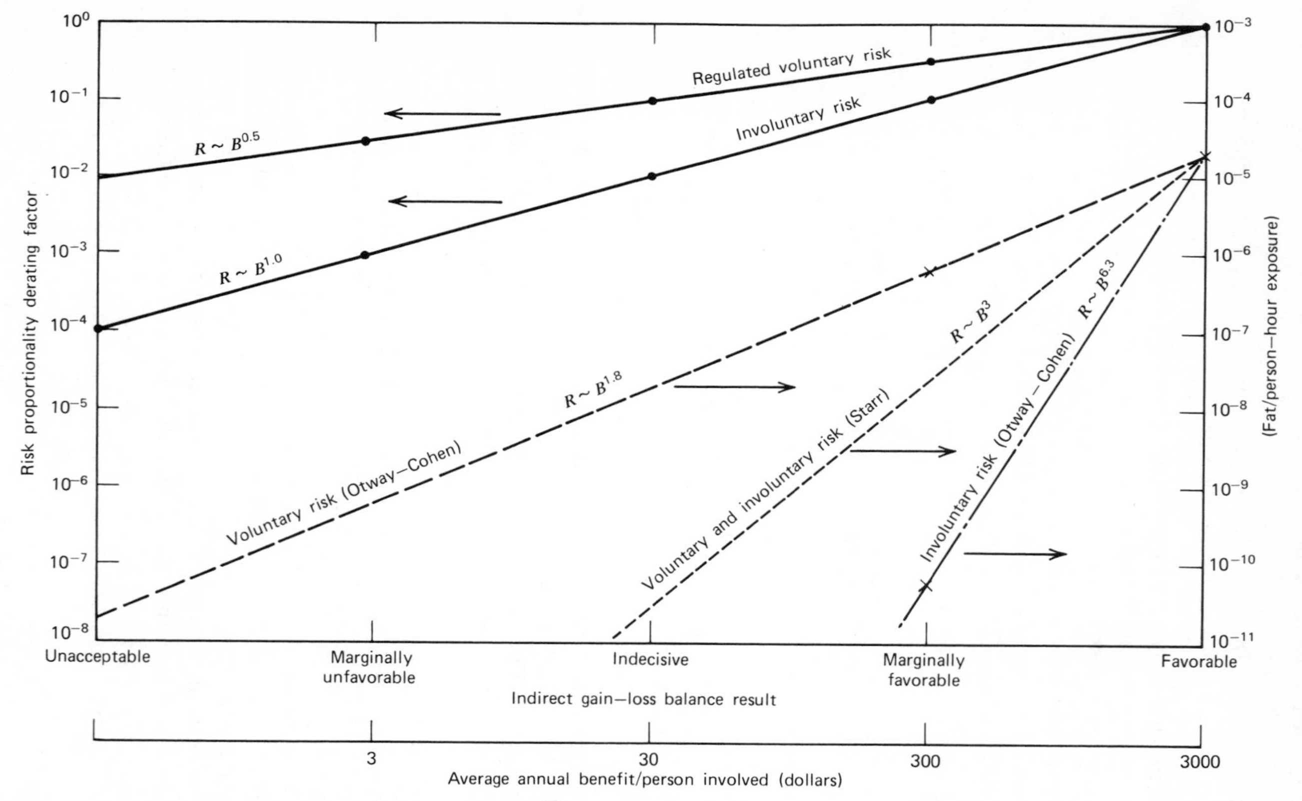

Figure 18.1. Slope of risk proportionality derating functions. Risk proportionality derating functions for involuntary and regulated voluntary risks in terms of the author's straw men, using the left-hand ordinate scale and the rank scale (upper scale) shown for the abscissa. The slopes of these functions are compared with those by Starr and those derived by Otway and Cohen using Starr's data. The comparison is made by assuming that a favorable balance is equivalent to a \$3000 benefit and a marginally favorable balance to a \$300 benefit, and so on, using an interval abscissa scale (lower scale) and the right-hand ordinate. The author's straw men are seen to be less pessimistic than either Starr's or Otway and Cohen's analysis; and, assuming that the choice of dollar values on the abscissa is valid, the involuntary risk function if $R \sim B^{1.0}$ and the regulated voluntary risk function is $R \sim B.^{0.5}$

sive factor must be taken into account as well; namely, the degree to which risks will be controlled in the future.

18.3.2.3 Determining Degree of Systemic Control

The level of risk that society is experiencing at any one time cannot necessarily be said to be acceptable to society. Society may be dissatisfied with the present level of risks, and in this case will want to reduce the risks, (never raise them unnecessarily). Although different expenditures can be more or less effective for different activities in reducing risks, the concept of degree of systemic controllability, as discussed in detail in Chapter 7 may be quantified in the form of a risk conversion factor as formulated in Chapter 16. The value of C_c derived in Section 16.2.4 is used to consider controllability. Several societal value judgments are involved.

> *Societal Value Judgment 5.* The degree of controllability of risks from a technological system is an important risk aversive factor in risk acceptance.
>
> *Societal Value Judgment 6.* If societal value judgment 5 is valid, develop a risk controllability derating function.

The author has used the values derived in Table 16.8 as straw men for the risk controllability derating factor C_c, which is the product of the four factors listed in Table 16.8 and reproduced in Table 18.1.

18.3.2.4 Derived Risk Referents

Risk referents for a given system are derived from the societal risk references $A(I)_{i,j,k}$ and $A(RV)_{i,j,k}$ by applying the system derating factors in terms of risk proportionality factors, risk proportionality derating values, and the degree of systemic control. The risk referents $R(I)_{i,j,k}$ and $R(RV)_{i,j,k}$ are of the form

$$R(I)_{i,j,k} = A(I)_{i,j,k} \times P(I) \times D(I) \times C_c \tag{18.1}$$

$$R(RV)_{i,j,k} = A(RV)_{i,j,k} \times P(RV) \times D(RV) \times C_c \tag{18.2}$$

$$R(V)_{i,j,k} = A(V)_{i,j,k} \times P(V) \times D(RV) \times C_c{}^{*} \tag{18.3}$$

These risk referents that take into account balancing of indirect gains and losses and risk aversion are the acceptable levels of the inequitable residue of societal risk. The actual levels of risk from a given system are compared against the risk referent for each type of risk to determine overall system acceptability for risk.

* The derating function $D(RV)$ is also used for voluntary risk.

Table 18.1 *Control Factors and Assignment of Cardinal Values of Desirability*[a]

Control Approach[b]		Degree of Control		State of Implementation		Control Effectiveness	
Factor	C_1	Factor	C_2	Factor	C_3	Factor	C_4
Systemic Control	1.0	Positive	1.0	Demonstrated	1.0	Absolute	1.0
Risk management system	0.8						
Special Design features	0.5			Proposed	0.5	Relative	0.5
Inspection and regulation	0.3	Level	0.3				
		Negative	0.2				
No control scheme	0.1	Uncontrolled	0.1	No Action	—[c]	None	—[c]

[a] This table is identical to Table 16.8 and reproduced here for convenience. See Chapter 16 for discussion.
[b] Each approach is in inclusive of those below it, except for "no control."
[c] These levels are included in "no control" and "uncontrolled" by definition.

18.3.2.5 Precision of Risk Referents and Risk Acceptance Comparisons

The precision of risk referents is limited as a result of the societal value judgments involved. Generally, risk referents should not be shown more precisely than one significant figure unless more are necessary to display differences between risk referents. In any case, a margin of one order of magnitude difference in the comparison between actual risks and risk referents should be considered acceptable. Conversely, if the actual levels of risk exceed the acceptance level by more than an order of magnitude, the activity is deemed unacceptable. In this case, two options remain. The first is to drop the activity and not implement it. The second is to apply higher degrees of control to the activity to reduce the risks to society. Of course this will change the gain-loss balances. In addition, the system may be forced to operate at levels that stretch technology to the limits. Actions, such as demonstrating rather than merely proposing positive control, can also be used to change the level of acceptability. However a new proposed activity may be limited to "proposed control." The point is, failure to achieve a level of acceptability does not necessarily mean that the activity should be discarded forever; further steps must be taken to reduce the risks or change the risk-cost-benefit balance, however. When this is accomplished, the system may then achieve an acceptable level of risk.

18.4 SUMMARY OF THE METHODOLOGY FOR RISK EVALUATION

The total methodology for risk evaluation envisions the following steps.

1. Direct gain-loss balance—made by the project sponsor or assumed to have been made. Involves voluntary risk only. Both variations of voluntary risk must be considered.
2. Indirect gain-loss balance—made by the governmental institution proposing the action, but review is needed to determine the gross level of balance (i.e., favorable, marginally favorable, etc.).
3. Determine the cost effectiveness of risk reduction for risk control alternatives for the project at hand.
4. Apply methodology for reconciling risk inequities to determine when cost effectiveness of risk reduction is adequate and feed back new direct and indirect costs into steps 1 and 2.
   a. Set up methodology for reconciling risk inequities and evaluate the six societal value judgments through formal review processes.

b. Develop absolute risk levels for different types of risk from historic data as societal risk references.

c. Derive risk referents for the specific project by applying societal value judgments for risk proportionality and degree of control.

d. Estimate the risks caused by the project, a technical analysis with technical value judgments.

e. Balance risks and probable gains of the same type to develop net risk levels.

f. Compare net risk levels for each type of risk with its appropriate risk referent.

i. If all risk levels, other than voluntary, do not exceed corresponding risk referents by more than one order of magnitude, the risks are in the acceptable range.

ii. For any risk exceeding its risk referent by more than an order of magnitude, apply further risk reduction in cost-effectiveness sequence. Apply incremental costs back into steps 1 and 2 and revise the balances as warranted.

Steps 4*a* and 4*b* must be implemented initially to set up the methodology, then used for all further cases in subsequent steps.

18.5 VALIDATION OF THE METHODOLOGY

One method of checking the validity of the methodology and the author's straw man value judgments as well has been suggested by Chauncey Starr.[17] This is to test the methodology on technological systems already accepted by the public, such as air travel and railroad transportation systems. These tests also serve to illustrate the application of the methodology.

One value judgment made for the present systems is that their existence implies a favorable indirect gain-loss balance and the risk proportionality derating value is unity. The risk references developed in Chapter 17 are also used for these tests.

18.5.1 Commercial Air Travel

Commercial passenger air travel in the United States is a closely regulated industry; the Federal Aviation Administration exercises demonstrated positive systemic control of risk, as shown previously in Figure 8.5. This control exists on both an absolute and a relative basis. The supersonic transport is excluded from this consideration of commercial aviation. Contribution to air pollution from ordinary jets, which occurs primarily during ground and low altitude operations, contributes less than 1% to ambient air levels and may be ignored. The main contribution to risk is from accidental events.

Crews on aircraft are primarily subject to regulated voluntary risk. Accidents, exclusive of passengers and crew, are considered involuntary when they do not involve occupational situations. This is due partly to the tight safety regulations maintained by the Federal Aviation Administration and partly to the absence of alternatives to rapid transportation.

18.5.1.1 Derivation of Risk Referents

All accidents for aircraft are considered to be man-originated. Pilot error is still the main cause of accidents in bad weather, with mechanical malfunction the second most prevalent cause. The appropriate risk references from Table 17.10 are as follows:

1. Man-originated catastrophic risk.

$$A(I)_{5,1} = 1 \times 10^{-7} \text{ fat/yr}$$
$$A(I)_{5,2} = 5 \times 10^{-7} \text{ he/yr}$$
$$A(I)_{5,3} = 2 \times 10^{-2} \text{ \$/yr}$$
$$A(V)_{5,1} = 2 \times 10^{-6} \text{ fat/yr}$$
$$A(V)_{5,2} = 2 \times 10^{-6} \text{ he/yr}$$
$$A(V)_{5,3} = 4 \times 10^{-1} \text{ \$/yr}$$
$$A(RV)_{5,1} = 3 \times 10^{-5} \text{ fat/yr}$$
$$A(RV)_{5,2} = 3 \times 10^{-6} \text{ he/yr}$$
$$A(RV)_{5,3} = 4 \times 10^{-1} \text{ \$/yr}$$

2. Man-originated ordinary risk.

$$A(I)_{6,1} = 5 \times 10^{-6} \text{ fat/yr}$$
$$A(I)_{6,2} = 3 \times 10^{-5} \text{ he/yr}$$
$$A(I)_{6,3} = 1 \times 10^{0} \text{ \$/yr}$$
$$A(RV)_{6,1} = 1 \times 10^{-4} \text{ fat/yr}$$
$$A(RV)_{6,2} = 6 \times 10^{-2} \text{ he/yr}$$
$$A(RV)_{6,3} = 3 \times 10^{1} \text{ \$/yr}$$
$$A(V)_{6,1} = 6 \times 10^{-4} \text{ fat/yr}$$
$$A(V)_{6,2} = 3 \times 10^{-1} \text{ he/yr}$$
$$A(V)_{6,3} = 3 \times 10^{1} \text{ \$/yr}$$

Since air travel is an existing system, it is assumed to have a favorable indirect gain-loss balance. Thus

$$P(I) = .1 \qquad D(I) = 1$$
$$P(RV) = 1.0 \qquad D(RV) = 1$$
$$P(V) = .1 \qquad D(RV) = D(V) = 1$$

Positive systemic control is demonstrated on both an absolute and a relative basis for air travel. Thus the factor for degree of controllability C_c is unity:

$$R(I)_{i,j} = 0.1 \times 1.0 \times 1.0 \times A(I)_{i,j} = .1\ A(I)_{i,j}$$
$$R(RV)_{i,j} = 1.0 \times 1.0 \times 1.0 \times A(RV)_{i,j} = A(RV)_{i,j}$$
$$R(V)_{i,j} = 0.1 \times 1.0 \times 1.0 \times A(V)_{i,j} = .1\ A(V)_{i,j}$$

The resulting risk referents for each case appear in the left-hand column of Table 18.2.

18.5.1.2 Derivation of Measured Risk Levels

There were two principal sources of data. First, data for 1973 and 1974 are taken directly from the National Transportation Safety Board (NTSB) Annual Review of Aircraft Accident Data[19, 20] for all cases except catastrophic involuntary risk. Since no on-the-ground fatalities or injuries involving the general population occurred for these two years, a 21-year coverage of catastrophic involuntary risk is taken from the CONSAD data summarized in Appendix A. The basic data (Table 18.3) include 42 accidents, of which 4 were catastrophic, for 1973, and 47 accidents, of which 5 were catastrophic, for 1974. The data include all U.S. carrier operations domestically or abroad, scheduled and unscheduled. There were 1,030,000 flight employees in the aircraft industry in 1973.

The CONSAD data indicate an average of 2.1 fat/yr, 12.43 he/yr, and \$30,000/yr associated with catastrophic involuntary risks from commercial aviation. A population of 2.1×10^6 people is the base for U.S. population. When both CONSAD and NTSB data are available, both are given. For measured risk levels, refer to the right-hand column of Table 8.2*

* Since no on-the-ground involuntary risks occurred for 1973 or 1974, a figure for involuntary risk was derived by taking all "other person" fatalities for the years 1968–1973, finding the average (5.5 fat/yr), assuming that these were involuntary risks, and subtracting the CONSAD average for involuntary catastrophic risks from this average. This resulted in 3.4 involuntary fatalities per year. This value is probably overstated, since many of the fatalities may be occupationally connected. Injuries were assumed to be identified in the same manner.

Table 18.2 *Risk Referents and Measured Risk Levels for Commercial Air Travel*

Risk Referents	Measured Risk Levels
Man-Originated Catastrophic Risk	
$R(I)_{5,1} = 1 \times 10^{-8}$ fat/yr	$M(I)_{5,1} = 1.0 \times 10^{-8}$ fat/yr[a]
$R(I)_{5,2} = 5 \times 10^{-8}$ he/yr	$M(I)_{5,2} = 5.9 \times 10^{-8}$ he/yr[a]
$R(I)_{5,3} = 2 \times 10^{-3}$ \$/yr	$M(I)_{5,3} = 1.4 \times 10^{-4}$ \$/yr[a]
$R(RV)_{5,1} = 3 \times 10^{-5}$ fat/yr	$M(RV)_{5,1} = 2.9 \times 10^{-5}$ fat/yr[b]
$R(RV)_{5,2} = 3 \times 10^{-6}$ he/yr	$M(RV)_{5,2} = 1.5 \times 10^{-6}$ he/yr[c]
	$= 1.9 \times 10^{-6}$ he/yr[d]
$R(RV)_{5,3} = 4 \times 10^{-1}$ \$/yr	$M(RV)_{5,3} =$ ND[e]
$R(V)_{5,1} = 2 \times 10^{-7}$ fat/yr	$M(V)_{5,1} = 1.0 \times 10^{-6}$ fat/yr[a]
	$= 1.5 \times 10^{-6}$ fat/yr[b]
$R(V)_{5,2} = 2 \times 10^{-7}$ he/yr	$M(V)_{5,2} = 2.6 \times 10^{-7}$ he/yr[a]
	$= 4.5 \times 10^{-7}$ he/yr[b]
$R(V)_{5,3} = 4 \times 10^{-2}$ \$/yr	$M(V)_{5,3} =$ ND[e]
Man-Originated Ordinary Risk	
$R(I)_{6,1} = 5 \times 10^{-7}$ fat/yr	$M(I)_{6,1} = 2.6 \times 10^{-8}$ fat/yr[b]
$R(I)_{6,2} = 3 \times 10^{-6}$ he/yr	$M(I)_{6,2} = 2.6 \times 10^{-8}$ he/yr[c]
$R(I)_{6,3} = 1 \times 10^{-1}$ \$/yr	$M(I)_{6,3} =$ ND[e]
$R(RV)_{6,1} = 1 \times 10^{-4}$ fat/yr	$M(RV)_{6,1} = 1.4 \times 10^{-4}$ fat/yr[f]
	$= 5.8 \times 10^{-6}$ fat/yr[b]
$R(RV)_{6,2} = 6 \times 10^{-2}$ he/yr	$M(RV)_{6,2} = 1.5 \times 10^{-5}$ he/yr[c]
	$= 2.6 \times 10^{-4}$ he/yr[d]
$R(RV)_{6,3} = 3 \times 10^{1}$ \$/yr	$M(RV)_{6,3} =$ ND[e]
$R(V)_{6,1} = 6 \times 10^{-5}$ fat/yr	$M(V)_{6,1} = 2.8 \times 10^{-8}$ fat/yr[b]
$R(V)_{6,2} = 3 \times 10^{-2}$ he/yr	$M(V)_{6,2} = 1.2 \times 10^{-7}$ he/yr[c]
	$= 1.4 \times 10^{-5}$ he/yr[d]
$R(V)_{6,3} = 3 \times 10^{0}$ \$/yr	$M(V)_{6,3} =$ ND[e]

[a] CONSAD data.
[b] NTSB data—Table 18.3.
[c] NTSB data—serious injuries only.
[d] NTSB data—all injuries.
[e] No data available.
[f] OSHA data.[21]

18.5.1.3 Comparision of Measured Risk Levels with Risk Referents

The concern for inequitable distribution of risk involves involuntary and regulated voluntary risks only. The voluntary risk data for passengers are shown merely to supply to passengers who are aware of these risk levels

Table 18.3 Accident Data for Air Travel

Accident Type		Fatalities	Serious Injury	Minor Injury
All				
Crew	1973	26	19	223
	1974	44	14	277
	avg	35	17	250
Passengers	1973	200	32	2574
	1974	421	40	3025
	avg	310	36	2756
Others	1973	1 (occ)	0	0
	1974	0	1 (occ)	0
	avg	0.5	0.5	0
Catastrophic				
Crew	1973	20	2	0
	1974	39	1	1
	avg	30	2.5	0.5
Passengers	1973	195	6	0
	1974	415	14	0
	avg	305	10	0
Others	1973	0	0	0
	1974	0	0	0
	avg.	0	0	0
Ordinary				
Crew	1973	6	17	223
	1974	5	13	276
	avg	6	15	249
Passengers	1973	5	26	2574
	1974	6	26	3025
	avg	6	26	2800
Others	1973	1 (occ)	0	0
	1974	0	1 (occ)	0
	avg	0.5	0.5	0

Source: NTSB *Annual Review of Aircraft Accidents*, 1973[19] and 1974.[20]

information permitting them to judge whether the risks are acceptable on a personal direct gain-loss balance.

In all cases the measured risk levels are no more than one order of magnitude greater than their associated risk referents. Involuntary risk levels compare very favorably with the risk referents, as do those for regulated voluntary risk. This should not be surprising for the case of catastrophic regulated voluntary risk, since the risk references used were derived in Chapter 17 from the same data from which the measured risk levels are generated here. Therefore this portion of the comparison should be ignored.

Catastrophic voluntary risks involving fatalities are very close to the margin, but within the limits of precision prescribed for the methodology, indicating the most sensitive area of air travel for passengers—namely, worry over a major air crash. Thus the methodology is not insensitive when identifying critical areas of concern. Aggregation to a single risk referent would mask these considerations.

18.5.2 Railroad Transportation

Railroad transportation consists of passenger travel and freight hauling. The railroad system, one of our oldest modern technologies, was used extensively throughout the nineteenth and twentieth centuries. Both passenger travel and freight hauling have declined over the past few years as a result of competing modes of travel. Nevertheless, the technology has been accepted by society and results in risk proportionality factors of $P(I) = .1$ and $P(RV) = 1.0$, with derating values of unity. The safety of passengers is also of interest. A value of $P(V) = .1$ has been used here, since many passengers would select rail travel as a perceived safe alternative to other modes of travel. As such, a passenger's decision is treated as separate decision from rail operator's direct gain-loss balance. Thus this is a customer gain-loss decision.

Control, at best, involves specific design features for safety. Between 1953 and 1973, however, total railroad miles have decreased from 1.3×10^9 to 7.8×10^8 miles, and passenger miles from 32×10^9 to 5×10^9 decreases of 40 and 84% during this period.[19] During the same period the total number of fatalities has decreased only by 35%, and passenger fatalities by 75%.[18] Employee fatalities have dropped by 60%. The control is positive on an absolute basis, but it is at best level or slightly negative on a relative basis.

The rating system for control shown in Table 18.1 is not particularly suitable for a declining technology. Two assumptions can be made in this case. First, positive control for specific design features on an absolute basis is actually demonstrated, since system use is declining. Second, what may have been systemic control in the past now indicates level or slightly negative control on a relative basis. Two different values of C_c can be derived.

$$\text{Absolute basis } C_c = .5 \times 1.0 \times 1.0 \times 1.0 = .5$$

$$\text{Relative basis } C_c = .5 \times .2 \times 1.0 \times .5 = .05$$

A value of .2 was chosen for C_2 in the relative case, since the degree of control is at best level.

18.5.2.1 Risk Proportionality Factors and Risk Referents

The risk proportionality factors for absolute control are

$$R(I)_{i,j} = .1 \times 1.0 \times 1.0 \times .5 = .05\ A(I)_{i,j}$$

$$R(RV)_{i,j} = 1.0 \times 1.0 \times 1.0 \times .5 = .5\ A(RV)_{i,j}$$

$$R(V)_{i,j} = .1 \times 1.0 \times 1.0 \times .5 = .05\ A(V)_{i,j}$$

The derived risk referents are given in the left-hand column of Table 18.4. The risk proportionality factors for relative control are:

$$R(I)_{i,j} = 0.005A(I)_{i,j}$$

$$R(RV)_{i,j} = 0.05A(RV)_{i,j}$$

$$R(V)_{i,j} = 0.005A(V)_{i,j}$$

The appropriate risk referents appear in the center column of Table 18.4.

18.5.2.2 Measured Risk Levels for Travel by Railroad

Catastrophic risk data are obtained directly from the CONSAD report using a U.S. population of 2.1×10^6 people. Involuntary risk data for catastrophes show 21-year averages of 1.1 fat/yr, 31.9 he/yr, and $880,000/yr. Voluntary risk data are used in the absence of regulated voluntary risk referents from catastrophic risk. These values are 24.6 fat/yr, 207.6 he/yr, and $1.02 million/yr.

Ordinary risk data for fatalities and health effects are obtained from *Accident Facts* (1974). Involuntary risk data are derived from the 5-year average (1968–1972) of all fatalities and health effects not associated with occupational, passenger, or motor vehicle accidents. Regulated voluntary risk data are derived from employee data for a worker population of 550,000 employees. The base data are summarized in Table 18.5. Voluntary risk levels are calculated for passengers on trains. Railroad grade crossing accidents are considered to be motor vehicle risks.

No data were found for economic costs of ordinary railroad accidents because separation from accidents involving freight was not feasible. Table 18.4 summarizes the measured risk levels in the right-hand column.

Table 18.4 *Risk Referents and Measured Risk Levels for Rail Transportation*

Risk Referents		
Absolute Control Basis	Relative Control Basis	Measured Risk Levels
	Man-Originated Catastrophic Risk	
$R(I)_{5,1} = 5 \times 10^{-9}$ fat/yr	5×10^{-10} fat/yr	$M(I)_{5,1} = 5.2 \times 10^{-9}$ fat/yr[a]
$R(I)_{5,2} = 3 \times 10^{-8}$ he/yr	3×10^{-9} he/yr	$M(I)_{5,2} = 1.5 \times 10^{-7}$ he/yr[a]
$R(I)_{5,3} = 5 \times 10^{-2}$ \$/yr	5×10^{-3} \$/yr	$M(I)_{5,3} = 4.2 \times 10^{-3}$ \$/yr[a]
$R(V)_{5,1} = 1 \times 10^{-7}$ fat/yr	1×10^{-8} fat/yr	$M(V)_{5,1} = 1.2 \times 10^{-7}$ fat/yr[a]
$R(V)_{5,2} = 1 \times 10^{-7}$ he/yr	1×10^{-8} he/yr	$M(V)_{5,2} = 9.8 \times 10^{-7}$ he/yr[a]
$R(V)_{5,3} = 2 \times 10^{-2}$ \$/yr	2×10^{-3} \$/yr	$M(V)_{5,3} = 4.8 \times 10^{-3}$ \$/yr[a]
	Man-Originated Ordinary Risk	
$R(I)_{6,1} = 3 \times 10^{-7}$ fat/yr	3×10^{-8} fat/yr	$M(I)_{6,1} = 2.0 \times 10^{-6}$ fat/yr[b]
$R(I)_{6,2} = 2 \times 10^{-6}$ he/yr	2×10^{-7} he/yr	$M(I)_{6,2} = 8.3 \times 10^{-6}$ he/yr[b]
$R(I)_{6,3} = 5 \times 10^{-2}$ \$/yr	5×10^{-3} \$/yr	ND[c]
$R(RV)_{6,1} = 5 \times 10^{-5}$ fat/yr	5×10^{-6} fat/yr	$M(RV)_{6,1} = 3.0 \times 10^{-4}$ fat/yr[b]
$R(RV)_{6,2} = 3 \times 10^{-2}$ he/yr	3×10^{-3} he/yr	$M(RV)_{6,2} = 2.7 \times 10^{-2}$ he/yr[b]
$R(RV)_{6,3} = 2 \times 10^{1}$ \$/yr	2×10^{0} \$/yr	ND[c]
$R(V)_{6,1} = 3 \times 10^{-6}$ fat/yr	3×10^{-7} fat/yr	$M(V)_{6,1} = 9.2 \times 10^{-7}$ fat/yr[b]
$R(V)_{6,2} = 2 \times 10^{-3}$ he/yr	2×10^{-4} he/yr	$M(V)_{6,2} = 3.6 \times 10^{-6}$ he/yr[b]
$R(V)_{6,3} = 1 \times 10^{-2}$ \$/yr	1×10^{-3} \$/yr	ND[c]

[a] CONSAD data.
[b] From Table 18.3.
[c] No data available.

Table 18.5 Ordinary Risk Data on Railroad Transportation

	fat/yr	he/yr
Totals	2273	21,131
Grade crossing	1488	3,410
Nonhighway	785	17,721
Passengers on trains	194	749
Employees on duty[a]	163	15,220
Involuntary risks	428	1,750

Source: Data from *Accident Facts,* 1974.
[a] Average number of employees, 555,000.

18.5.2.3 Comparison of Measured Risk Levels with Risk Referents

In all cases for railroad travel, the measured risk levels on an absolute control basis are within one order of magnitude of the risk referents where such referents are exceeded. Within this margin railroad travel would be deemed acceptable to society, and especially to individual passengers, since voluntary risk levels are well below equivalent referents. Occupational risks and involuntary injuries are areas for consideration for increased safety.

On a relative control basis, however, railroad transportation becomes an unacceptable system for fatalities and injuries in all cases except ordinary voluntary risks. This implies that although railroad transportation has been an acceptable system in the past, its acceptability now and in the future may be seriously questioned. This is not an unreasonable conclusion, since lack of investment in railroads has been evident over the past 20 years. Deterioration of rights-of-way and rolling stock has resulted with an apparent decrease in safety. Consolidation of rail lines, abandonment of old, deteriorated routes, and the establishment of Amtrak may provide the mechanism to reverse these trends.

From the point of view of the passenger, the risks are acceptable in either case. It is only the inequitable risk that causes concern. It would seem that regulatory action to improve safety for involuntary risks and occupational conditions is warranted. Since passengers will not be inhibited, there is little incentive for the railroads to improve safety on their own.

The methodology has provided a means to show that rail travel is acceptable on an absolute basis but is in trouble on a relative basis. Apparently, then, the methodology, the straw man value judgments, and the risk reference data are not totally invalid, and the methodology has utility.

18.5.3 Methodology Validation

Two existing transportation systems have been used to test the reasonableness of the methodology, the value judgments used as straw men, and the absolute risk levels developed in Chapter 17. Both cases seem to result in situations that are not anti-intuitive. Such tests do not validate the methodology, but only indicate two cases in which it is not unreasonable.

There are probably other systems not tested that fail to meet the conditions. Thus further effort toward validation is necessary. At the same time, value judgments must be refined, data bases for determining absolute risk levels must be establisted, and estimation of risks must be improved.

Our purpose is not to prove that the specific methodology is the right one, but that the utility of one such methodology will demonstrate that a methodology for risk assessment is not only possible, but useful.

The next chapter undertakes the next step: to test the utility of the method on some relatively new technological systems.

REFERENCES

1. Robert W. Kates, "Risk Assessment of Environmental Hazard," SCOPE Report No. 8, International Council of Scientific Unions, Scientific Committee on Problems of the Environment, Paris, 1976.
2. Jacob Marschak, "Rational Behavior, Uncertain Prospects and Measureable Utility," *Economtrica,* Vol. 18, No. 2, 1950.
3. Anthony Downs, *Inside Bureaucracy.* Boston: Little, Brown, 1967.
4. Louis M. Kohlmaier, Jr. *The Regulators.* New York: Harper & Row, 1969.
5. John C. Chicken, *Hazard Control Policy in Britain.* Oxford, England: Pergamon Press, 1975.
6. Hubert A. Simon, "A Behavioral Model of Rational Choice," *Quarterly Journal of Economics,* Vol. 69, No. 1, 1955.
7. Charles E. Lindblom, "The Science of 'Muddling Through,'" *Public Administration Review,* Vol. 19, 1959.
8. Henry C. Hart, "Crisis, Community and Consent in Water Politics," *Law and Contemporary Problems,* School of Law, Duke University, Vol. 22, No. 3, 1957.
9. George A. Peters, "Systematic Safety," *National Safety News,* September 1975, pp. 83–90.
10. Joanne Linnerooth, *The Evaluation of Life-Saving: A Survey,* Laxenburg, Austria: International Institute for Applied Systems Analysis, 1975.
11. Gerald J. Rausa, "Some Estimates of the Worth of a Human Life," EPA Memorandum, July 2, 1973.
12. Kendall Moll, Sanford Baum, Erwin Carpanar, Francis Dresch, and Rose M. Wright, "Methods for Determining Acceptable Risks from Cadmium, Asbestos, and Other Hazardous Wastes." Menlo Park, Calif.: Stanford Research Institute, July 1975.

13. E. G. Altouney, *The Role of Uncertainties in the Economic Evaluation of Water Resource Projects,* Report EEP-7. Palo Alto, Calif.: Stanford University Institute in Engineering-Economic Systems, 1963.
14. T. D. Crocker, "Benefit-Cost Analysis of Benefit-Cost Analysis." Presented at the Environmental Protection Agency Symposium on Cost-Benefit Analysis, Washington, D.C., September 1973.
15. Harry J. Otway and J. J. Cohen, *Revealed Preferences: Comments on the Starr Benefit-Risk Relationships,* Laxenburg, Austria: International Institute for Applied Systems Analysis, 1975.
16. Chauncey Starr, "General Philosophy of Risk-Benefit Analysis." Presented at Electric Power Research Institute (EPRI), Stanford IES Seminar, Stanford, Calif., September 30, 1974.
17. Personal communication from Chauncey Starr, May 9, 1975.
18. *Accident Facts.* Chicago: National Safety Council, 1973, 1974.
19. National Transportation Safety Board, "Annual Review of Aircraft Accident Data," Report NTSB-ARC-74-2. Washington, D.C.: Air Carrier Operations, 1973.
20. National Transportation Safety Board, "Annual Review of Aircraft Accident Data, Report NTSB-ARC-76-1. Washington, D.C.: Air Carrier Operations, 1974.
21. U.S. Department of Labor, *Occupational Injuries and Illnesses by Industry, 1972*: Bulletin 1830. Washington, D.C.: Bureau of Labor Statistics, 1974.

19

Application of the Risk Evaluation Methodology

The validity and utility of this or any other methodology for risk evaluation can be demonstrated only by meaningful application to technological systems of interest involving difficult societal questions to be clarified and successfully resolved. Four technological systems, all involved in electrical energy production, are evaluated by the proposed methodology. The use of energy systems relying on coal, oil, natural gas, including liquefied natural gas (LNG) and liquefied propane gas (LPG), and light-water nuclear reactors, allows intersystem comparison and makes possible balanced perspectives.

All systems generate electrical energy, and only the part of the system employed for generation is considered. Substitution among systems for energy production precludes consideration of the ability to meet energy demand in this analysis.

19.1 BACKGROUND

Together with hydroelectric, geothermal, and perhaps solar energy, the four energy sources to be examined provide the basic methods of conversion of fuel for generation of electricity. Table 19.1 provides some summary statistics on the four fuel systems in question. The data are based on 1972 production (i.e., prior to the 1973 oil embargo) and are taken from the Joint Economic Committee Congressional hearing record[1] and the Project Independence report.[2] Oil and gas, including LNG and LPG, are primarily used for purposes other than electrical power production. Coal and nuclear (uranium) fuel are primarily aimed at power production. Industry risks independent of use must be prorated for each intended use. Gas and oil have very limited domestic reserves, whereas coal and nuclear energy have greater domestic availability.

Since 1973, dependence on oil and gas imports has been a major concern. Minimizing the dependence on imports is a national goal, but based on the

Table 19.1 *Basic Data on Energy Production by Fuel Source*

Statistic	Petroleum	Natural Gas	Coal	Uranium
Uses (1972)[a]				
Generation of electricity	9%	18%	66%	96%
Household and commercial	17%	33%	3%	—
Industrial	21%	46%	31%	—
Transportation	53%	3%	0.1%	—
Miscellaneous	—	—	—	4%
Total dollar value estimate (1972)[a]	$26 billion	$14 billion	$5.5 billion	$153 million
Number of employees (1972)[a]	263,000 Production 151,000 Refining		148,000	7000
Consumption by quads (1972)[b]	22.4 Domestic 11.7 Imports	22.1	12.5	0.6
Consumption by quads (1985 estimate)[b]	31.3 Domestic 6.5 Imports	24.8	22.9	12.5
Years of proven reserves at 1972 [b] consumption levels	8 yr	11 yr	823 yr	40–1000 yr depending on Pu, recycling, etc.

[a] Data from Joint Economic Committee Report.[1]

[b] Data from Project Independence report,[2] where one quad is equal to a quadrillion Btu = 10^{15} Btu.

Table 19.2 *Environmental Comparisons for All Energy Uses—1972–1985*

Pollutant Category	1972 Level (bbl)	1985 Level $7/bbl	1985 Level $11/bbl
Water pollution			
Dissolved solids (tons/day)	37,000	5,200	5,800
Suspended solids (tons/day)	7,600	240	300
Thermal discharges (billion Btu/day)	19,500	24,500	24,000
Air pollution			
Particulates (tons/day)	1,800	2,200	2,300
Nitrogen oxides (tons/day)	38,000	41,800	46,700
Sulfur oxides (tons/day)	58,900	47,100	53,700
Hydrocarbons (tons/day)	33,200	18,800	18,800
Carbon monoxide (tons/day)	7,900	1,000	1,400

Source: Project Independence report (p. 33).[2]

relatively small proportions of oil and gas used for electrical energy, perhaps this use of gas and oil should be a small component of such concern. However, substitution among fuel sources is most easily accomplished in the area of electrical generation. It should be noted that coal and nuclear energy are particularly amenable to base-load operation, whereas natural gas is used for peaking-loads.

Like low-sulfur oil, natural gas is a very clean fuel when used in electrical generation. Coal and high-sulfur oils contribute significantly to air pollution, nuclear energy does not, but it involves radiation hazards and potential radioactive contamination of the environment. All methods involve thermal discharges to air and water, but size of plant and concentration of thermal sources are of most concern for nuclear plants and very large coal-fired plants. Table 19.2 estimates pollutant discharges from all energy uses for 1972 and 1985, based on the price of a barrel of oil. Radioactive discharges are not shown.

The risks of each energy source for generation of electricity arise from different parts of the energy production cycle. A general overview of these follows.

19.1.1 Coal

Coal is mined by open-pit or deep-shaft methods. Risks to miners during the extraction operation include both ordinary and catastrophic risks due to accidents, and delayed chronic effects from black lung disease, silicosis, and pneumoconiosis. Subsidence of old mines entails some involuntary financial risk to landowners above mines. Environmental damage from unrestored open-pit mines takes the form of loss of productive land and unsightly gouges in the earth. Waste disposal piles are dangerous because culm pile slides can occur. Coal refuse dam failures of this nature, such as that which occurred in Saunders, West Virginia, in February 1972, involve involuntary risks. Occupational risks are regulated voluntary risks.

The normal risks of ordinary processing of materials and freight handling are associated with preparation and transportation. When coal is burned to produce energy, air-polluting sulfur dioxide, sulfates, and particulates are generated. These are controlled by regulating the sulfur content of the incoming coal or by cleaning off-gases. Solid wastes in the form of ash and pollution control residues must be disposed of and water pollution controlled. There are ordinary accidental risks for steam-electric plants for both workers and the surrounding population. Air pollution, a delayed chronic risk, is the primary involuntary exposure.

19.1.2 Oil

Oil is produced by on- and off-shore drilling, and the accidental risks that accompany production are ordinary and occupational. Transportation of crude and refined petroleum products can result in fires and spills. Fires subject people to involuntary risks. Spills involve environmental risks and direct costs to the transporter under environmental regulations that establish penalties for spills. Oil storage and pipeline transmission involve the risks of fire and explosion. Pipeline breaks can result in environmentally damaging spills. There are no direct waste, land, or refuse problems associated with the production and refining of petroleum. Environmental pollution occurs to some extent during refining, but it is controlled.

Air pollution from sulfur dioxide and sulfates is controlled by use of low sulfur fuels or stack gas cleaning. Risks of oil-fired power generation are similar to those for coal.

19.1.3 Natural Gas

Natural gas is produced in conjunction with oil production. Transportation is by pipeline to a processor for separation of butane and propane and/or liquification, or directly to users with intermediate storage at production sites, processors, or distributors. LNG is usually shipped by tanker, and LPG by tank truck. All forms of transportation and storage involve the possibility of fire and explosion.

There is virtually no pollution from the burning of natural gas,* no wastes, and only minimal ordinary risks are incurred in power generation. Since electrical energy production is not involved in home use, explosions and fires in the home are not of concern.

The major risks of natural or synthetic gases and their by-products (LNG and LPG) are accidental risks in transportation and storage. Risks are both ordinary and catastrophic and are either involuntary or regulated voluntary (occupational).

19.1.4 Nuclear Energy

Uranium is mined by deep-shaft or open-pit methods. Occupational risks include the ordinary and catastrophic accidental risks of general metallic ore production, as well as similar chronic risks except for inhalation of radon gas. The latter poses an increased risk of lung cancer and is presently regulated by the Mine Enforcement Safety Administration to EPA guidelines. Uranium mine wastes lead to the same environmental problems associated with coal or any other mining activity, but there are additional risks from radioactivity due to radon emanations, radium in particulate form, and runoff of radioactive alpha emitters into surface and groundwaters. Similar problems are found in the milling operation, including the presence of mill tailings, a major source of radioactive effluents.

Preparation and transportation of new fuel bear no special risks other than those involved in ordinary fabrication and transportation situations.

The generation of electrical energy results in the planned releases of radioactive effluents to air and water, exposures to workers, and thermal discharges. Unplanned accidental releases can occur, producing both ordinary and catastrophic consequences. A major core meltdown and breach of containment would result in serious, large consequences to the general population. Reprocessing of spent fuel involves both planned and unplanned releases. The large inventory of radioactivity at fuel reprocessing plants make accidents a matter of major concern, but the magnitude of consequences is expected to be much less than those possible at a reactor. Air and water pollution from ordinary chemicals is extremely small.

The need to isolate waste materials from the environment until they have decayed to innocuous levels is a major requirement of nuclear energy. Risks occur from planned and unplanned exposures during the transportation of spent fuel and processed wastes to a reprocessor or repository. Long-term risks from improper waste management are a large unknown factor.

The existence of plutonium in the fuel cycle causes a unique type risk.

* Production of carbon dioxide is a potential future problem.

Though all fuel systems are subject to sabotage and terrorist attack, the possibility that stolen plutonium would be used for construction of nuclear weapons or other forms of blackmail demands special consideration. The adequacy of safeguards to prevent such events has yet to be demonstrated for commercial systems.

Table 19.3 summarizes qualitatively the particular sources of risk from each fuel cycle. Risks common to all forms of steam-electric generation are nonparticular and are omitted. Risks to the population (involuntary) and occupational risks (regulated voluntary) are shown separately. The magnitude of these risks is addressed later.

19.2 DIRECT GAIN–LOSS BALANCE

Direct gain-loss balances are made by electric utility companies, either privately or publicly owned, which must supply power to all customers at reasonable rates (based on the efficiency of a single noncompetitive supplier of service and a regulated rate structure*). Each utility will make a direct gain-loss balance involving alternative fuel systems, based on capital costs, operational and fuel costs, availability of capital, assurance of supply, cost of meeting environmental and safety regulations, and other factors. Employing this analysis, a utility will make the choice that is best for itself and, presumably, its customers.

This voluntary weighing of gains and losses, assuming that a favorable balance is found for at least one system, is not of concern in this risk evaluation. It is assumed that cases can be found for favorable direct gain-loss balances for all four systems. It should be stated that the nuclear option involves relatively high initial investment, but substantially lower fuel and operating costs, than the other alternatives. Thus nuclear energy is a particularly sought-after option for large base-load plants by utilities, assuming capital is available.

19.3 INDIRECT GAIN–LOSS BALANCES

An indirect, gross gain-loss balance must be made for each of the energy systems considered. However these are part of a larger gain-loss balance, namely, the need to assure adequate supplies of electrical energy to meet conservative demand estimates in the light of competing uses of resources and global environmental impact. For present purposes, the author has

* Which may or may not serve to encourage efficiency.

made the judgment that having adequate energy to meet reasonable demand projections results in a favorable indirect gain-loss balance; that is, a major energy shortfall would be worse than reasonable environmental and health impacts incurred in assuring adequate energy. This judgment is debatable, but it provides a means of viewing the four energy systems as substitutable electrical energy sources along with hydroelectric, wind, solar, and geothermal renewable energy sources, as well as new forms in development, involving fusion, gasification and liquification of fossil fuels, and hydrogen production by thermal processes. In this manner, each system is initially judged as having a favorable balance; then particular indirect losses for each system are examined, which alter the individual system balance.

The value judgments used are those of the author, who has a bias toward protection of public health and the environment. It is not expected that all will agree with the value judgments, but they are made visibly and the reader is welcome to substitute his own.

In making these gain-loss balances, only gross consideration of risk as a loss is involved. The detailed evaluation of risks is a separate step, taken after the indirect gain-loss balance has been made. Furthermore, all four systems are in use; but ground rules have changed since 1973, and new, additional plants are being proposed. All new installations must meet regulatory requirements for health, safety, and the environment.

19.3.1 Oil Indirect Gain-Loss Balance

We must import petroleum on a massive scale because of shortfalls in domestic oil production, the dependence of transportation systems on petroleum fuels, the need for feedstocks in the chemical industry, and the lack of near-term substitutes for these applications. Dependence on foreign imports produces problems in strategic foreign policy and balance of payments. In eventually achieving energy and material independence, the administration hopes to reduce the proportion and volume of oil imports. Since electrical generation has substitutable alternatives, expansion of use of oil as a fuel in this area is divergent from the objective of energy independence.

The advantages of oil are ease of transportation and storage and the ability to treat crude oil to remove impurities such as sulfur. Oil is used for both base-load and peak-load applications, and conversion of existing base-load plants to coal has occurred under Federal Energy Administration policy.

Air pollution from oil-burning plants has been a problem in the past, but low sulfur fuels and stack-cleaning technologies are available at some direct cost premium. The Clean Air Act of 1970 provides regulatory environmental controls.

Table 19.3 *A Qualitative Comparison of Risks in the Four Fuel Cycles*[a]

Cycle Part	Oil	Gas	Coal	Nuclear
Production of fuel				
Occupational				
Accident	Fire	Explosion	Mine explosion or collapse	Mine collapse
Chronic	—	—	Lung diseases: Dust	Lung cancer: radon
Population				
Environmental	—	—	Waste: Scars	Waste: Scars
Environmental			Drainage and	Drainage
Health			Refuse dams	Radon
Transportation and storage				
Occupational				
Accident	Fire	Explosion	—	Spent fuel
Chronic	—	—	—	Spent fuel exposure

Population				
Accident	Fire	Explosion	—	Spent fuel, wastes, sabotage and terrorism
Chronic	—	—	—	Spent fuel exposure
Environmental	Spills and leaks			Sabotage and terrorism
Power generation				
Occupational				
Accident	—	—	—	Small and large accidents
Chronic				Radioactive exposure
Population				
Accident	—	—	—	Small and large accidents
Chronic	SO_2, SO_3	—	SO_2, SO_3, particulates	Radioactive effluents
Wastes and treatment of wastes				
Occupational	—	—	—	Radioactive exposure
Population-environmental	—	—	Ash and residues	Long-term containment
				Reprocessing exposures
				Safeguards

[a] Risks common to generic steam electric generation are not included.

New sources of oil from oil shale, tar sands, and other materials have not yet been developed, and gains and losses have not been totally identified. They are not considered in this balance.

The use of oil would have a favorable balance for production of electric energy if it were not for the strategic problems that dependence on imports entails. This overriding concern has made the future use of oil as a fuel unclear.

The indirect gain-loss balance is marginally favorable to indecisive. On this basis interpolation between the two may be meaningful.

19.3.2 Natural Gas Indirect Gain-Loss Balance

Natural gas, including LNG and LPG, is an environmentally clean fuel that is relatively easy to use, transport, and store. It is uniquely suitable for peak-load plants.

Its disadvantages lie solely in its shortage, perhaps due in part to price regulation, and future availability of this fuel source is in doubt. Commercial, industrial, and home use of gas as a fuel is substitutable in some cases by electrical energy, such as in home heating, but conversion costs are extremely high for homes. Industrial uses as a fuel and as feedstock are already on an interruptible basis.

The use of natural gas as a fuel would be very favorable if the substance were readily available. Since it is not, a marginally favorable balance results. Methods for stimulating increased outputs from existing wells only result in incremental extension of supplies.

19.3.3 Coal Indirect Gain-Loss Balance

Coal is in plentiful supply in the United States, at least in the ground. Removal involves strip-mining and new transportation routes. The mined coal has many impurities, including sulfur, which when burned results in air pollution, fly ash, and slag. These cause environmental problems that can be controlled—at a substantial premium for the direct gain-loss balance—and are regulated by the Clean Air Act of 1970 and the Water Pollution Control Act of 1972. Reclamation of strip mines is not yet under federal regulation, although such legislation is expected.

If there were no environmental and transportation problems, the short-term* indirect gain-loss balance would be quite favorable. However even with existing environmental regulations and new regulations for reclamation of strip-mined land, coal has a marginally favorable balance as a result of

* Short-term since all fossil fuels are nonrenewable resources.

environmental impact. Without regulation for reclamation before opening extensive operations for stripping western coal, this author would even place coal at an indecisive balance level. It is assumed that legislation regulating strip mines will be forthcoming before extensive extraction takes place. However, both cases, with and without strip mine restoration, are retained for examination.

19.3.4 Nuclear Energy Indirect Gain-Loss Balance

Nuclear energy is essentially clean. There is an abundance of natural uranium, although extraction from low grade sources implies high prices and will affect the direct gain-loss balance. The ore can be taken from the earth by deep-shaft and open-pit mining; and milling operations involve large quantities of wastes, many containing natural radioactivity. Regulations for control of these wastes, which are presently inadequate, are under review by both the Nuclear Regulatory Commission and the Environmental Protection Agency. The use of plutonium as a fuel in conjunction with enriched uranium to extend the supply of uranium and the use of breeder reactors to generate plutonium are options under development and evaluation.

Two problem areas, the planned releases of radioactivity and nuclear accidents, are directly amenable to risk evaluation. Two other problems call for special consideration: management of radioactive wastes, and adequate institutional safeguards. The management of radioactive wastes, is a presently unsolved problem that can have an impact on future generations if adequate solutions are not found. All types of radioactive waste associated with nuclear energy production are included. The second problem involves the theft and use of plutonium to make weapons for terroristic and nationalistic purposes.

No institutional safeguards have yet been demonstrated for the use of mixed-oxide fuels and breeding of plutonium in this country and abroad. Export policy on reprocessing capability is a major factor in the safeguards area. The proliferation of weapons and the opportunity for blackmail and terrorism makes the use of these fuel options and exports of reprocessed fuel unacceptable without implemented and demonstrated safeguards systems. Mixed-oxide fuels and reactor-bred plutonium will not be considered as a part of this energy option; only the light-water reactor, enriched uranium fuel system is included in this analysis.

Planned releases from the uranium fuel cycle are assumed to meet existing environmental standards as promulgated by the Environmental Protection Agency. Accidental releases will be viewed as part of the risk evaluation step. The major detrimental problem accompanying nuclear energy is the absence of acceptable methods for the final disposal of radioactive waste of all types, and it will be 7 to 10 years before acceptable methods can be demonstrated. On the basis of this waste disposal problem alone, the indirect gain-loss balance is at best marginally favorable. A proved waste disposal capability for all wastes would result in a favorable balance.

19.4 COST EFFECTIVENESS OF RISK REDUCTION

The cost effectiveness of risk reduction has not been practiced in setting government regulations for these systems up to this time, except for nuclear energy.* It is assumed, however, that existing regulations or those which are imminent will result in cost-effective reduction of risks. Examples are the Water Pollution Control Act of 1972, which regulates chemical and thermal discharges to water from power plants and controls oil spills involved in the storage and transportation of petroleum and hazardous materials, the Clean Air Act of 1970, which regulates chemical effluents from power plants, regulations for mine health and safety by the Mine Enforcement Safety Administration, occupational standards set by OHSA, radiation standards for planned releases of the uranium fuel cycle by the EPA, and continued effort by Congress to pass a strip-mining reclamation act.

19.5 DEVELOPMENT OF RISK REFERENTS

Risk referents for each system are developed by establishing P and D values from the indirect gain-loss balance as well as values for the controllability factor, and are used in conjunction with risk references to form a set of risk referents. Risk referents are restricted for involuntary and regulated voluntary risks, since voluntary risks are of concern only to those involved in the direct gain-loss balance. Involuntary risks are interpreted as risks to the general population, and regulated voluntary risks as occupational risks.

For coal and nuclear energy, two cases for evaluation for each system are considered. The coal system is shown with and without a reclamation program for strip mining. Nuclear energy, which is restricted to enriched uranium fuel light-water reactors,† is shown for the present case (i.e., waste disposal is an unsolved problem) and for a future condition (i.e., an acceptable waste disposal program is implemented).

* The concept of "as low as practicable" is used by EPA in its Uranium Fuel Cycle Standards and by NRC in their Appendix I to 10CRF50 for reactor design regulations.

† Plutonium recycling, breeder reactors, and export of fuel reprocessing systems are omitted. A different indirect balance would result for these systems until adequate international safeguard programs were assured.

19.5.1 Risk proportionality Factors and Derating Factors

The risk proportionality factors P and the risk proportionality derating factors D for each system (Table 19.4) are derived directly from the indirect gain-loss balance on the basis of values given in Chapter 18.

19.5.2 Controllability Factors

The controllability factors for each system involve aspects of control specific to each particular system. These are developed as follows.

19.5.2.1 Oil Controllability Factors

Although there are accidents and fires at well sites and offshore rigs, refineries, and storage facilities, the transportation of oil involves the most hazards—both insults to people from accidents and insults to the environment from oil spills. The Oceanographic Commission of Washington state has undertaken studies on oil transportation, including a risk analysis. One study shows a direct proportional relationship between the number of vessel casualties and the number of ship movements in major U.S. ports.[3] The report further indicates that:

> Throughout the years, in an attempt to reduce the occurrence of accidents, the maritime industry has developed rules, regulations, navigation aids, navigation equipment, and systems. Unfortunately, hidebound by tradition, economic pressures and controls, these have not been exemplary of progress and innovation. . . .
>
> It is a fact that oil spillage will continue to occur as long as we transport it. The cost of cleanup, the cost of paying for the damage and/or the destruction of beaches, private property and marine life have to be weighed against the expense of prevention.
>
> Passing strong laws and writing good contingency plans help in establishing an oil spill protection plan, but the actual cleanup effort can be no better than is permitted by the state of the art of oil spill cleanup technology.[3]

The Water Pollution Control Act of 1972 covers prevention and cleanup of oil spills, and the Oceanographic Commission of Washington has recommended even stronger regulations and technological design for oil couplings to the State of Washington. There is no question that inspection and regulation are implemented, as well as special design features. The degree of control is negative on an absolute basis, but is level or even slightly positive on a relative basis involving the volume of traffic. Oil spills have been reduced over the past few years under the Water Pollution Control Act; thus a positive control factor has been demonstrated here on a relative basis.

$$C_c = .5 \times 1.0 \times 1.0 \times .5 = .25$$

Table 19.4 *Risk Proportionality and Controllability Factors for Electric Generation Systems Evaluated*

			Coal		Nuclear, ^{238}U Only	
Factor	Oil	Gas	No Reclamation	Reclamation Control	Waste Problem	Waste Contained
Indirect Gain-Loss Balance	Indeterminate	Marginally Favorable	Indeterminate	Marginally Favorable	Marginally Favorable	Favorable
$P(I)$	.1	.1	.1	.1	.1	.1
$P(RV)$	1.0	1.0	1.0	1.0	1.0	1.0
$D(I)$	.03	.1	.01	.1	.1	1.0
$D(RV)$	.1	.25	.1	.3	.3	1.0
C_c	.9	.5	.9	.9	.4	.4
$R(I)$	2.7×10^{-3}	5×10^{-3}	9×10^{-4}	9×10^{-3}	4×10^{-3}	4×10^{-2}
$R(RV)$	9×10^{-2}	1.3×10^{-2}	9×10^{-2}	2.7×10^{-1}	1.2×10^{-1}	4×10^{-1}

This consideration of controllability has taken into account only accidental risks. Those casused by air pollution are considered to be of a much higher magnitude than risks from accidents; these effects are often delayed. The Clean Air Act of 1970 has been effective in reducing air pollution from all sources in spite of increasing use of fossil fuels. Table 19.5 reveals a reduction in all pollutants except nitrogen oxides between 1970 and 1974. This demonstrates positive systemic control on an absolute basis that results in controllability value of unity.

$$C_c = 1.0 \times 1.0 \times 1.0 \times 1.0 = 1.0$$

Assuming a weighted average of accidental risk and pollution control, a value of .9 seems appropriate for C_c for oil.

19.5.2.2 Gas Controllability Factors

For natural gas and LNG and LPG, we examine only accidents involving production, refining, transportation, and storage relating to power generation. Risks in the home, commercial establishments, and other industries are not of concern here.

From 1970 on, the volume of natural gas used has been diminishing, while shipments of LNG and LPG have been increasing. A significant number of explosions and fires occur in the home, but there have been relatively few accidents in the production and transportation of natural gas. Catastrophic risk in tanker transport and in truck and railroad car transport are associated with LNG and LPG, respectively.

Table 19.5 *Estimates Amounts of Air Pollutants Nationwide*

	Estimates (millions of tons)	
Pollutant	1970	1974
Carbon monoxide	107.3	94.6
Particulates	27.5	19.5
Sulfur dioxide	34.3	31.4
Hydrocarbons	32.1	30.4
Nitrogen oxides	20.4	22.5
Totals	221.6	198.4

Source: "Clean air, the Breath of Life." U.S. Environmental Protection Agency, April 1976.

Specific design features exist for all three forms of gas. Level control on a relative basis is well demonstrated.

$$C_c = .5 \times .3 \times 1.0 \times .5 = .075$$

The same air pollution argument holds for gas as well as oil, although the contribution of air pollution from natural gas is much less than that of oil or coal. Nevertheless, demonstrated positive systemic control on an absolute basis exists. Since air pollution effects are a smaller proportion of overall risk, however, a value of .5 has been assigned to the gas fuel cycle.

19.5.2.3 Coal Controllability Factors

Except for ordinary transportation accidents, the major type of accident connected with the use of coal comes under the heading, occupational risks from mining. These risks are now regulated by the Mine Enforcement and Safety Administration under a number of safety laws. Specific design features of coal mines and mining equipment to reduce hazards date back to the Davy safety lamp.

This program has been effective, and as shown in Table 19.6, the frequency of fatalities has been decreasing at a more rapid rate than the number of miners has decreased. Thus positive control has been demonstrated on an absolute basis.

$$C_c = 1.0 \times 1.0 \times 1.0 \times 1.0 = 1.0$$

Sulfur and particulate emissions from coal-fired power plants are a major source of pollution. The control demonstrated by the Clean Air Act is absolute positive systemic control, and a value of .9 is assigned to controllability for the coal fuel cycle.

19.5.2.4 Nuclear Controllability Factors

The long-term concern for accidental releases of nuclear material, both in weapons programs and in energy production, has made safety a major concern of the nuclear industry. The Nuclear Regulatory Commission institutes a risk management system; however the level of quality control in commercial electric generation is well short of that in military nuclear propulsion systems. Thus nuclear fuel does not qualify for total systemic control, but is not far from it.

There have been no fatalities from accidents attributable to fissionable materials or their wastes with about 60 nuclear plants operating. However the Reactor Safety Study[5] indicates an expectation of an increase in risk for each additional reactor system. Thus, positive control has been demonstrated on a relative basis, at least up to this time.

$$C_c = .8 \times 1.0 \times 1.0 \times .5 = .4$$

Table 19.6 *Statistics for the Coal Industry, United States, 1955–1973*

Year	Men Employed	Fatal Injuries	Fatality Frequency Rate per 10^6 hours
1955	260,089	420	1.00
1956	260,285	448	1.03
1957	254,725	478	1.17
1958	224,890	358	1.11
1959	203,597	293	0.99
1960	189,679	325	1.15
1961	167,568	294	1.15
1962	161,286	289	1.16
1963	157,126	284	1.12
1964	150,765	242	0.96
1965	148,734	259	1.04
1966	145,244	233	0.96
1967	139,312	222	0.92
1968	134,467	311	1.33
1969	133,302	203	0.85
1970	144,480	260	1.00
1971	142,098	181	0.71
1972	129,378	156	0.53
1973	133,867	132	0.45

Source: U.S. Department of the Interior, Bureau of Mines, "Fuels," in *Minerals Yearbook,* 1953–1973. Washington, D.C.: Government Printing Office.

19.5.2.5 Risk Referent Levels

The risk proportionality computation results are the last two entries of Table 19.4. Referents for ordinary and catastrophic involuntary and regulated voluntary risk for each system (Table 19.7) are used for comparison with measured values for each system to be evaluated.

19.6 MEASURED RISK LEVELS

To make equivalent comparisons among fuel cycles, we assume that each fuel cycle involves installation of 100 gigawatts* of new electrical capacity

* 100 gigawatts (GW) = 100×10^9 W.

Table 19.7 *Risk Referents for Electrical Generation Systems Evaluated*

			Coal		Nuclear	
Risk type	Oil	Gas	No Reclamation	Reclamation	Waste Problem	Waste Solved
Catastrophic						
$R(I)_{5,1}$	3×10^{-10}	5×10^{-10}	9×10^{-11}	9×10^{-10}	4×10^{-10}	4×10^{-9}
$R(I)_{5,2}$	2×10^{-9}	3×10^{-9}	5×10^{-10}	5×10^{-9}	2×10^{-9}	2×10^{-8}
$R(I)_{5,3}$	3×10^{-5}	5×10^{-6}	9×10^{-6}	9×10^{-5}	8×10^{-5}	8×10^{-4}
$R(RV)_{5,1}$	3×10^{-6}	4×10^{-6}	3×10^{-6}	8×10^{-6}	4×10^{-6}	1×10^{-5}
$R(RV)_{5,2}$	3×10^{-7}	4×10^{-7}	3×10^{-7}	8×10^{-7}	4×10^{-5}	1×10^{-4}
$R(RV)_{5,3}$	4×10^{-2}	5×10^{-2}	4×10^{-2}	1×10^{-1}	5×10^{-2}	2×10^{-1}
Ordinary						
$R(I)_{6,1}$	2×10^{-8}	3×10^{-8}	5×10^{-9}	5×10^{-8}	2×10^{-8}	2×10^{-7}
$R(I)_{6,2}$	9×10^{-7}	2×10^{-7}	3×10^{-8}	3×10^{-7}	1×10^{-7}	1×10^{-6}
$R(I)_{6,3}$	3×10^{-3}	5×10^{-3}	9×10^{-4}	9×10^{-3}	4×10^{-3}	4×10^{-2}
$R(RV)_{6,1}$	9×10^{-6}	1×10^{-5}	9×10^{-6}	3×10^{-5}	1×10^{-5}	4×10^{-5}
$R(RV)_{6,2}$	5×10^{-5}	8×10^{-3}	5×10^{-5}	2×10^{-2}	7×10^{-3}	2×10^{-2}
$R(RV)_{6,3}$	3	4	3	8	4	12

about year 1985. It is assumed that all existing environmental and safety regulations, as well as those expected to be implemented prior to 1980, are met. A 1-GW electric power plant is estimated to supply 7×10^9 kilowatt-hours (kWh) at an 80% load factor, which is equivalent to 0.07 quads (10^{15} Btu). Thus 100 GW is equivalent to 7 quads.

Historic risk data up to 1972 are based on total usage of fuel systems for all purposes. Only that portion of these risks equivalent to 7 quads can be assigned to the fuel system. An apportionment based on 1972 statistics from Table 19.1 follows:

Fuel Cycle	1972 Consumption (quads)	7 Quad Equivalent (%)
Oil	34.1	21
Gas	22.1	32
Coal	12.5	56

Table 19.8 compares various environmental and health factors for the fuel cycle in question. Annual occupational injury data from the same source—the Council on Environmental Quality[6] (CEQ)—are as follows:

Fuel	Accidental Fatalities per year (occupational)
Coal	
Deep-mined	112.3
Surface-mined	41.2
Oil	
Domestic	32.3
Import	5.7
Gas	18.3
Nuclear	5.4

These data are for accidents covering extraction, processing, transport, and conversion. The occupational health data are used for determining total regulated voluntary risk for each of the 100 plants involved. Catastrophic risks and delayed fatalities are separated from ordinary risks using CONSAD report data and other sources.

Other assumptions are required—namely, that 40% of oil used will be imported and that 28% of gas used will be LNG shipped by tanker. Risks for LNG and nuclear energy will be determined from existing models; other systems will use historic data.

Ordinary involuntary risks arise from accidents and delayed effects due to release of pollutants. Accidents for oil and natural gas associated with home heating, use of fuel in transportation, and cooking, and are not related to the production of electrical energy. Accidents relating to transportation of fuels, but not involving fires, spills, or radiation exposure, are considered as regular transportation accidents not peculiar to energy production. On this basis, ordinary involuntary risks from oil, gas, and coal are basically from chemical effluents and from nuclear energy radioactive exposure. These risks usually involve delays between exposure and onset of consequences.

Table 19.9 shows the impact of introducing additional, available environmental controls to the systems listed in Table 19.8. These data do not include modification of existing systems, but they indicate the extent to which required controls can be effective. These factors are assumed to be implemented for the systems covered.

The problem of health effects from pollutant emissions is the most difficult aspect of risk estimation, especially for the oil and coal fuel cycles, which involve sulfur dioxide, sulfates, and particulate releases to the atmosphere. The health effect that can be attributed to these pollutants is a function of the selection of either a threshold or a nonthreshold dose-effect model for these pollutants. A best judgment of short-term exposure thresholds appears in Table 19.10. Conversely, experimental data in the

Table 19.8 *Comparative Environmental Impacts of 1000-MW Electric Energy Systems Operating at a 0.75 Load Factor with Low Levels of Environmental Controls or with Generally Prevailing Controls*

	Air Emissions			Water Discharges				Solid Waste			Land Use		Occupational Health		
System	Tons ($\times 10^3$)	Curies ($\times 10^3$)	Severity[a]	Tons ($\times 10^3$)	Curies ($\times 10^3$)	Btu ($\times 10^{13}$)	Severity[a]	Tons ($\times 10^3$)	Curies ($\times 10^8$)	Severity[a]	Acres ($\times 10^3$)	Severity[a]	Deaths	Workdays Lost ($\times 10^3$)	Potential for Large-Scale Disaster
Coal															
Deep-mined	383	—	5	7.33	—	3.05	5	602	—	3	29.4	3	4.00	8.77	Sudden subsidence in urban areas, mine accidents
Surface-mined	383	—	5	40.5	—	3.05	5	3,267	—	5	34.3	5	2.64	3.09	Landslides
Oil															
Onshore	158.4	—	3	5.99	—	3.05	3	NA	—	1	20.7	2	0.35	3.61	Massive spill on land from blowout or pipeline rupture
Offshore	158.4	—	3	6.07	—	3.05	4	NA	—	1	17.8	1	0.35	3.61	Massive spill on water from blowout or pipeline rupture
Imports	70.6	—	2	2.52	—	3.05	4	NA	—	1	17.4	1	0.06	0.69	Massive oil spill from tanker accident
Natural gas	24.1	—	1	0.81	—	3.05	2	—	—	0	20.8	2	0.20	1.99	Pipeline explosion
Nuclear	—	489	1	21.3	2.68	5.29	3	2,620	1.4	4	19.1	2	0.15	0.27	Core meltdown, radiological health accidents

Source: "Energy and the Environment, Electric Power."[6]

[a] Severity rating key: 5 = serious, 4 = significant, 3 = moderate, 2 = small, 1 = negligible, 0 = none.

Table 19.9 *Cost (%) of Controls and Changes in Environmental Impacts of 1000-MW Electric Energy Systems Operating at 0.75 Load Factor with a High Level of Environmental Controls*

	Air		Water[a]		Land		Solid Waste[b]	
System	Tonnage Change	Cost Increase	Tonnage Change	Cost Increase	Acreage Change[c]	Cost Increase	Tonnage Change	Total Cost Increase[d]
Coal								
Deep-mined	−81.3	23	−96.2	4	+1	0	+159	28
Surface-mined	−81.3	23	−92.4	4	−37	4	+29	31
Oil								
Onshore	−73.0	31	−38.8	5	0	0	0	36
Offshore	−73.0	31	−38.8	5	0	0	0	36
Imports	−39.4	28	−77.2	5	0	0	0	34
Natural gas	0	0	0	5	0	0	0	5
Nuclear	−29.3[f]	1	0	4	0	0	0	5

Source: "Energy and the Environment, Electric Power."[6]

[a] Costs of cooling tower construction and operation are considered with water controls.

[b] Solid waste costs are included in air and water pollution controls.

[c] Land impacts are reduced by reclamation but are increased for solid waste disposal.

[d] Total percentage cost increase may differ from sum of media percentage cost increases because of rounding.

[e] In the case of oil-fired systems, price increases for desulfurized oil are the primary cost effect.

[f] Radioactive emissions from the nuclear system are reduced by controls at the powerplant.

Table 19.10 *Best Judgment Exposure Thresholds for Adverse Effects*

Effects	24-hour Threshold ($\mu g/m^3$)		
	Sulfur Dioxide	Total Suspended Particulates	Particulate Sulfate
Short term			
Mortality harvest	300–400	250–300	—[a]
Aggravation of symptoms in elderly	365	80–100	8–10
Aggravation of asthma	180–250	100	8–10
Acute irritation symptoms	340	170	—[a]
Present standard	365	260	—[b]
	Annual Threshold ($\mu g/m^3$)		
Long term			
Decreased lung function of children	200	100	11
Increased acute lower respiratory disease in families	90–100	80–100	9
Increased prevalence of chronic bronchitis	95	100	14
Present standard	80	75 (Geometric)	No Standard

Source: Environmental Protection Agency, CHESS Program.[7]
[a] No data.
[b] No standard.

New York area show a linear dose relationship in the range of 20 to 1000 micrograms per cubic meter ($\mu g/m^3$). This relationship (Table 19.11) represents a 1% increase in mortality rate for each incremental change of 300 $\mu g/m^3$. Since these are observed changes and sulfur dioxide is accompanied by sulfates and particulates, this represents to some extent total increased mortality. If the figures are applied nationally, the increased mortality rate is 1% of the U.S. mortality rate of 9.42×10^{-3} fat/yr from all causes (1974) or 9.42×10^{-5} fat/yr for a 300 $\mu g/m^3$ incremental dose. This is equivalent to 3.1×10^{-7} fat/yr for a 1 $\mu g/m^3$ increment.

North[8] has undertaken a study of the health effects from sulfur emission from coal plants, using a diffusion model for different locations and types of coal-fired plants affecting 50 million people in the New York area. Based on this model, two types of plants are of interest because they meet maximum environmental requirements. A plant located in Appalachia burning low sulfur (0.9%) coal puts out 48,200 tons of sulfur dioxide for a 1000-MW power plant operating at 80% load factor. This results in an increases in ambient sulfate and sulfur dioxide of 0.07 and 0.17 $\mu g/m^3$, respectively. A similar size plant located in the New York area, burning high sulfur coal but with flue gas desulfurization (FGS), discharges 17,100 tons of sulfur dioxide per year and yields increases in New York City ambient sulfate and sulfur dioxide of 0.33 and 0.80 $\mu g/m^3$. Assuming that half of new plants are located remotely and are equivalent to the first case, and the remaining half are located near population centers as in the second case, the average change in ambient for a single average 1000-MW plant is 0.5 $\mu g/m^3$, representing an increased risk rate of 1.5×10^{-7} fat/yr using the linear relationship. The effects would not be additive for 100 such plants because they would not be all colocated nor would their effects overlap. An estimate of a maximum of four total overlaps is assumed, providing an overall rate of 6×10^{-7} fat/yr.

North[9] also estimates that chronic respiratory disease occurs at a rate 1850 times higher than fatalities. This is a rate of 1×10^{-3} he/yr. Property damage from materials damage by sulfur oxides and sulfates are

Table 19.11 *Relationship of Daily Sulfur Dioxide Levels and Daily Mortality Rates in the New York City Metropolitan Area*

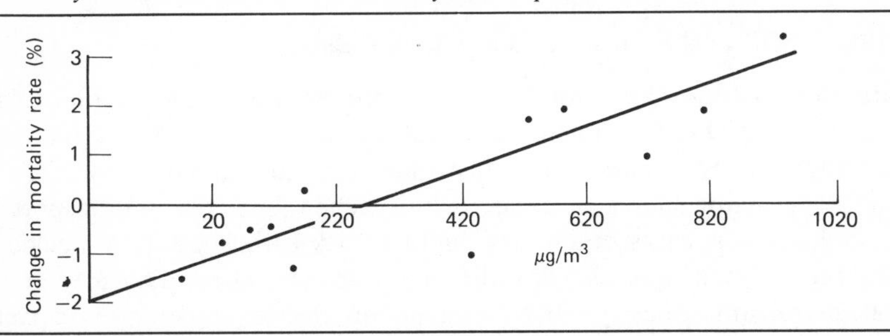

Source: A. Hershaft and G. Shea, "Assessment of the Interpretative Utility of Air Pollution Damage Functions," prepared by Enviro Control, Inc., for the Council of Environmental Quality under contract E05AC012, 1975; based on R. W. Buechley et al., "SO_2 Levels and Perturbations in Mortality," *Archives of Environmental Health,* Vol. 27, September 1973, p. 134, and M. Glasser and L. Greenberg, "Air Pollution Mortality and Weather: New York City 1960–1964." Presented at the annual meeting of the American Public Health Association, Philadelphia, November 11, 1969.

equivalently at a rate of 1×10^{-1} dollars per year, and damage from acid rain is 6×10^{-2} dollars per year for a total (rounded) of 2×10^{-1} \$/yr.

All these figures are based on assumptions of no threshold and are invalid if a threshold exists and the ambient is below these levels. Furthermore, a remote plant burning high sulfur coal with flue gas desulfurization would reduce all the figures by a factor of 3, for a release of 17,000 tons of sulfur dioxide per year.

Radiation health effects assume a linear, nonthreshold relationship with dose. Values of dose-effect relations are obtained from the BEIR report.[10] These effects, involving cancer and genetic risks, have not been observed directly but were extrapolated from dose-effect data taken at higher exposure levels in man and animals. Even though not directly observable, the radiation data represent a confidence in the information several orders of magnitude greater than similar dose-effect data for air pollutants. The health effects for sulfur oxides and particulates are, indeed, observed, but dose-response and threshold questions are not easily quantified with information presently available.

Table 19.8 shows other environmental effects on land use, aesthetics, water purity, and so on, that do not threaten life directly and are not a part of the risk acceptance decision. They should be considered as part of the indirect gain-loss balance; however such a balance would have to be finer grain and more precise than attempted here. The factors represent societal losses but are not directly considered as risks.

19.6.1 Oil Fuel System Measured Risk Levels

Data on catastrophic risk levels are obtained from an examination of data in the CONSAD report for the oil industry and are averaged over the years 1953 to 1973. The sources of catastrophic risk are summarized in Table 19.12. The major source of catastrophic occupational risk is transport by oil tanker. For catastrophic involuntary risk, transportation accidents involving railroad tank cars are the major source. These data are for the total petroleum industry, and 7 quads of electric generation capacity represent only 21% of the values shown.

Total occupational risks are obtained from the Council on Environmental Quality data[6] and Table 19.8. Based on a 60/40 domestic-import mix for 100 GW, these risks involve 23.4 fat/yr and 2166 he/yr, all accidental.

Oil spills result in property damage and cleanup expenditures. These costs are usually passed on to the general population and are the major contributor to involuntary financial losses. In 1974 spillage of petroleum products in U.S. waters amounted to 15.7 million gallons.[11] Cost for cleanup average about \$1 to \$2 per gallon but can go as high as \$25 per gallon, as

Table 19.12 *Risk Data for the Petroleum Industry: Average Rates, 1953–1973*

	Risk Rates (fat/yr)		Risk Rates (he/yr)		Risk Rates ($/yr)	
	Voluntary	Involuntary	Voluntary	Involuntary	Voluntary	Involuntary
Catastrophic risks: CONSAD data — (avg. 1953–1973)						
Transportation						
Tanker	11.43	—	10.38	—	2,170,000	—
Pipeline	0.81	—	0.43	—	—	—
Truck	1.86	0.10	4.19	1.40	—	—
Railroad tank car	0.05	0.71	1.71	15.95	—	900,000
Refining	1.00	0.10	2.71	1.90	5,170,000	760,000
Drilling	—	0.19	—	0.71	160,000	240,000
Storage	1.00	—	9.43	—	200,000	—
Spills	—	—	—	—	—	1,000,000
Total	16.15	1.10	28.05	19.96	7,700,000	2,000,000
7 quads at 21%	3.39	0.23	5.89	4.19	1,620,000	420,000
Total occupational risk: CEQ Data	23.40		2166			
Total ordinary occupational risk	20.01		2160			
Risk rates						
Catastrophic, individual[a]		1×10^{-9}		2×10^{-8}	8×10^{-3}	2×10^{-3}
Catastrophic, occupational[b]	8×10^{-6}		1×10^{-5}			
Ordinary, individual[a]		4×10^{-7}		7×10^{-4}		1×10^{-2}
Ordinary, occupational[b]	2×10^{-4}		2×10^{-2}			

Source: CONSAD report.

[a] Based on a population of 2.1×10^8.

[b] Based on 21% of a work force of 4.14×10^5.

experienced on the St. Lawrence River in 1976.[12] The average annual cost is about \$20 million, and no single spill has exceeded \$3 million in the data for 1973 and 1974. Catastrophic spills are estimated to be about 5% of the cost of all spills. Only 21% of these are attributable to electric power generation. Ordinary involuntary property damage is \$4 million/yr and catastrophic involuntary property damage is \$200,000/yr (Table 19.12). The ordinary involuntary risk rate is 2×10^{-2} \$/yr, and the catastrophic risk is added to other property damage in the table.

Sulfur oxides and sulfates produce the major deleterious health effects from air pollution. Although the preconversion processing of petroleum is a prime source of sulfur release, these activities are particularly subject to environmental controls, which are better than 96% efficient.[6] Similarly, at the conversion step environmental control requiring the use of low sulfur oil (0.6%) and/or fuel gas desulfurization results in emissions that are about 85% of those from an equivalent coal-fired plant. Using the North[9] estimates, this gives a level of 5.1×10^{-7} fat/yr for 100 plants. These premature deaths are the result of exposure over 20 to 30 years and can be discounted at a 0.5% effective discount rate for involuntary risks. For 25 years a present value factor of .88 of the future value is obtained (see Table 17.5). Using both factors, we obtain a risk level of 4×10^{-7} fat/yr for 100-GW oil-fired plants. This is equivalent to a rate of 7×10^{-4} he/yr and a property damage of 1×10^{-2} \$/yr. These estimates must be considered speculative and extremely soft.

The measured risk values are summarized and compared against risk referents in Table 19.13. Examination of measured risk rates more than an order of magnitude greater than the risk referents shows that risks from 100 GW of oil-fired electrical generation are unacceptable for regulated voluntary risks, both ordinary and catastrophic, that arise primarily in transportation (which is a high risk occupation). On an involuntary risk basis, the respiratory problems of sulfur oxide pollution are major problems, along with property damage from catastrophic accidents, including major oil spills.

19.6.2 Gas Fuel System Measured Risk Levels

Since natural gas is in limited domestic supply, importation of LNG will increase substantially. By 1980 to 1985, it is expected that the United States will be importing 24×10^{9} gallons of LNG, which is equivalent to 2.07×10^{12} cubic feet of natural gas or about 2 quads. It will be assumed that of 7 quads of gas per year for 100 GW electrical power, 2 quads will be from LNG. Even if LNG is not used for electrical generation, substitution for other uses of natural gas will take place.

Table 19.13 *Comparison of Measured Risk Rates for Oil Fuel System Against Referents*

Risk Referents		Measured Risk Rates	
	Catastrophic Risks		
$R(I)_{5,1}$	3×10^{-10} fat/yr	$M(I)_{5,1}$	1×10^{-9} fat/yr
$R(I)_{5,2}$	2×10^{-9} he/yr	$M(I)_{5,2}$	2×10^{-8} he/yr
$R(I)_{5,3}$	3×10^{-5} \$/yr[a]	$M(I)_{5,3}$	2×10^{-3} \$/yr
$R(RV)_{5,1}$	3×10^{-6} fat/yr	$M(RV)_{5,1}$	8×10^{-6} fat/yr
$R(RV)_{5,2}$	3×10^{-7} he/yr[a]	$M(RV)_{5,2}$	1×10^{-5} he/yr
$R(RV)_{5,3}$	4×10^{-2} \$/yr	$M(RV)_{5,3}$	8×10^{-3} \$/yr
	Ordinary Risks		
$R(I)_{6,1}$	2×10^{-8} fat/yr[a]	$M(I)_{6,1}$	4×10^{-7} fat/yr
$R(I)_{6,2}$	9×10^{-7} he/yr[a]	$M(I)_{6,2}$	7×10^{-4} he/yr
$R(I)_{6,3}$	3×10^{-3} \$/yr	$M(I)_{6,3}$	1×10^{-2} \$/yr
$R(RV)_{6,1}$	9×10^{-6} fat/yr[a]	$M(RV)_{6,1}$	2×10^{-4} fat/yr
$R(RV)_{6,2}$	5×10^{-5} he/yr[a]	$M(RV)_{6,2}$	2×10^{-2} he/yr
$R(RV)_{6,3}$	3 \$/yr	$M(RV)_{6,3}$	—[b]

[a] Measured level exceeds the referent by more than one order of magnitude.
[b] Data not available.

Catastrophic risk data for 5 quads of natural gas use are obtained from the CONSAD report and are summarized in Table 19.14 along with occupational risk data from Table 19.8. Risk for 2 quads of LNG and LPG are obtained from a study made by John A. Simmons for the Environmental Protection Agency.[13] This study provides an event tree analysis of the risks involved in these transportation modes, estimates the fatalities that might be expected in the industry now, and offers some perspective on future risks. In using the data, no attempt is made to ascertain the validity of the Simmons estimates, which are reviewed and argued on their own merits elsewhere.

Both LNG and LPG are considered in this study because of the similarity of their hazards and the large volume transported (or planned to be transported). Both are highly volatile liquids, and spills may create extensive flammable plumes. The plumes are negatively buoyant and tend to lie flat on the ground or water. On the other hand, there are important differences. LNG is a cryogenic liquid consisting primarily of methane and is stored at atmospheric pressure. LPG is a compressed gas consisting primarily of propane and is stored at ambient temperature at a pressure of about 100 psig. Because of this, an LPG spill results in the immediate flashing of 30 to 35% into vapor.

Table 19.14 *Risk Data for Natural Gas Industry*

	Risk Rates (fat/yr)		Risk Rates (he/yr)		Risk Rates ($/yr)	
	Voluntary	Involuntary	Voluntary	Involuntary	Voluntary	Involuntary
Catastrophic risks						
CONSAD data						
Storage	1.9	—	—	—	620,000	—
LPG RR car	—	1.2	—	7.0	—	—
LPG truck	—	0.5	—	0.4	—	—
Total—less LPG truck	1.9	1.2	—	7.0	620,000	—
5 quad equivalent @ 23%	0.4	0.3	—	1.6	140,000	—
Simmons data						
LNG tanker	0.1	0.3				
LPG truck	0.3	0.8	3.0	7.6	—	—
Total catastrophic risk	0.8	1.4	3.0	9.2	140,000	—
Total occupational risk: CEQ data	20.0		1830			
Ordinary (difference)	19.2	—	1827	—		—
Simmons data						
LNG tanker	4×10^{-3}	1×10^{-2}				
LPG truck	3×10^{-2}	0.1		1.0		
Risk rates						
Catastrophic, individual,[a]	—	3.5×10^{-8}	—	4×10^{-8}		7×10^{-4}
LNG portion		3.0×10^{-8}				
Catastrophic, occupational,[b]	2×10^{-6}	—	7×10^{-6}	—	4×10^{-1}	—
Ordinary, Individual,[a]	—	1.5×10^{-9}	—	5×10^{-9}		3×10^{-2}
LNG portion		1.0×10^{-9}				
Ordinary, occupational[b]	5×10^{-5}	—	4×10^{-3}	—		—

[a] Population 2.1×10^8 except for LNG tankers, which use a population of 10^7.
[b] Occupational population of 414,000.

About 90% of the LPG (20×10^9 gallons in 1973) is transported by truck or a combination of pipeline and truck. The average truckload is 4370 gallons. LNG is transported in tank ships carrying approximately 32.5×10^6 gallons each. Experiments with a few thousand gallons of LNG suggest that the accidental rupture of a single cargo tank could result in an LNG pool in water 1500 feet in diameter and a vapor air cloud that might remain flammable for a distance of several miles downwind. Examination of accidents involving spills of LPG and other volatile fuels indicates that a major fraction of deaths, injuries, and property damage is caused by flash fire in a large, flammable vapor plume formed before ignition. In some cases explosions and even detonations result when a portion of the plume has infiltrated a building or other confined region.

Other fires, such as liquid pool fires, often cause extensive property damage, but only infrequently cause fatalities and injuries. The explosive rupture of LPG tanks because of overheating in a fire is another mechanism that may produce fatalities. This type of accident is unlikely for LNG, however, because the fuel is stored at ambient pressure.

The accidents considered in the Simmons[13] study are limited to events that lead to the formation of a flammable plume. Table 19.15 reproduces the results of the study for leaks and tank ruptures for LPG tank trucks and LNG tank ships. Based on the definition for catastrophic accidents of 10 or more fatalities without consideration of injury or property damage, we see that these data may be divided into two categories—one for failure rate for catastrophic accidents and another for failure rate for accidents of other types.

For LNG tank or transportation, the catastrophic rate (fat/yr) is 0.4 and for the other lesser accidents it is 0.015. For LPG truck and truck pipeline transportation, the catastrophic failure rate is 0.1 fat/yr; that for other types of accidents is 1.1 fat/yr.

Of these fatalities, Simmons estimates that 76% do not involve employees of the company and are in the area of involuntary risk. The total U.S. population is subject to risk from LPG tank trucks, since liquid propane is shipped by truck throughout the country on almost all our roads. The population exposed to risks from LNG tankers is harder to estimate, since it is primarily located at and near LNG-equipped seaports. A population at risk of 10 million people, chosen for this example, has not been verified, but is seems to be a reasonable estimate, arrived at from examination of the number of seaports and the people living within reasonable distances of those seaports at any given time.

For LPG, Simmons has provided a "soft" estimate on injuries.[13] For the 36-year period 1938–1973, there were 453 reported injuries from LPG fires and explosions, giving a mean number of 10.6 injuries per year with a stan-

Table 19.15 *LNG and LPG Risk Estimates*

	Accident Frequency (per year)			
	LPG Tank Trucks		LNG Tank Ships	
Fatalities per Accident	Leak	Tank Rupture	Leak	Tank Rupture
0.001–0.003	3.5	1.30	—	—
0.003–0.01	4.6	1.7	—	—
0.01–0.03	2.9	1.1	3.5×10^{-3}	—
0.03–0.1	1.4	0.86	4.9×10^{-3}	—
0.1–0.3	0.50	0.54	1.8×10^{-3}	2.0×10^{-3}
0.3–1.0	0.15	0.17	1.2×10^{-3}	1.6×10^{-3}
1.0–3.0	0.046	0.047	7.3×10^{-4}	1.3×10^{-3}
3.0–10	0.011	0.021	4.3×10^{-4}	9.0×10^{-4}
10–30	1.2×10^{-3}	3.3×10^{-3}	1.7×10^{-4}	6.2×10^{-4}
30–100	4.2×10^{-3}	1.2×10^{-4}	3.3×10^{-5}	3.8×10^{-4}
100–300	—	—	1.2×10^{-6}	2.2×10^{-4}
$300–1 \times 10^3$	—	—	—	1.1×10^{-4}
$1 \times 10^3–3 \times 10^3$	—	—	—	5.6×10^{-5}
$3 \times 10^3–1 \times 10^4$	—	—	—	1.8×10^{-5}
$1 \times 10^4–3 \times 10^4$	—	—	—	7.4×10^{-7}

Source: Simmons[13] (pp. 52–53).

dard deviation of 46.5. No data were found for LNG for injuries, and data on property damage for both are even more suspect.*

Table 19.14 summarizes these risk estimates and the involuntary risk rates to individuals. Risk rates for LNG tankers are based on an exposed population of 10 million, whereas other involuntary risks are based on national exposure to a population of 210 million. Occupational risk rates are based on petroleum industry employment of 414,000. Property damage data are calculated using national and occupational populations for the same dollar estimate, assuming that these costs will be passed on to consumers. The data should be considered soft.

Ordinary risk occupational data are obtained from the CEQ report[6] after catastrophic risks have been subtracted from totals. The only ordinary involuntary risks are from Simmons data,[13] since there are no significant sulfur oxide emissions from combustion of gas. These figures are derived

* Data for property damage for both LNG and LPG were never requested before 1971. Data for 1972 and 1973 exist, but are too soft to include.

using 76% of the total risks for involuntary risk and 24% for voluntary risk. The risk rates for LNG are shown separately from the total in Table 19.14 because values of 3×10^{-8} fat/yr for involuntary catastrophic risk and 1×10^{-9} fat/yr for ordinary involuntary risk dominate these categories.

Table 19.16 compares measured risk rates with risk referents for natural gas. Catastrophic involuntary risks are unacceptably high; all other are acceptable. If the LNG tanker estimate of Simmons is removed (i.e., if only natural gas or LPG is used), $M(I)_{6,1}$ would be 5×10^{-9} fat/yr (for 7 quads) and would be just acceptable. For injuries, two sets of LPG involuntary catastrophic accident values are shown: for CONSAD, 0.4 he/yr based on historic data for 5 quads equivalent; and for Simmons,' 7.6 he/yr based on a 2 quads modeled estimate of equivalent. Simmons "soft" estimate may be too high. If his estimate were just 10 times higher than historic data, this value would be acceptable. The property damage value is a totally suspect number.

In summary, natural gas, if available, would be a totally acceptable method of producing electric energy in terms of risk. However if 2 quads were dependent on LNG, the catastrophic risks from LNG tankers would be unacceptable under present practice.

Table 19.16 *Comparison of Measured Risk Rates for Natural Gas Fuel System Against Referents*

Risk Referents		Measured Risk Rates	
	Catastrophic Risks		
$R(I)_{5,1}$	5×10^{-10} fat/yr[a]	$M(I)_{5,1}$	4×10^{-8} fat/yr
$R(I)_{5,2}$	3×10^{-9} he/yr[a]	$M(I)_{5,2}$	4×10^{-8} he/yr
$R(I)_{5,3}$	5×10^{-6} \$/yr	$M(I)_{5,3}$	7×10^{-7} \$/yr
$R(RV)_{5,1}$	4×10^{-6} fat/yr	$M(RV)_{5,1}$	2×10^{-6} fat/yr
$R(RV)_{5,2}$	4×10^{-3} hr/yr	$M(RV)_{5,2}$	7×10^{-6} he/yr
$R(RV)_{5,3}$	5×10^{-2} \$/yr	$M(RV_{5,3}$	4×10^{-1} \$/yr
	Ordinary Risks		
$R(I)_{6,1}$	3×10^{-8} fat/yr	$M(I)_{6,1}$	2×10^{-9} fat/yr
$R(I)_{6,2}$	2×10^{-7} he/yr	$M(I)_{6,2}$	5×10^{-9} he/yr
$R(I)_{6,3}$	5×10^{-3} \$/yr	$M(I)_{6,3}$	3×10^{-2} \$/yr
$R(RV)_{6,1}$	1×10^{-5} fat/yr	$M(RV)_{6,1}$	5×10^{-5} fat/yr
$R(RV)_{6,2}$	8×10^{-3} he/yr	$M(RV)_{6,2}$	4×10^{-3} he/yr
$R(RV)_{6,3}$	4	$M(RV)_{6,3}$	—[b]

[a] Measured level exceeds the referent by more than one order of magnitude.

[b] Data not available.

19.6.3 Coal Fuel Cycle Measured Risk Rates

Regulated voluntary risks in the coal industry include both mine accidents and respiratory disease. Table 19.6 shows accidental fatalities by year for mining for an average employment of 173,000 in the mines over this period. CONSAD data are summarized along with several other sources of data in Table 19.17. The only catastrophic accident involving involuntary catastrophic risk was a single incident of a refuse dam failure in 1972. This is a man-triggered risk and the risk referent is adjusted accordingly* in Table 19.18. The CONSAD data are adjusted for electrical generation of 7 quads by a factor of 56% of the total historic industry data.

Total risk data for a 1000 MW/yr plant is derived from Tables 19.8 and 19.9 using the CEQ study.[6] These data cover accidents only and do not include respiratory diseases. Based on tonnage supplied by each source, 42% of coal is deep mined and 58% surface mined. This mix results in a total of 320 fatalities and 7100 injuries. These figures, divided by the occupational population of 173,000, represent ordinary occupational risks after subtracting catastrophic risks from the total.

Health effects from sulfur emissions are determined by using the North data,[9] as previously described, for oil, except that the 85% factor for oil is omitted, and a .88 discount factor is used. A premature death rate of 5×10^{-7} fat/yr, a chronic respiratory disease rate of 9×10^{-4} he/yr, and a property damage rate of 2×10^{-2} \$/yr result. Again these results must be considered speculative and extremely questionable.

Well-known occupational risks for miners are black lung disease and other respiratory ailments. Based on the 1950 Public Health Service Study of respiratory disease in American coal miners, observed deaths from respiratory disease exceed expected deaths by nearly a factor of 5.[14] The average death rate for respiratory disease in the United States in 1971 was 14.5 deaths per 100,000 population. Thus the increased death rate of miners is equivalent to 6×10^{-4} fat/yr. The latency period is about 20 years, and a regulated voluntary discount rate of 2.5% is assumed.† This results in a discount factor of .67, and the discounted risk rate for respiratory disease for miners is 4×10^{-4} fat/yr.

Property damage values have not been included, since the data presented in the CONSAD report and elsewhere are incomplete. Basically, there is very little property damage from coal except that due to mining. These are direct voluntary risks to the mine owners. However North[9] has indicated material damage costs from sulfates and sulfur oxides to be about \$3.4 million for a populated area involving 50 million people. For a 5 million

* Increased by a factor of 2 over that shown in Table 19.7.

† Five times the involuntary risk rate.

Table 19.17 *Risk Data for Coal Industry*

	Risk Rates (fat/yr)		Risk Rates (he/yr)		Risk Rates ($/yr)	
	Voluntary	Involuntary	Voluntary	Involuntary	Voluntary	Involuntary
Catastrophic risks						
CONSAD data (1953–1973)						
Coal mine (avg)	18.3	—	1.7	—	—	—
Refuse dams (avg)	—	5.5	—	0.5	—	—
Total electric generation, 56%	10.3	3.1	1.0			
Total coal mine risk						
Bureau of Mines data	208	1.7				
Total electric generation, 56%	116	1.0				
Total risks: CEQ data						
42% deep, 58% surface	320		7100			
Ordinary risks						
Mines	106	1.0				—
Other	204					
Total	310		7100	—		—
Risk rates						
Catastrophic, individual[a]		1×10^{-8}		5×10^{-9}		
Catastrophic, occupational[b]	6×10^{-5}		5×10^{-9}			
Ordinary, individual[a]		5×10^{-7}		9×10^{-4}		2×10^{-2}
Accident portion		5×10^{-9}				
Ordinary, occupational risk[b]	2×10^{-3}		4×10^{-2}			
Respiratory disease portion	4×10^{-4}					

[a] Population, 2.1×10^8.

[b] Occupational population, 173,000.

average population for each of 100 plants throughout the United States, this is about $17 million and represents a value of 8×10^{-2} $/yr.

The measured risk values are summarized and compared against two sets of risk referents in Table 19.18. When strip mine reclamation takes place, coal is unacceptable in two areas—involuntary health effects and occupational fatalities for ordinary risks. When strip mine reclamation is not implemented, all risks are unacceptable except for catastrophic health effects for involuntary and regulated voluntary risks (miners either are killed or survive mine disasters; few injuries result). The major problem areas are risk to miners and health effects from sulfur oxides.

19.6.4 Nuclear Fuel Cycle Measured Risk Rates

Since there have been very few identifiable deaths or injuries from nuclear energy, modeled risk data must be used to estimate risk levels from planned and accidental releases of radioactivity from radioactive effluents and exposure, as well as for ordinary accidents in the fuel cycle. Catastrophic risk data are derived from the Reactor Safety Study[4] modified by an EPA

Table 19.18 *Comparison of Measured Risk Rates for Coal Fuel System Against Referents*

	Risk Referents			
	No Reclamation	Reclamation	Measured Risk Rates	
		Catastrophic Risks		
$R(I)_{7,1}$	2×10^{-10a}	2×10^{-9} fat/yr	$M(I)_{7,1}$	1×10^{-8} fat/yr
$R(I)_{7,2}$	1×10^{-9}	1×10^{-8} he/yr	$M(I)_{7,2}$	2×10^{-9} he/yr
$R(I)_{7,3}$	2×10^{-5}	2×10^{-4} $/yr	$M(I)_{7,3}$	—
$R(RV)_{5,1}$	3×10^{-6a}	8×10^{-6} fat/yr	$M(RV)_{5,1}$	6×10^{-5} fat/yr
$R(RV)_{5,2}$	3×10^{-7}	8×10^{-7} he/yr	$M(RV)_{5,2}$	5×10^{-9} he/yr
$R(RV)_{5,3}$	4×10^{-2}	1×10^{-1} $/yr	$M(RV)_{5,3}$	—
		Ordinary Risks		
$R(I)_{6,1}$	5×10^{-9a}	5×10^{-8} fat/yr	$M(I)_{6,1}$	5×10^{-7} fat/yr
$R(I)_{6,2}$	3×10^{-8a}	3×10^{-7a} he/yr	$M(I)_{6,2}$	9×10^{-4} he/yr
$R(I)_{6,3}$	9×10^{-4a}	9×10^{-3} $/yr	$M(I)_{6,3}$	2×10^{-2} $/yr
$R(RV)_{6,1}$	9×10^{-6a}	3×10^{-5a} fat/yr	$M(RV)_{6,1}$	2×10^{-3} fat/yr
$R(RV)_{6,2}$	5×10^{-5a}	2×10^{-2} he/yr	$M(RV)_{6,2}$	4×10^{-2} he/yr
$R(RV)_{6,3}$	3	8	$M(RV)_{6,3}$	—[b]

[a] Measured level exceeds the referent by more than one order of magnitude.
[b] Data not available.

Table 19.19 *Approximate Average Societal and Individual Risk Probabilities per Year from Potential Nuclear Plant Accidents*[a]

Consequence	Risk to Society	Risk to Individual
Early fatalities[b]	3×10^{-3}	2×10^{-10}
Early illness[b]	2×10^{-1}	1×10^{-8}
Latent cancer fatalities[c]	7×10^{-2}/yr	3×10^{-10}/yr
Thyroid nodules[c]	7×10^{-1}/yr	3×10^{-9}/yr
Genetic effects[d]	1×10^{-2}/yr	7×10^{-11}/yr
Property damage ($)	2×10^{6}	—

Source: Data from Reactor Safety Study.[4]
[a] Based on 100 reactors at 68 current sites.
[b] Individual risk value based on the 15 million people living in the general vicinity of the first 100 nuclear power plants.
[c] This value is the rate of occurrence per year for about 30 years following a potential accident. The individual rate is based on the total U.S. population.
[d] This value is the rate of occurrence per year for the first generation born after a potential accident; subsequent generations would experience effects at a lower rate. Individual rate based on total U.S. population.

factor. Ordinary risk data are derived from a paper by Ellett and Richardson[15] using BEIR estimates.[10]

Involuntary catastrophic risks are derived from the Reactor Safety Study for 100 reactors as shown in Table 19.19. It is estimated by the EPA that these factors are low by at least a factor of 4, if not more, for consequences, and by a factor of 2 for probability.[16] These are within the error ranges of the Reactor Safety study.* A factor of 8 is used here, along with an effective discount factor of .86 to discount latent cancers and thyroid nodules over a 30-year delay of onset for involuntary risks. These result in catastrophic voluntary risk factors of 4.6×10^{-9} fat/year and 1.04×10^{-7} he/yr. The fatality rate covers early fatalities, latent cancer fatalities discounted over 30 years, and genetic effects. The health effect rate includes early illnesses and thyroid nodules discounted over 30 years. Property damage is calculated by taking the societal risk number and spreading it over the entire population of 210 million. This results in a rate of 4×10^{-2} \$/yr.

* The Reactor Safety Study uncertainties are represented by factors of $1/4$ and 4 on consequence magnitudes and by factors of $1/5$ and 5 on probabilities.

Catastrophic occupational risks have not been calculated to the author's knowledge. One rough estimate based on Reactor Safety Study probabilities is to assume that an accident that kills 100 people in the general population might kill 10 of the 20 to 40 work personnel at a reactor during a single shift. On this basis an accident with a probability of 1×10^{-7} per reactor year would result in 10 fatalities, and an accident with a probability of 1×10^{-8} per reactor year would produce 40 fatalities. This is about 4×10^{-8} fat/yr for a single plant and 4×10^{-6} fat/yr for 100 plants. The same argument holds for health effects, giving a risk rate of 1×10^{-4} he/yr.

Ordinary occupational risks are calculated by Bunger[17] for an exposure of 5 rems per year between the ages of 18 and 65 and are equivalent through life tables to 3.8×10^{-4} fat/yr for accidental fatalities. To this the electric utility occupational accident of 2.2×10^{-4} fat/yr must be added, to yield an ordinary occupational risk rate of 6×10^{-4} fat/yr. This compares favorably with the data for deaths in Table 19.8, based on 100 reactors and a work force of 75,000, which results in a rate of 6.12×10^{-4} fat/yr. Injuries can also be obtained from the CEQ data[6] for 5.37 injuries per year per 1000 MW electricity. This results in a rate of 2×10^{-2} he/yr.

For ordinary involuntary risk, Ellett and Richardson[15] estimate that releases from 1 GW/yr of operation of the fuel cycle will result in

	0.5	fatality from uranium mills
	0.1	fatality from reactors
	0.3	fatality from fuel reprocessing
	0.001	fatality from transportation[18]
	0.1	fatality from noncatastrophic accidents*
Total	1.0	(This is about one fatality per GW/yr)

In addition, about 0.9 additional health effects involving nonfatal cancers and nonfatal genetic effects is estimated. For the same population of 15 million people used for the Reactor Safety Study, this results in ordinary risk rates for 100 reactors of 6×10^{-6} fat/yr and 5×10^{-6} he/yr. These rates have been discounted by a factor of .86 in their derivation.

These risk rates are compared against two sets of risk referents in Table 19.20, where one set of referents is for the case of an unsolved waste disposal problem and one is for the case of ultimate disposal of wastes having been demonstrated. For the first case (the present situation for nuclear energy), catastrophic involuntary health effects and property damage

* Accidents of Nuclear Regulatory Commission classes 1 to 8 as opposed to class 9 used in the Reactor Safety Study.

Table 19.20 *Comparison of Measured Risk Rates for Nuclear Fuel System Against Referents*

	Risk Referents			Measured Risk Rates (Modeled)
	Waste Problem	No Waste Problem		
		Catastrophic Risks		
$R(I)_{5,1}$	4×10^{-10}	4×10^{-9} fat/yr	$M(I)_{5,1}$	4×10^{-9} fat/yr
$R(I)_{5,2}$	2×10^{-9}[a]	2×10^{-8} he/yr	$M(I)_{5,2}$	1×10^{-7} he/yr
$R(I)_{5,3}$	8×10^{-5}[a]	8×10^{-4}[a] \$/yr	$M(I)_{5,3}$	4×10^{-2} \$/yr
$R(RV)_{5,1}$	4×10^{-6}	1×10^{-5} fat/yr	$M(RV)_{5,1}$	5×10^{-6} fat/yr
$R(RV)_{5,2}$	4×10^{-5}	1×10^{-4} he/yr	$M(RV)_{5,2}$	1×10^{-4} he/yr
$R(RV)_{5,3}$	5×10^{-2}	2×10^{-1} \$/yr	$M(RV)_{5,3}$	—[b]
		Ordinary Risks		
$R(I)_{6,1}$	2×10^{-8}[a]	2×10^{-7}[a] fat/yr	$M(I)_{6,1}$	6×10^{-6} fat/yr
$R(I)_{6,2}$	1×10^{-7}[a]	1×10^{-6} he/yr	$M(I)_{6,2}$	5×10^{-6} he/yr
$R(I)_{6,3}$	4×10^{-3}	4×10^{-2} \$/yr	$M(I)_{6,3}$	—[b]
$R(RV)_{6,1}$	1×10^{-5}[a]	4×10^{-5}[a] fat/yr	$M(RV)_{6,1}$	6×10^{-4} fat/yr
$R(RV)_{6,2}$	7×10^{-3}	2×10^{-2} he/yr	$M(RV)_{6,2}$	2×10^{-2} he/yr
$R(RV)_{6,3}$	4	12 \$/yr	$M(RV)_{6,3}$	—[b]

[a] Measured level exceeds the referent by more than one order of magnitude.
[b] Data not available.

exceed the referents by more than a factor of 10. The health effects would even exceed the referents if the less conservative Reactor Safety Study estimates were used without modification. Catastrophic property damage exceeds the referent for both cases, but insurance up to the level of the limit set by Price-Anderson legislation* is supposed to offset this problem.

For ordinary risks, fatalities for involuntary and regulated voluntary are too high in all cases. Health effects for the first case are also too high for involuntary risks. If a total U.S. population of 210 million is used instead of 15 million, $M(I)_{6,1} = 4 \times 10^{-7}$ fat/yr and $M(I)_{6,2} = 4 \times 10^{-7}$ he/yr. Here fatalities for the initial case are still too high, but the fatalities for the second case, as well as health effects for the first case, become acceptable. Regardless of whether use of the larger population is valid, it does provide a better comparison with the other fuel cycles, since the larger population estimates were used in these cases.

* Commercial insurance companies indemnify utility companies up to a limit of $125 million, and the federal government is responsible for liability over this amount up to a level of $560 million. The Price-Anderson Act limits liability above the latter amount.

19.7 INTERPRETATION AND SUMMARY

All four of the fuel systems examined present unacceptable risks in some respects. This should not be surprising, since installation of 100 GW/yr of new electrical generation capacity is a major technological undertaking. The reasons for unacceptability, and the steps possible for reconciliation, are best separated into two parts—those for involuntary risks and those for regulated voluntary risks.

19.7.1 Involuntary Risk Inequities

For catastrophic risks for the oil fuel system, inequities arise from large oil spills and property damage from railroad tank car fires and explosions. Oil spills come under the Water Pollution Control Act, and fines and damages can be assessed to the spiller, so that costs are not necessarily passed on to the consumer or property owner. Safety of tank cars is not under such control.

Ordinary risks for the oil fuel system are unacceptable for both fatalities and health effects based on a linear nonthreshold model for air pollution from sulfur oxides and sulfate. If there were a threshold above the present standards as cited in Table 19.10, this problem might not exist. In this case, the major unacceptable risk would be property damage from railroad tank car catastrophic accidents, and so on.

For the natural gas fuel the only unacceptable risk is from catastrophic explosion and fires of LNG tankers in populated areas around ports or rivers. If LNG was not used, natural gas would be acceptable in all respects. Locating LNG loading and unloading sites in remote areas and prohibition of water traffic in populated areas would reduce risks from LNG tanker transportation to acceptable levels.

For the coal fuel system with reclamation of strip-mined land, which is highly probable in the next few years, the only unacceptable involuntary risk is from chronic respiratory disease caused by sulfur oxides and sulfates. This is based on a linear nonthreshold model; and, if a threshold exists above the present standards, this problem disappears. Coal would be acceptable on this basis, assuming strict pollution controls.

Without reclamation, the coal fuel system is virtually unacceptable in all categories of ordinary risk and catastrophic fatalities. The methodology is particularly sensitive to this reclamation question.

Demonstrated nuclear waste disposal methods will not be available before 1985 for present Energy Research and Development Administration programs. Thus the waste problem will exist for some time. On this basis, health effects and property damage for catastrophic accidents and fatalities and health effects from planned releases of radioactivity are too high. In the latter case, use of the total population risk leading to a fatality and health effect risk rate of 4×10^{-7} still yields too many fatalities.

If waste disposal methods existed, property damage from large accidents, covered by the Price-Anderson Act, would be the only problem, assuming that the larger population base leading to a risk rate of 4×10^{-7} fat/yr is used. On this basis, nuclear power would be acceptable.

19.7.2 Regulated Voluntary Risk Inequities

Risks associated with the transportation of oil, especially by tankers, are great and are not regulated. Other risks (e.g., fires and explosions at oil storage locations) involve the total petroleum industry and are not peculiar to electrical generation. The systems exist, but it may be necessary to make them safer.

The natural gas system is acceptable in all cases, including LNG and LPG transportation. This is also true for the nuclear industry except for ordinary risk to radiation workers for fatalities. However these fatality figures are based on the assumption that all workers receive the maximum regulatory dose limit of 5 rems per year. This is possible but unlikely, and the average dose to workers is less than 1 rem per year.[19] On this basis occupational risks would be acceptable. Perhaps the standard is too high.

Coal involves unacceptable risks to miners, primarily from black lung disease for the case with reclamation, and from accidents, in addition, if there is no reclamation. New occupational health and safety requirements may make these risks acceptable, but data showing improvement are not yet available. Modeled estimates could provide enough confidence if they were available.

19.7.3 SUMMARY

The methodology has been very sensitive in pointing out inequities. Amelioration of many of these inequities is possible, since extra safety measures can be taken. This may involve going back to the "as-low-as-practicable" step of cost effectiveness of risk reduction and iterating the four-step process. Even using gross values and imprecise risk estimates, critical risk inequities have been illuminated and possible approaches to reducing the inequities identified.

Again, the purpose is not to make the "final" analysis of risk acceptability of these power systems, but to present an "initial" analysis to demonstrate the value of the methodology and the need to implement it or some other equivalent approach on a wider, more precise scale. The object is to promote formal risk analysis by showing that it is posssible to do, not that the methods used here are more valid than any others.

REFERENCES

1. Joint Economic Committee, Congress of the United States, *Energy Statistics*. Hearings before the Subcommittee on Priorities and Economy in Government, 93rd Congress. Washington, D.C.: Government Printing Office, 1974.
2. Federal Energy Administration, *Project Independence Report*, November 1974.
3. "Offshore Petroleum Transfer Systems for Washington State," Oceanographic Commission of Washington, Seattle, 1974.
4. U.S. Environmental Protection Agency, "Clean Air—The Breath of Life." Washington, D.C., April 1976.
5. "Reactor Safety Study: An Assessment of Accident Risks in U.S. Commercial Nuclear Power Plants," WASH-1400 (NUREG-75/014). Nuclear Regulatory Commission, October 1975.
6. "Energy and the Environment, Electric Power," Washington, D.C.: President's Council on Environmental Quality, August 1973.
7. U.S. Environmental Protection Agency, CHESS Program. Reproduced from *Energy and Environmental Standards*, Subcommittee on Energy of the Committee on Science and Astronautics, House of Representatives, 93rd Congress, 1st and 2nd Sessions, September 25, 27; October 4, 18, 1973.
8. Warner North, "Methodology of Decision Analysis as Applied to Coal and Nuclear Fuel Cycles," *Proceedings of Quantitative Environmental Comparison of Coal and Nuclear Generation and Their Associated Fuel Cycles Workshop*, MTR-7010, Vol. 1. McLean, Va.: The MITRE Corporation, August 1975, pp. 106–136.
9. Derived from an incremental health effect assessment by Warner North for an ambient of 16 $\mu g/m^3$ as cited previously.
10. National Academy of Sciences, "The Effects on Populations of Exposure to Low Levels of Ionizing Radiation." Report of the Advisory Committee on the Biological Effects of Ionizing Radiation (BEIR). Washington, D.C.: Division of Medical Sciences, National Academy of Sciences, National Research Council, November 1972.
11. "Polluting Incidents in and Around U.S. Water, Calendar Year 1974," Washington, D.C.: Department of Transportation, U.S. Coast Guard.
12. Environmental Protection Agency, "Oil Spills and Spills of Hazardous Substances." Washington, D.C.: Environmental Protection Agency, Division of Oil and Special Materials Control, March 2, 1975.
13. John A. Simmons, *Risk Assessment and Transport of LNG and LPG*. Final Report for Contract 68,01.2695. Washington, D.C.: Environmental Protection Agency, November 25, 1974.
14. W. Baldewicz, G. Haddock, Y. Lee, Prajoto, R. Whitley, and V. Denny, "Historical Perspectives on Risk for Large-Scale Technological Systems," UCLA-ENG-7485. Los Angeles: UCLA School of Engineering and Applied Science, November 1974, p. 192.

15. W. Ellett and A. Richardson, "Estimates of the Cancer Risk Due to Nuclear Power Generation," Cold Spring Harbor Laboratory Conference on Origins of Human Cancer, Cold Spring Harbor, N.Y., September 7–14, 1976.
16. "Reactor Safety Study (WASH-1400): A Review of the Final Report," EPA-520/3-76-009. Washington, D.C.: Environmental Protection Agency, June 1976.
17. Byron Bunger and John R. Cook, "Use of Life Table Methods to Compare Occupational Radiation Exposure Risks and Other industrial Risks." Presented at the meeting of the Health Physics Society, Buffalo, N.Y., June 30, 1976.
18. "Transportation Accidents in the Nuclear Power Industry, 1975–2000," EPA-520/3-75-023. Washington, D.C.: Environmental Protection Agency, March 1975.
19. "Radiological Quality of the Environment," EPA-520/1-76-010. Washington, D.C.: Environmental Protection Agency, May 1976, p. 153.

20

Some Thoughts on the Implementation of Formal Assessment of Risks

The objective of *An Anatomy of Risk* has been to demonstrate that in spite of the complexity and imprecision inherent in risk analysis, formal assessment of risks is both desirable and possible. The development of the methodology of risk assessment undertaken is in the form of an existence theorem to demonstrate that it is possible to evaluate risks formally. No other claim is made for the particular methodology presented; its validity will be determined by its utility. It is hoped that improved methods will be developed as further work progresses.

However it is not only desirable but necessary that formal assessment of risks be carried out. The increased scope and complexity of technology and the scale of man's endeavors have made all society aware of risks that heretofore were not perceived. There is no possibility of retreat; society will continue to demand that all risks be assessed and that assessment be performed on behalf of society as a whole and in a credible, visible manner.

In the past, risk assessment has been dominated by two opposing, extreme groups. Technologists who typify Kates' "Count the Bodies" school often do society a real disservice in risk analysis because they restrict their analysis to objective risks they can measure and deny the existence of subjective risk perception. A subjective society can become antagonized by "scientific experts" pronouncements, suspecting their motives and becoming antitechnology oriented. The shortcoming is not society's; it lies with the objectively trained technocrats who are unable to admit the existence of anything they cannot count. The situation is often exacerbated by industry propaganda which is frequently counterproductive.

On the other side of the fence are those representing Kates' "Tip of the Iceberg" school. These people base the opposition to many projects entirely on emotions and consider only the extreme crises that might occur as a result of a technological undertaking. These individuals may initially serve society well by making people aware of risks they would otherwise not have considered. A few, however, have become addicted to the process of antiestablishment action, as opposed to treating the issues, and these few prevent the rest of society from making reasonable decisions. Neither all technologists nor all highly risk aversive people are included in the groups just described. However the loudest "squawks" come from these two extremes, and society is asea.

A formal approach to risk assessment can take the inputs from each of these groups along with interests of the rest of society and put them into proper perspective. Such an approach can filter out the signal from the noise.

If risk assessment is to be formalized, a number of aspects must be considered. First, risk is indeed a complex matter with many factors and variables. Too often risks are compared with one another erroneously, in disregard of the different risk factors involved. Essentially, "comparing apples with oranges" misrepresents the underlying processes. This study is an effort to indicate the existence of these factors, to estimate their sensitivity and impact on risk decisions, to quantify the effect of the different factors, and as a result, to supply a basis for comparing different types of risks. However this has only been an initial attempt with limited input. Considerably more effort is needed.

Second, the difficulties in obtaining data on risks and risk factors are considerable. For example, few data have been taken on cancer fatalities to determine which cases are involuntary or voluntary (such as workers taking risks knowingly). Nevertheless, pertinent data can be obtained (or synthesized for sensitivity analysis) to provide useful insight and to aid in decision making. The risk factors associated with man-originated involuntary, catastrophic accidents constitute an illustrative case of reasonable data available for some types of risk. Data bases and collection of data must be expanded and restructured to provide a base for examining the many risk factors identified. Both historic and experimental data are required.

Such data can be used to assure that new risks can be compared with existing levels of risk of the same type. These comparisons allow methodologies, such as the one prescribed to be formulated to determine acceptable levels of societal risk. Value judgments made in a visible, repeatable manner are an inherent part of such methodologies, imparting a subjective element to their application. However the impact of these value judgments must not be masked by improper data comparison due to oversimplification of the problem.

Several major conclusions can be stated as a result of this effort: (1) the factors affecting risk valuation can be identified and studied in detail to improve our understanding of this extremely complex individual and societal problem; (2) oversimplification of the risk problem can lead to misrepresentation of risk conditions and levels of risk acceptance; (3) the effect of risk factors and individual propensity for risk taking can be measured, but detailed data in forms suitable for such analysis are indeed difficult to obtain because the existing data acquisation base seldom permits the identified factors to be analyzed easily.

Acceptable levels of risk for society can be obtained through examination of historic societal reactions to existing risks (when risks are known) as an external referent, and comparisons of new risks against existing societal behavior for similar risks by preestablished methodologies. The methodologies involve value judgments but, if made in a visible manner, can be argued and agreements and disagreements made specific.

Two types of societal value judgments are involved: gross level, nontechnical judgments relating to the burden society will accept for an activity, and technical judgments about the degree to which given systems meet different levels of indirect benefit-cost balances and degrees of control. The separation of these value judgments allows all risk takers to participate in the key judgments made at gross levels. These gross value judgments do not imply simplification; quite the contrary, they indicate the level of precision that is meaningful to risk agents, demonstrating that further precision is fruitless. Clearly much more effort must be expended in this area.

It is hoped that this categorization and identification of risk factors and the demonstration of at least one risk acceptance methodology will stimulate further efforts in the field.

September 1976
WDR.

A

Data Bases

A.1 CONSAD REPORT DATA ANALYSIS

A.1.1 Data Analysis

A set of data developed for the Environmental Protection Agency through a contract arrangement with CONSAD Research Corporation, which resulted in a report entitled "The Consequences and Frequency of Selected Man-Originated Accident Events,"[1] is reported here. This exhaustive investigation into man-made catastrophic accidents covers the years 1953–1973, inclusive. The selection criteria consisted of the three conditions for a catastrophic accident used previously.* The original data consisted of 814 distinct entries. However, many of these entries represent events that resulted in deaths among populations outside the United States. To assure that consistent population or risk data were used, 53 events involving foreign military air crashes, 14 foreign commercial aircraft accidents, 3 events in foreign buildings and structures, and 15 foreign marine events were deleted. Table A1.1 summarizes the basic data for fatalities, along with the number of events in Table A1.2. The total number of accidents and events, the mean number of each per year, and the standard deviation for the means are also shown. A derived statistic of interest is that the mean number of fatalities per event for the total data set is 16.0 deaths per catastrophic event. This includes events that have made the data set without fatalities, since their entry may be based on the two other nonfatal conditions defining a catastrophe.

Table A1.3 gives the numbers of involuntary fatalities from the data set, which are highly dependent on the definition of involuntary risk and the availability of data as stated in the CONSAD report. This definition is given in the next section.

The CONSAD report also presents data on injuries and property damage from man-made catastrophic events. The total number of injuries from man-made catastrophic events in the United States and the number of involuntary injuries appear in Tables A1.4 and A1.5, respectively. It would

* Ten or more fatalities, 30 or more injuries, or $3 million in damages.

Table A1.1 *Number of Fatalities by Year and Source for Catastrophic Events in United States, for 1953–1973*

	Airlines										
Year	All	Military	Commercial	Railroads	Building and Structures	Marine	Mines	Bus, Auto, Truck	Miscel-laneous	Total	Total Non-military
1953	381	194	187	32	133	82	0	0	0	628	434
1954	154	68	86	4	40	140	16	0	0	354	286
1955	421	178	248	15	48	0	0	0	0	489	311
1956	282	244	38	89	59	67	0	0	12	509	265
1957	124	76	48	12	99	20	48	45	0	348	272
1958	300	180	120	50	157	51	35	27	32	652	472
1959	252	21	231	34	59	8	12	52	0	417	396
1960	534	74	460	14	22	61	18	0	0	649	575
1961	311	14	297	26	91	3	22	13	0	466	452
1962	144	44	100	20	43	38	48	10	0	298	254
1963	226	0	226	30	225	168	40	11	0	700	700
1964	295	19	276	0	62	21	0	8	0	386	367
1965	447	181	226	0	119	91	0	16	0	633	452
1966	211	59	152	31	88	56	0	10	0	396	337
1967	374	36	338	8	192	13	21	15	0	639	603
1968	377	113	259	11	101	30	99	20	16	633	520
1969	451	181	270	22	115	65	9	18	0	680	499
1970	303	76	227	3	129	31	38	9	0	513	437
1971	279	34	245	13	117	35	17	41	0	502	468
1972	232	24	200	88	211	59	91	61	0	742	718
1973	185	45	140	14	190	105	0	34	0	528	483
Total	6283	1861	4374	516	2300	1144	514	390	60	11,162	9301
$\bar{X}$/yr	299.2	88.6	208.3	24.6	109.5	54.5	24.5	18.6	2.86	531.5	442.9
σ	110.2	73.2	100.2	24.7	58.9	44.0	28.6	18.2	7.91	131.3	133.2

Source: Data from CONSAD report.[1]

Table A1.2 *Number of Catastrophic Events in United States by Source, 1953–1973*

Year	Airlines			Railroads	Building and Structures	Marine	Mines	Bus, Auto, Truck	Miscellaneous	Total	Total Non-Military
	All	Military	Commercial								
1953	20	13	7	4	12	5	0	0	0	41	28
1954	17	13	4	4	5	2	1	0	0	29	16
1955	21	12	9	2	7	0	0	0	0	30	18
1956	16	14	2	7	10	3	0	0	3	39	25
1957	14	9	5	1	7	2	2	3	1	30	21
1958	17	14	3	4	10	2	2	1	2	38	24
1959	18	8	10	2	5	1	1	4	0	31	23
1960	22	13	9	1	5	2	1	0	2	33	20
1961	9	3	6	2	8	1	1	1	0	23	20
1962	5	1	4	2	10	2	2	1	0	22	21
1963	6	3	3	1	10	2	2	1	0	22	19
1964	8	1	7	1	12	1	0	1	0	23	22
1965	16	7	9	0	9	1	0	2	0	28	21
1966	9	5	4	3	11	3	0	1	0	27	22
1967	9	2	7	1	15	1	1	2	1	31	29
1968	16	7	9	5	13	2	3	2	0	41	34
1969	20	8	12	7	21	3	1	2	0	54	46
1970	13	4	9	5	25	3	1	2	0	49	45
1971	9	2	7	3	19	2	1	4	0	38	36
1972	8	2	6	5	20	4	1	8	0	46	44
1973	9	3	6.	6	28	5	0	4	0	52	49
Total	282	144	138	66	262	47	20	39	9	727	583
$\bar{X}$/yr	13.43	6.86	6.57	3.14	12.48	2.24	0.95	1.86	0.43	34.62	27.76
σ	5.33	4.71	2.66	2.10	6.57	1.30	0.86	1.93	0.87	9.83	10.32

Source: Data from CONSAD report.[1]

Table A1.3 *Involuntary Fatalities by Year and Source for Catastrophic Events in United States, 1953–1973 (Number of Events)*

Year	Airlines			Railroads	Building and Structures	Marine	Mines	Bus, Auto, Truck	Miscel-laneous	Total	Total Non-military
	All	Military	Commercial								
1953	—	—	—	—	5(2)	—	—	—	—	5(1)	5(1)
1954	—	—	—	—	—	—	—	—	—	—	—
1955	14(1)	14(1)	—	—	1(1)	—	—	—	—	15(2)	1(1)
1956	—	—	—	—	—	—	—	—	—	—	—
1957	2(1)	2(1)	—	—	—	—	—	—	—	2(1)	2(1)
1958	1(1)	1(1)	—	2(1)	—	—	—	—	—	3(2)	2(1)
1959	8(1)	8(1)	—	20(1)	24(2)	—	—	—	—	52(4)	44(2)
1960	8(2)	2(1)	6(1)	—	2(1)	—	—	—	—	10(3)	8(2)
1961	—	—	—	—	—	—	—	—	—	—	—
1962	—	—	—	—	—	—	—	—	—	—	—
1963	—	—	—	—	68(1)	—	—	—	—	68(1)	68(1)
1964	—	—	—	—	—	—	—	—	—	—	—
1965	43(2)	43(2)	—	—	28(2)	—	—	—	—	71(4)	28(2)
1966	14(2)	14(2)	—	—	—	—	—	—	—	14(2)	—
1967	13(1)	—	13(1)	—	46(1)	—	—	—	—	59(2)	59(2)
1968	—	—	—	—	41(1)	—	—	—	—	41(1)	41(1)
1969	6(1)	6(1)	—	—	—	—	—	—	—	6(1)	—
1970	—	—	—	—	—	—	—	—	—	—	—
1971	—	—	—	—	—	—	—	36(2)	—	36(2)	36(2)
1972	25(3)	—	25(3)	—	116(1)	—	—	10(1)	—	151(5)	151(5)
1973	14(1)	14(1)	—	—	—	—	—	—	—	14(1)	—
Total	148(16)	104(11)	44(5)	22(2)	331(12)	0	0	46(3)		547(32)	445(21)
$\bar{X}$	7.0	5.0	2.1	1.1	15.8	0	0	2.2		26.1	21.2
σ	10.8	10.1	6.1	4.8	30.0	0	0	8.1		37.5	36.9

Source: Data from CONSAD report.[1]

Table A1.4 *Number of Injuries by Year and Source for Catastrophic Events in United States, 1953–1973*

Year	Airlines			Railroads	Building and Structures	Marine	Mines	Bus, Auto, Truck	Miscel-laneous	Total	Total Non-military
	A l	Military	Commercial								
1953	28	28	—	263	292	43	—	—	—	626	598
1954	12	4	8	198	175	200	—	—	—	585	581
1955	99	8	91	81	121	—	—	—	—	301	293
1956	21	21	—	446+	379	—	—	—	—	846+	825+
1957	146	70	76	5	60	43	—	17	—	271	201
1958	9	—	9	290	218	33	—	—	137	687	687
1959	16	13	3	—	35	41	—	167	—	259	246
1960	37	—	37	63	140	336	—	—	—	576	576
1961	—	—	—	90+	—	—	—	—	—	90	90
1962	26	—	26	243	355	—	—	17	—	641	641
1963	—	—	—	—	392	—	—	—	—	392	392
1964	14	14	—	32	149	24	—	60	—	279	265
1965	108	23	85	—	114	10	—	64	—	296	273
1966	62	45	17	87	301	72	—	—	—	522	477
1967	213	13	200	73	147	—	—	60	—	493	480
1968	49	—	49	340	110	47	155	34	—	725	725
1969	29	23	6	656+	107	110	31	80	—	1,013+	990
1970	151+	—	151+	228	161	67	—	92	—	699+	699+
1971	8	—	8	173	184	15	—	117	—	497	497
1972	303	—	303	618	249	12	—	223	—	1,405	1,405
1973	79	40	39	473	276	—	—	40	—	871	831
Total	1410+	302	1108+	4359+	3965	1053	186	971	137	12,074+	11,772
$\bar{X}$	67.7	14.4	52.8	207.6	188.8	50.1	8.9	46.2	6.5	575.0	304.0
σ	78.9	18.8	78.7	201.5	110.9	81.1	34.2	61.6	29.9	301.4	304.0

Source: Data from CONSAD report.[1]

Table A1.5 *Number of Involuntary Injuries by Year and Source for Catastrophic Events in United States, 1953–1973*

Year	Airlines All	Airlines Military	Airlines Commercial	Railroads	Building and Structures	Marine	Mines	Bus, Auto, Truck	Miscel-laneous	Total	Total Non-military
1953	—	—	—	—	159	—	—	—	—	159	159
1954	—	—	—	—	87	—	—	—	—	87	87
1955	8	8	—	—	40	—	—	—	—	48	40
1956	—	—	—	—	—	—	—	—	—	—	—
1957	70	70	—	—	—	—	—	—	—	70	—
1958	—	—	—	240	—	—	—	—	—	240	240
1959	13	13	—	—	—	—	—	135	—	148	135
1960	—	—	—	—	70	—	—	—	—	70	70
1961	—	—	—	—	—	—	—	—	—	—	—
1962	—	—	—	—	—	—	—	17	—	17	17
1963	—	—	—	—	340	—	—	—	—	340	340
1964	14	14	—	—	—	—	—	—	—	14	—
1965	23	23	—	—	67	—	—	—	—	90	67
1966	56	56	—	—	—	—	—	—	—	56	—
1967	40	—	40	—	5	—	—	—	—	45	45
1968	—	—	—	—	110	—	150	—	—	260	260
1969	12	12	—	—	76	—	—	—	—	88	12
1970	—	—	—	70	15	—	—	40	—	125	125
1971	—	—	—	—	33	—	—	33	—	66	66
1972	221	—	221	230	—	8	—	—	—	459	459
1973	40	40	—	130	—	—	—	—	—	170	130
Total	497	236	261	670	1002	8	150	225	—	2552	2252
$\bar{X}$	23.67	11.24	12.43	31.90	47.71	.38	7.14	10.71	—	121.52	107.24
σ	49.71	20.11	48.58	74.47	80.88	1.75	32.73	30.65	—	118.02	124.78

Source: Data from CONSAD report.[1]

seem that the data are less firm than for fatalities, since the reporting of incidents with fewer than 10 fatalities, but more than 30 injuries, may not be as complete as that for large numbers of deaths.

The data for property damage for all U.S. man-made catastrophic incidents (Table A1.6) are still more suspect, since the reporting of incidents with fewer than 10 fatalities and/or 30 injuries, but more than $30 million damage, is even more doubtful. Table A1.7 shows the property damage for incidents associated with risks that had involuntary fatalities or injuries. The property damage figures could not be separated in terms of values for voluntary and involuntary risks from the source data.

A.1.2 Comment on Classification of Involuntary Risk*

Classifying each accident within the chronology—as involving voluntary or involuntary risk—was problematic. Accidents whose description specifically noted casualties to bystanders or other victims who could not reasonably be expected to have anticipated the possibility of such an event, were classified as having an involuntary risk factor; in the absence of such specific mention, accidents were classified as involving voluntary risk only. Undoubtedly this type of arbitrary classification scheme, and its apparent lack of specificity, pose some very real problems that should be recognized before attempting to use these figures in any specific context.

An example of the weaknesses in this scheme can be found in the classification of accidents involving oil refineries and chemical plants. In most cases, available information was very sketchy—sketchy in the sense that no clear indication was given regarding what segment (i.e., employees or residents of the surrounding area) suffered the casualties. For example, an explosion at a chemical plant could involve injuries to employees, local residents, or both. If the explosion caused injuries to employees only, the risk factor was considered voluntary on the grounds that those who work at the plant consciously accept the possibility of being involved in an accident when they make the decision to work there. If the explosion resulted in injuries to residents of the surrounding area, the risk factor was involuntary—those who reside in the vicinity of the plant do not consciously accept the possibility that an accident at the chemical plant will directly affect them.

Obviously, this interpretation is only one of many interpretations possible—another being that those residing in the vicinity of the plant consciously accept the risk of an accident at the plant involving them when they decide to establish residence near the plant—making the risk factor to them a voluntary one. A case can be made for the viability of either instance.

* Directly quoted from CONSAD report.

Table A1.6 *Property Damage (millions of dollars) Reported by Year and Source for Catastrophic Events in United States, 1953–1973*

	Airlines				Building			Bus,			Total
					and			Auto,	Miscel-		Non-
Year	All	Military	Commercial	Railroads	Structures	Marine	Mines	Truck	laneous	Total	military
1953	18.25	14.75	3.50	—	130.9	12.6	—	—	6.9	168.65	153.90
1954	42.00	40.15	1.85	—	7.4	2.0	—	—	—	51.40	11.25
1955	25.00	22.00	3.00	—	31.0	—	0.25	—	—	56.25	34.25
1956	32.15	29.50	2.65	—	36.2	3.2	—	—	10.8	82.35	52.85
1957	46.45	33.45	13.00	—	18.2	4.0	—	—	24.3	92.95	59.50
1958	54.25	53.75	0.50	2.0	14.3	1.3	—	—	—	71.85	18.10
1959	80.20	63.90	16.3	—	6.7	6.0	—	10.0	—	102.90	39.00
1960	68.1	63.1	5.0	—	11.1	48.0	—	—	22.0	149.20	86.10
1961	31.2	21.0	10.2	—	32.2	4.0	—	—	—	67.40	46.40
1962	—	—	—	—	23.7	—	—	—	—	23.70	23.70
1963	18.4	18.4	—	—	16.2	—	—	—	—	34.60	16.20
1964	—	—	—	—	59.3	2.5	—	—	—	61.80	61.80
1965	10.8	5.5	5.3	—	52.0	—	—	—	—	62.80	57.30
1966	10.0	10.0	—	—	13.2	—	—	—	—	23.20	13.20
1967	0.6	—	0.6	—	261.3	—	—	—	—	261.90	261.90
1968	15.3	3.0	12.3	—	38.7	—	0.6	—	—	54.60	51.60
1969	11.0	—	11.0	—	107.3	0.9	—	—	—	119.20	112.2
1970	57.7	27.7	30.0	3.0	118.45	5.25	—	—	—	184.40	156.70
1971	—	—	—	—	60.5	20.0	—	—	—	80.50	80.50
1972	—	—	—	7.5	48.1	15.2	—	—	—	70.80	70.80
1973	13.9	4.9	9.0	9.0	125.7	—	—	—	—	148.60	143.70
Total	535.3	411.1	124.2	21.5	1212.45	124.95	0.85	10.0	64.0	1969.05	1550.95
$\bar{X}$	25.7	19.6	6.0	1.02	57.7	6.0	0.04	.48	3.1	93.8	73.9
σ	24.1	21.1	7.6	2.53	61.5	11.1	0.14	2.2	7.2	56.6	62.0

Source: Data from CONSAD report.[1]

Table A1.7 *Property Damage (millions of dollars) Reported in Association with Catastrophic Events Involving Involuntary Death or Injuries in United States by Year and Source 1953–1973*

Year	Airlines All	Airlines Military	Airlines Commercial	Railroads	Building and Structures	Marine	Mines	Bus, Auto, Truck	Miscellaneous	Total	Total Non-Military
1953	—	—	—	—	8.0	—	—	—	—	8.0	8.0
1954	—	—	—	—	—	—	—	—	—	—	—
1955	1.0	1.0	—	—	16.0	—	—	—	—	17.0	16.0
1956	—	—	—	—	—	—	—	—	—	—	—
1957	—	—	—	—	—	—	—	—	—	—	—
1958	—	—	—	2.0	—	—	—	—	—	2.0	2.0
1959	—	—	—	—	—	—	—	—	—	—	—
1960	3.2	3.2	—	—	—	—	—	10	—	10.0	10.0
1961	—	—	—	—	—	—	—	—	—	3 2	—
1962	—	—	—	—	—	—	—	—	—	—	—
1963	—	—	—	—	—	—	—	—	—	—	—
1964	—	—	—	—	—	—	—	—	—	—	—
1965	—	—	—	—	10.0	—	—	—	—	10.0	10.0
1966	—	—	—	—	—	—	—	—	—	—	—
1967	.65	—	.65	—	—	—	—	—	—	.65	.65
1968	—	—	—	—	2.0	—	—	—	—	2.0	2.0
1969	—	—	—	—	—	—	—	—	—	—	—
1970	—	—	—	—	5.0	—	—	—	—	5.0	5.0
1971	—	—	—	—	—	—	—	—	—	—	—
1972	—	—	—	7.5	—	10.1	—	—	—	17.6	17.6
1973	4.9	4.9	—	9.0	—	—	—	—	—	13.9	9.0
Total	9.75	9.10	.65	18.5	41.0	10.1	—	10	—	89.35	80.25
$\bar{X}$	.46	.43	.03	.88	1.95	.48	—	.48	—	4.25	3.82
σ	1.25	1.25	.14	2.50	4.30	2.20	—	2.18	—	6.00	5.65

Source: Data from CONSAD report.[1]

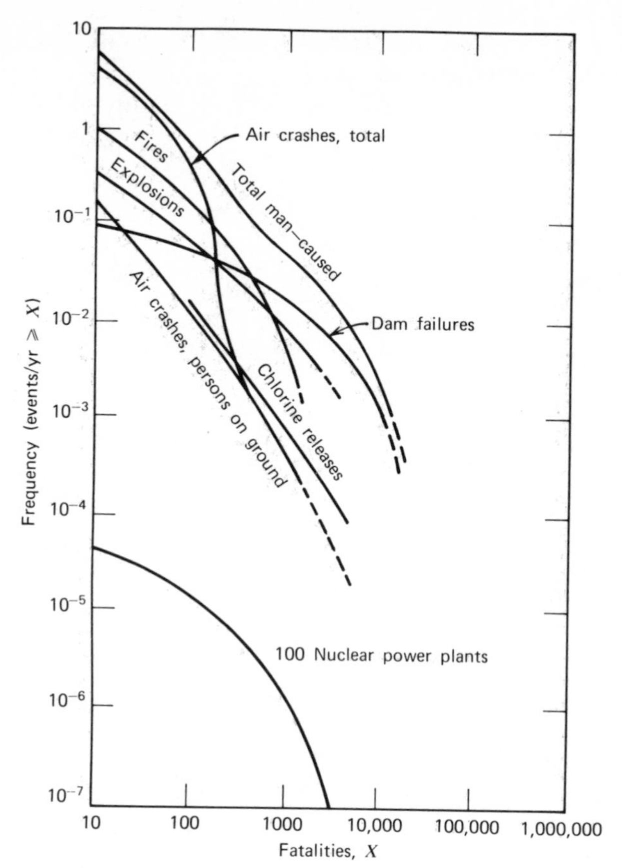

Figure A3.1. Frequency of man-caused events involving fatalities.
NOTE: Fatalities due to auto accidents are not shown because data are not available for large consequence accidents. Auto accidents cause about 50,000 fatalities per year. *Source:* Reactor Safety Study, p. 119.

Keeping the feasibility of those interpretations in mind, one could then deduce that there are very few instances of accidents involving involuntary risk—again, classification is very difficult and not entirely reliable.

We fully recommend this classification—voluntary or involuntary risk factor—should be used always, keeping in mind the apparent weaknesses of the scheme.

A.2 DATA ON NATURAL DISASTERS

Kates has compiled information on natural disasters using the data of the Disaster Research Unit Study and his own Natural Hazard Research

Group. Table A2.1 compares disaster numbers, type, and region of occurrence. The first two columns are attempts at total compilation; the remaining estimates are more incomplete having been compiled from records of selective disaster aid.

Sources are standard English language international newspapers and reference works; the period chosen, 1947–73, designed to include only the period of modern, post-World War II, expanded communication. The compilation suffers from biases of definition from under-representation of Northern Europe, Africa, U.S.S.R and China, and overrepresentation of the United States; and from the omission of drought in order to avoid the obvious under-reporting of occurrence and confusions over drought timing and territorial extent.

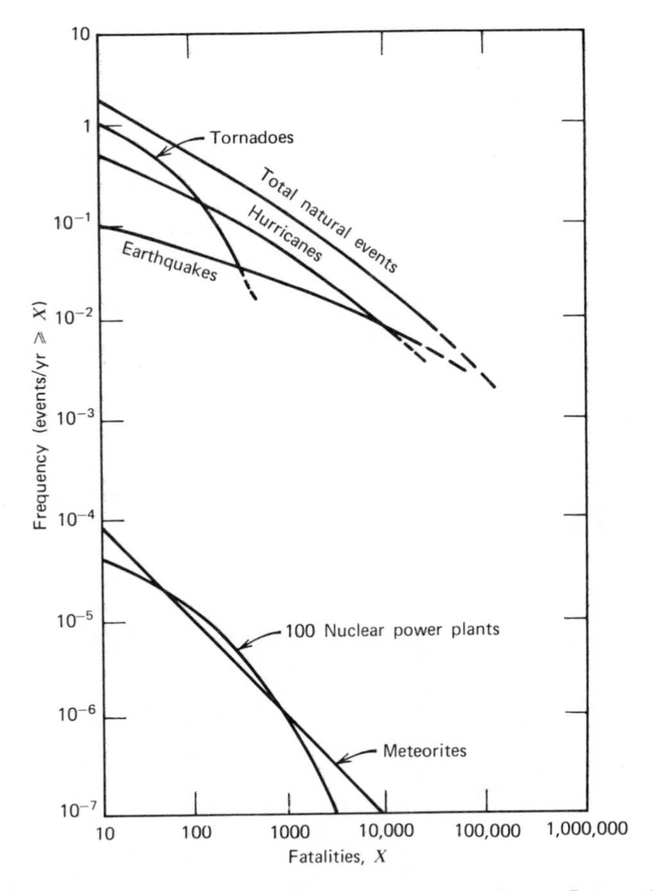

Figure A3.2. Frequency of natural events involving fatalities. *Source:* Reactor Safety Study, p. 120.

Table A2.1 *Global Disasters and Deaths: Available Compilations by Number, Type, and Region*

	Recorded Disasters		Disasters in Which Aid Was Given				Recorded Deaths
Years of Record	(1) Natural Hazard Research Group, 1947–1973	(2) UN Disaster Relief Office, 1919–1971	(3) League of Red Cross Societies' Major Appeal, 1961–1970	(4) Catholic Relief Services, 1970–1972	(5) US Agency for International Development, 1968–1971	(6) (UK) Office of Disaster Monitoring, 1974–1975 <1 yr.	(7) Natural Hazards Research Group, 1947–1973
Total	833	251	112	110	104	13	848,815
per year	31	5	11	55	26	13	27
1968–1971 average	28	13	NA	NA	26	NA	NA
Disaster type							
Flood	269	NA	58	51	41	5	182,270
Tropical cyclone	169	NA	18	15	24	3	467,508
Earthquake	115	NA	19	7	19	1	129,076
Drought and famine	NA	NA	8	34	18	2	NA
Volcano	13	NA	3	3	—	—	7,220
Epidemic	NA	NA	6	—	2	—	NA
Other	267	NA	—	—	—	2	62,741
Region of occurrence							
Asia	{325	89	46	{45	32	{5	{706,566
Middle East		38	20		5		
Africa	23	33	22	31	19	6	18,994
Americas	—	42	11	—	41	1	—
United States/Canada	287	—	—	—	—	—	9,177
Latin America	121	—	—	31	—	—	88,963
Europe	102	49	13	3	7	1	20,698
Australia	13	—	—	—	—	—	4,417

Sources: Data from Kates.[2] Column 1, Dworkin,[4] columns 2–6, Baird et al.,[5] column 7, Dworkin.[4]

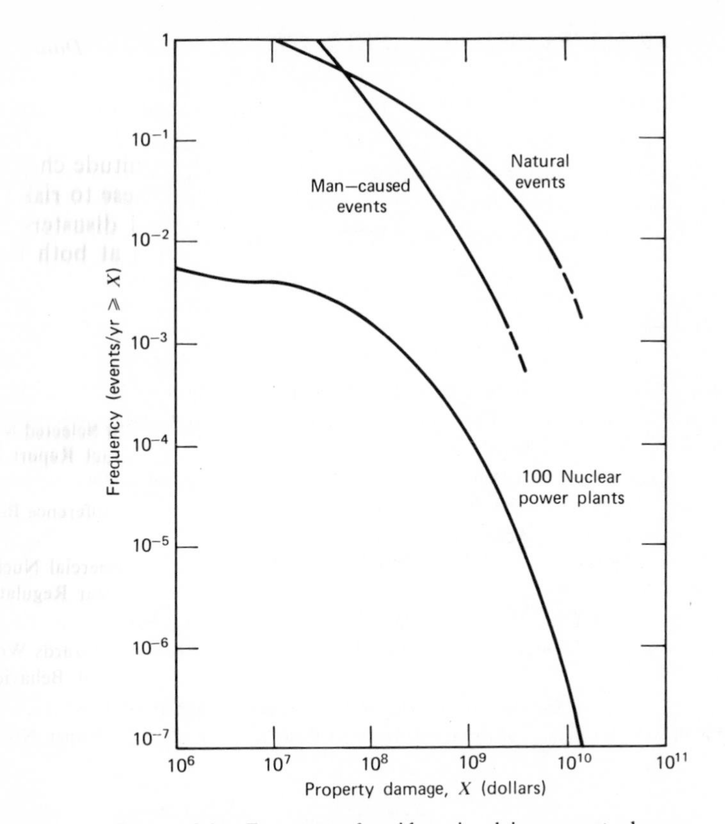

Figure A3.3. Frequency of accidents involving property damage.
NOTE: Property damage due to auto accidents is not included because data are not available for low probability events. Auto accidents cause about $15 billion damage each year. *Source:* Reactor Safety Study, p. 121.

If the number of disasters and their death tolls are sketchy at best, estimates of the global losses of material wealth and productive activity and of the costs of disaster aid or prevention are non-existent. With the organization of the United Nations Disaster Relief Office in 1972, better compilations may be hoped for. But these will be always approximations at best, because of the distortions in making damage estimates.

Earthquakes are relatively over-estimated because of the ease of their detection and drought under-estimated because of difficulties in its definition, timing, siting and in distinguishing drought impacts from perennial seasonal hunger or malnutrition. And the proportion of deaths accounted for between floods and tropical cyclones is probably reversed—floods more frequent and doing the greater damage; tropical cyclones are more deadly.[2]

A.3 DATA FROM REACTOR SAFETY STUDY

The Reactor Safety Study[3] uses frequency-consequence magnitude charts to present data. Appendix B.2 provides means for converting these to risk rate format. Figures A3.1 and A3.2, respectively, cover natural disasters and man-made disasters, based on fatalities. Figure A3.3 looks at both man-made and natural disasters based on property damage.

REFERENCES

1. "CONSAD Research Corporation, "The Consequences and Frequency of Selected Man-Originated Accident Events." Environmental Protection Agency Contract Report No. 68-01-0492. Washington, D.C., 1975.
2. Robert W. Kates, "Natural Disasters and Development." Wingspread Conference Background Paper, Racine, Wis., October 19–22, 1975.
3. Reactor Safety Study: An Assessment of Accident Risks in U.S. Commercial Nuclear Power Plants, WASH-1400 (NUREG-75/014). Washington, D.C.: Nuclear Regulatory Commission, October 1975.
4. J. Dworkin, "Global Trends in Natural Disasters, 1947–1973," Natural Hazards Working Paper No. 26. Boulder, Colo.: University of Colorado, Institute of Behavioral Science, 1974.
5. A. Baird, "Toward an Explanation of Disaster Proneness," Occasional Paper No. 10. United Kingdom: University of Bradford, Disaster Research Unit, June 1975.

B

Problems of Data Analysis

B.1 AN EXAMPLE OF DIFFICULTY IN ANALYZING RISK DATA

There is a wealth of data available concerning voluntary risks in the coal mining industry. Annual reports from the U.S. Department of Interior, Bureau of Mines, provide information on in-mine and work-related fatalities in the form of Mineral Industry Surveys. The data go back at least as far as 1941.[1] The overall industry trend in total fatalities per year is downward, and the frequency rates of fatalities per million man-hours show a slight downward trend, indicating positive systemic control. Analysis of different types of mining operations and accident situations reveals a variety of processes, some positively controlled, some not.[2] The decline of the number of fatalities is even greater if it is not measured in fatalities per year, but in fatalities per tons of coal mined. Because of mechanization, the efficiency or productivity of the industry per mine worker has increased manyfold in the last 20 to 30 years. Hence for the coal mined, fewer workers are employed. In 1973 the total individual voluntary risk rate was about 1.1×10^{-3} fat/yr for 132 fatalities for about 120,000 mine workers.[3]

On the other hand, involuntary risks associated with coal mining are derived from subsidence of structures, refuse pile movement, refuse dam failure, and attractive nuisances from abandoned mines and refuse piles. The 1972 refuse dam failure at Saunders, West Virginia, is well documented,[4] and 125 fatalities of an involuntary nature occurred. However failures of this kind occur infrequently, and although workers in the field remember others, there is little documentation of frequency and average number of fatalities. More important, no record of the failures or failure rates of such dams* or the populations at risk as a result of a dam's existence was found even after exhaustive search and direct communication with the Mining Enforcement and Safety Administration. In other words, there are inadequate data to compute the involuntary risk rate. This is also

* There is only a record in West Virginia of the most important of such dams as a result of the 1972 disaster.

true of subsidence fatalities (if any exist) or involuntary fatalities from refuse pile movement in the United States. Likewise, the Bureau of Mines keeps records of involuntary fatalities only for government-controlled mines. For example, some fatalities occur in old abandoned mines where workers or children fall into shafts, and so on, but there are no official records of such accidents.

The problem is exacerbated because if the events and magnitude of consequences are well documented, estimation of the actual population at risk is usually lacking. The inescapable conclusion is that society in the past has been so unconcerned with involuntary risks from industrial activity that we now lack data in this area. Existing labor contracts of the United Mine Workers have no provisions for involuntary fatalities from coal mining activities, although they offer protection to the miners. It may be that the possibility of a large number of involuntary fatalities from the nuclear power industry is, for the first time, making the problem of involuntary risk from man-made activities a central issue.

B.2 DIFFERENCES IN PRESENTATION OF DATA

The validity of conclusions drawn from analysis of data depends wholly on the quality of the data base and validity of statistical inferences. Valid comparison of results among different data bases is even more difficult to achieve. Problems that must be considered involve the comprehensiveness

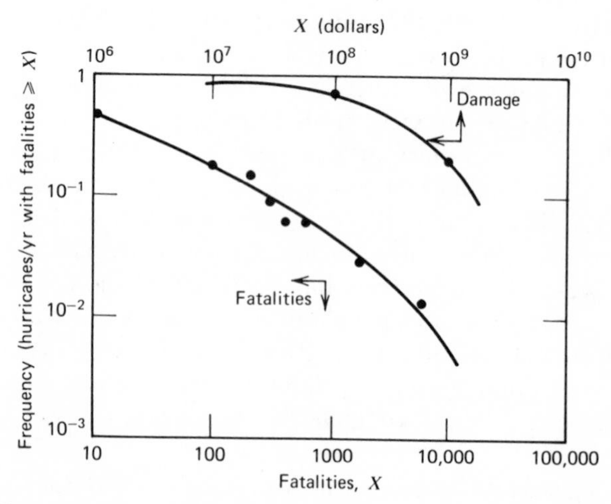

Figure B2.1. Frequency of hurricane consequences. *Source:* Reactor Safety Study.

of the data reporting system, the definition of reportable events and resulting ambiguity in report generation, along with valid use of statistical presentations.

An example of the need to resolve statistical presentation methods is provided by the Reactor Safety Study.[5] Data are given for 51 events associated with major hurricanes in the United States between 1900 and 1972. The consequence ranges from 6000 fatalities to none. Of these, 46 specific events having fatalities are individually listed, along with a single grouping of five events with no fatalities, but damages exceeding $5 million.

A cumulative distribution of the frequency of hurricanes with consequences greater than N is given and reproduced as Figure B2.1. Figure B2.2 is a histogram of the number of events occurring in decade intervals for the 46 events involving fatalities. On the basis of this histogram, the mode of the distribution is between 10 and 100. The mean value of the 46 events is 273.11, with a standard deviation of 915.95, and the total number of fatalities is 12,577. When a log-normal distribution is used for the data, however, the

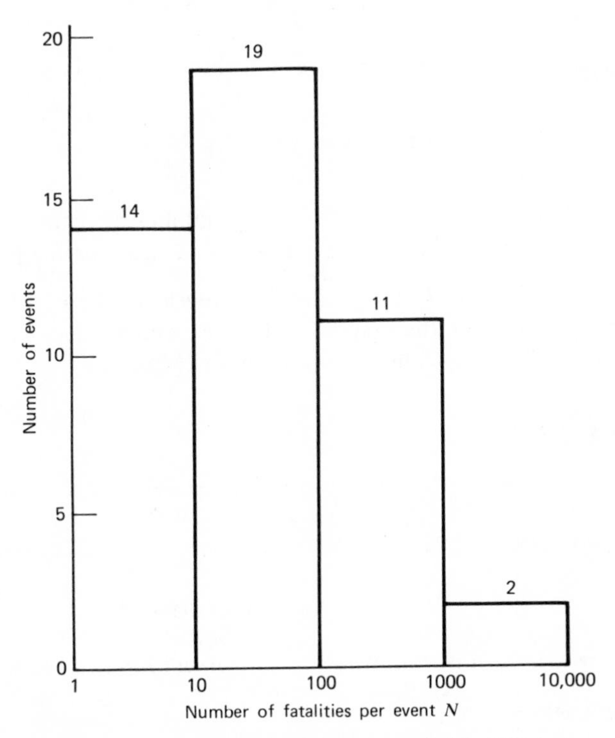

Figure B2.2. Histogram of magnitude of US. hurricanes (1900–1972) in terms of number of fatalities N on a logarithmic scale.

mean number of fatalities is 34.24, with a standard deviation of 7.76. The mean frequency of occurrence is 0.64 event per year, and probability of occurrence per year is

$$\bar{N} \text{ (normal distribution)} = 0.64 \times 273.11 = 174 \text{ fat/yr} \quad \text{(B.1)}$$

$$\bar{N} \text{ (log-normal distribution)} = 0.64 \times 34.24 = 22 \text{ fat/yr} \quad \text{(B.2)}$$

To gain the same information from the cumulative distribution in Figure B2.1 the curve in Figure B2.2 must be integrated over the number of fatalities per event. The expected value (EV) can be defined as follows:

$$EV = p(1) \times 1 + p(2) \times 2 + \cdots + p(n)_{\max} \times n_{\max} \quad \text{(B.3)}$$

where $p(n) \times n$ = probability of exactly n fatalities. However,

$$p(N \geq 1) = p(1) + p(2) + p(3) + \cdots + p(n)_{\max} \quad \text{(B.4)}$$

and

$$p(N \geq n) = p(n+1) + p(n+2) + \cdots + p(n_{\max}) \quad \text{(B.5)}$$

Therefore

$$EV = p(\mathrm{N} \geq 1) + p(\mathrm{N} \geq 2) + P(N \geq 3) + \cdots + p(N \geq n_{\max}) \quad \text{(B.6)}$$

$$EV = \sum_{n=1}^{n_{\max}} p(N \geq n)$$

where $p(\mathrm{N} \geq n)$ = (number of events $N \geq n$)/(number of years measured)
= probability of an event with $N \geq n$ occurring per year

The expected value is 255.72 fat/yr for the integral of the curve in Figure B2.1, and is, by definition, the best estimate of expected value. This value is 46% greater than the mean value for a normal distribution.

REFERENCES

1. "Coal-Mine Fatalities in 1972," U.S. Department of Interior, Bureau of Mines, January 14, 1972.
2. "Coal-Mine Fatalities in 1970," U.S. Department of Interior, Bureau of Mines, January 10, 1971.
3. "Coal-Mine Fatalities in 1973," U.S. Department of Interior, Bureau of Mines, January 1974.
4. "Preliminary Analysis of the Coal Refuse Dam Failure at Saunders, West Virginia, February 26, 1972," U.S. Department of Interior Task Force to Study Coal Waste, *HAZARDS*, March 12, 1972.
5. "Reactor Safety Study: An Assessment of Accident Risks in U.S. Commercial Nuclear Power Plants," WASH-1400 (NUREG-75/014). Washington, D.C.: Nuclear Regulatory Commission, October 1975.

C

Crisis Management

Reprinted by permission from Robert H. Kupperman, Richard H. Wilcox, and Harvey A. Smith, "Crisis Management: Some Opportunities" (*Science,* Vol. 187, February 7, 1975, pp. 404–410). Copyright 1975 by the American Association for the Advancement of Science.

C.1 WHAT ARE CRISES?

Crises are matters of degree, being emotionally linked to such subjective terms as calamity and emergency. In fact it is not necessary to define crises in order to discuss problems generally common to their management, including the paucity of accurate information, the communications difficulties that persist, and the changing character of the players as the negotiations for relief leave one or more parties dissatisfied.

In a sense, crises are unto the beholder. What is a crisis to one individual or group may not be to another. However, crises are generally distinguished from routine situations by a sense of urgency and a concern that problems will become worse in the absence of action. Vulnerability to the effects of crises lies in an inability to manage available resources in a way that will alleviate the perceived problems tolerably. Crisis management, then, requires that timely action be taken both to avoid or mitigate undesirable developments and to bring about a desirable resolution of the problems.

Crises may arise from natural causes or may be induced by human adversaries, and the nature of the management required in response differs accordingly.

C.2 PRIORITIES FOR CRISIS MANAGEMENT

The problems a crisis manager faces in establishing priorities for action seem to follow the hierarchy of needs presented in Chapters 2 and 4 according to the authors.

The order in which his [the manager's] objectives are placed may be quite controversial, and his priorities may change as he learns of the success or failure of his prior actions and the sharply vocalized opinions of his supporters and detractors.

There is generally no single acceptable course of events for him to follow; rather, there is a wide selection which can be ordered more or less by their relative economy and political difficulty of accomplishment.

To illustrate this phenomenon, consider the plight of the Federal Disaster Coordinator during a major natural disaster. There is limited manpower, equipment, and time. Few would disagree that the saving of lives and providing of food and initial shelter is the first order of business; and few would disagree that reconstruction of public parks can be postponed. But when one considers the multiple and relatively incommensurate questions of the restoration of utilities, the securing of mobile homes for temporary housing, and the clearance of major debris, the manager has his hands full in setting priorities. The widespread effects of Hurricane Agnes of 1972 provide an example: extensive damage, severe resource shortages, and a lot of political heat.

C.3 CRISIS MANAGEMENT DEFINED

We will use the term "crisis management" to mean any process which a manager exercises to meet his goals within a potentially deteriorating situation at an acceptable cost to him, persuading those with whom he is interacting that the costs of opposing him are greater than the cost of allowing him to attain his objectives.

This definition of crisis management is broad and admittedly something of a straw man. It implicitly involves resource allocation, urgency, and various forms of communications. It admits the possibilities of programmed conflict as well as of situations offering minimal feedback. The management of economic problems, disasters, famines, and the termination or avoidance of war are includable within the definition.

In international relations, the process of crisis management can be regarded as having the objective of convincing an opponent that the immediate and long-run costs of opting for, continuing, or increasing hostilities exceed the immediate costs of accepting an offer to resolve conflict that minimally promises face-saving opportunities to the leaderships of the more powerful nations.

C.4 THE PROCESS OF MANAGING CRISES

Negotiations with an adversary in an international crisis rest, at bottom, on conveying to him the costs of opposition in terms of his own value system. (In many cases these may be opportunity costs, that is, costs attributable to lost or deferred opportunities.) Moreover, for our position to be credible to an opponent he must perceive that the costs we propose to meet in gaining our objectives are actually acceptable to us in terms of our value system. There are a number of hazards in this task. One of the hazards implicit in conveying costs is that the tactics themselves are a part of the value system. An opponent may, for instance, have a cultural bias toward noncooperative behavior, such as playing zero-sum games—that is, treating all bargaining as though a gain for one side must represent a loss for the other.

Another hazard is connected with the conceptual problems of exploring fully an adversary's measures of utility (quantitative indicators of desire or goodness), as well as our own. Nations foster economic and political institutions that are imperfect images of differing implicit value systems. At best we can discern preferences among alternatives, and these preferences need not be internally consistent nor temporally constant.

C.5 DIFFICULTIES IN CRISIS MANAGEMENT

A number of difficulties have been noted here as being common to the management of crises: uncertainty, poor data-handling methods, too little data, too much data, inadequate communications, differing value systems, changing management objectives, political harassment, little planning, and insufficient time in which to learn. To this long list of conceptual difficulties must be added the psychological and physical problems of confusion and fatigue. Clearly, the successful crisis manager must be a versatile person—a resource allocator, a communicator, and an artful negotiator with a tough hide.

Of all the planning concerns that must be addressed, the most difficult is the measurement of utility, especially among comparative value systems. Here we face what might appear to be limiting conceptual problems. How can two nations with radically different outlooks and perceptions engage in a meaningful dialogue involving respective value systems not well understood by either?

The often intangible problems of measuring utility are at the root of the conceptual difficulties of socioeconomic modeling. Nevertheless, analytic tools can at least help in establishing relative preferences among alternatives.

Glossary

A number of terms used in this work are listed here, with their definitions as specified by the author. Whenever possible, existing definitions from other sources have been used, to minimize the proliferation of terms. Two sources that were used extensively are John N. Warfield and J. Douglas Hill, "A Unified Systems Engineering Concept," Battelle Memorial Institute, Columbus, Ohio, June 1972, and J. Watson, R. Kuehnel, and J. Golden, "A Preliminary Study of a Concept for Categorizing Benefits," MTR-6569, MITRE Corp., McLean, Virginia, January 1974. A number of other sources have served to a lesser degree.

Accuracy: The quality of being free from error. The degree of accuracy is a measure of the uncertainty in identifying the true measure of a quantity at the level of precision of the scale used for the quantity.

Algorithm: A standard set of procedures for solving a mathematical problem (as of determining the greatest common divisor) that frequently involves repetition of an operation. An algorithm is an expression utilizing factors (some of which may be intangibles) that can be assigned value (i.e., can be quantified).

Alternative: One member of a set of options associated with a decision, the decision being limited to a choice of one and only one.

Axiology: The study of the nature of types and criteria of values and of value judgments, especially in ethics.

Baseline: A known reference used as a guide for further development activities.

Bayesian statistics. "Bayes rule" (Thomas Bayes, a nineteenth century English mathematician and clergyman) states that the probability that both of two events will occur is the probability of the first multiplied by the probability that if the first has occurred the second will also occur. Baysian statistics is a way of making quantity of information substitute for quality of information. There are two kinds of probability: the classical type derived from empirical information, and subjective probability. Bayesian statistics is based on these "subjective probabilities." It involves the joint probability of A and B. The probability of the second event occurring if the first has occurred is called the conditional probability of the second, given the first. Stated another way, the probability of any event $P(A)$ is always positive but never greater than 1. Symbolically, $0 \leq P(A) \leq 1$. If $P(A) = 0$, the occurrence of the event B is considered impossible. If $P(A) = 1$, the occurrence of the event B is considered to occur with $P(B)$.

Behavior: The observable manifestations of performance.

Benefit: (*a*) An axiological concept representing anything received that causes a net improvement to accrue to the recipient.
(*b*) A result of a specific action that constitutes an *increase* in the production possibilities or welfare level of society.

Benefit-cost ratio. The ratio of total social benefit to total social costs related to a specific activity.

Candidate set(structured value analysis): A particular system to be evaluated against particular value sets.

Capability: A measure of the degree to which a system is able to satisfy its performance objectives.

Cardinal (interval) Scale: A continuous scale between two end points, neither of which is necessarily fixed.

Causative event: The beginning in time of an activity that results in particular outcomes of the activity.

Classify: To group or segregate in classes.

Cognition: The act or process of receiving and storing information, with intent to use the information to influence future behavior of the recipient.

Cognitive system: A system that is capable of cognition.

Cognitive dissonance: Behavior toward minimization of internal mental conflicts by rationalizing mistakes in decisions already made.

Communication: The activity of transferring information between two or more entities that share a common language.

Component utility function: The utility assigned to a subgoal.

Compound utility function: The utility function formed by combining a multiplicity of component utility functions by some mathematical or logical rule.

Consequence: See *Risk Consequence.*

Consequence value: The importance a risk agent subjectively attaches to the undesirability of a specific risk consequence.

Consensus: Group solidarity in sentiment and belief . . . general agreement.

Consumers' surplus: The excess of the consumer's willingness to pay over the sacrifice he is required to make to obtain goods or services.

Continuing process: A process in which the variables are uninterrupted in their values.

Control: Given a standard of comparisons or means of verification, that which affords a means of directing performance in the direction of the standard of comparison.

Controllability: A performance measure of a system that describes whether the system will perform as it is directed.

Cost: A result of a specific action that constitutes a *decrease* in the production possibilities or welfare level of society. Also see *Loss*.

Cost-benefit analysis: An attempt to delineate and compare in terms of society as a whole the significant effects, both positive and negative, of a specific action. Generally a number of alternative actions are analyzed, resulting in the selection of the alternative that provides either the largest benefit-cost ratio (total benefit/total cost) or one with a positive ratio at least. If an alternative results in a net benefit less than zero or a benefit-cost ratio less than 1, it is deemed socially inefficient and is not carried out.

Cost-effectiveness analysis: A term less specific than cost-benefit analysis, usually meaning the selection of the lowest cost alternative that achieves a predetermined level of benefits. Alternatively, the analysis and selection of the path that yields the largest social benefit for a predetermined specified level of social costs.

Cost function: A mathematical function defining the cost of an entity in terms of the values of the variables of entity. Also see *Loss function*.

Creativity: The gift of imaginatively recombining known things into something new.

Data set (structured, value analysis): The set of values obtained for a particular candidate set. The data sets are composed of a set of single values for each parameter of the value set.

Decision analysis: A methodology of decomposition of the decision-making process into parts, whereby the appropriate data can be associated with the parts, to provide a rational basis for decision making.

Decision making: A dynamic process of interaction, involving information and judgment among participants who determine a particular policy choice. Decision models are either models of the decision-making process itself, or analytical models (e.g., decision trees, decision matrices) used as aids in arriving at the decisions. Decision theories usually are in relation to the process itself.

Decision matrices: Matrices whose elements exhibit quantitative relationships (cardinal or ordinal) among sets of factors coming into play in the decision-making process.

Decision tree: A device used to portray alternative courses of action and relate them to alternative decisions showing all consequences of the decision. The tree represents alternative courses or series of actions related to a previous decision.

Decisive decision conditions: Conditions in which the preference between values on a utility scale is clearly discernible because ranges of uncertainty of the two values do not overlap (in the case of uniform distributions of uncertainty) or are below a certain error level (for normal distributions of uncertainty).

Degree of uncertainty: That proportion of information about a total system that is unknown in relation to the total information about the system.

Delphi technique: An iterative method designed to produce a consensus by repeated queries of an individual with feedback of group responses. Members of the group do not interact directly.

Descriptive uncertainty: The absence of information about the completeness of the description of the degrees of freedom of a system.

Deterministic system: A system with specific outputs, collectively called "system behavior" or "response," such that the outputs are completely determined by the past and present inputs.

Diagnosis (hazards): An assessment of hazards in terms of their symptoms or consequence in relation to possible causes.

Disaggregation: Reducing an entity to its component parts.

Discount rate: Based on the principle that a dollar received today is valued more than a dollar received in the future, an interest rate that reflects either the social opportunity cost rate or the social time preference rate, permitting the conversion of all benefits and costs into present values for analytical purposes.

Discounted cash flow: The use of an appropriate interest rate to discount all cash flows to the same point in time.

Dissonant behavior: An increased display of dissatisfaction with a situation as a result of disconfirmation of a strong expectancy.

Ecology: The study of the interrelationship of organisms and their environment . . . the totality or pattern of relations between organisms and their environment.

Empirical: Originating in or based on observation or experience.

Endogenous: Occurring or originating within a system.

Endogenous risk imposition: Choice of risk exposure is under control of the risk agent alone.

Entity: Anything capable of unique description.

Equitable risk: A risk agent receives direct benefits as a result of exposure to a risk, and the knowledge of the risk is not purposely withheld from the risk agent.

Equity: A social standard delineating the manner in which individuals are to be dealt with in social relationships. Equity considerations are generally deemed to be subjective, although two basic principles are almost universally accepted; that is, equals are to be treated equally, and unequals are to be treated unequally.

Estimation: The assignment of probability measures to a postulated future event.

Estimator uncertainty: Uncertainty in measurement resulting from deliberate use of less complex measures such as central value estimates of dispersion and smoothing functions for time-dependent parameters.

Evaluation: Comparison of performance of an activity with the objectives of the activity and assignment of a success measure to that performance.

Event: A particular point in time associated with the beginning or completion of an activity, and possibly accompanied by a statement of the benefit or result attained or to be attained because of the completion of an activity.

Exogenous: External to a system (part of the environment of the system).

Exogenous risk imposition: Choice of risk exposure is not under control of the risk agent alone.

Expected value, use of: Valuation of an uncertain numerical event by weighting all possible events by their probability of occurrence and averaging.

Expert judgment: Designating the relevance of opinions of persons well informed in an area for estimates (e.g., forecasts of economic activity).

Exposure (to risk): The condition of being vulnerable to some degree to a particular outcome of an activity, if that outcome occurs.

Externalities: The result of an activity that affects someone who is not a direct party in the activity; sometimes referred to as "third-party effects" or "spillovers."

Extrapolation/projection: The technique of estimating the future by a continuation of past trends without attempts to understand the underlying phenomena.

Extrinsic parameter: A variable whose value can be determined empirically by direct physical measurement.

Feasible: That which is possible to do, realistically.

Feedback: The return of performance data to a point permitting comparision with objective data, normally for the purpose of improving performance (goal-seeking feedback), but occasionally to modify the objectives (goal-changing feedback).

Figure of merit (worth): An index used to provide a measure of the degree of performance and value of a system.

Forecast: A part of a scenario in which the future of some entity is described, based on an arbitrary set of assumptions about the future environment of that entity.

Fuzzy: Ill-defined, thereby frustrating efforts to arrive at either an intentional or extentional definition.

Generational effects: Effects that occur during the lifetime of persons involved in making a decision, or within 30 years of that decision.

Group resiliency: The perception by a community or group of its ability to control or deal with a risk consequence. The perception of control decreases with the size of the consequence in a highly nonlinear manner.

Heuristic: Describing an operational maxim derived from experience and intuition.

Hierarcial clustering: The organized grouping of related elements.

Hierarchy: A partially ordered structure of entities in which every entity but one is successor to at least one other entity, and every entity except the basic entities is a predecessor to at least one other entity.

Homeostasis: An equilibrium condition attributable to the existence of a set of self-regulating properties of a living process.

Imprecision: Inexactness (in the measurement of a quantity).

Inaccuracy: Error in identifying the true measure of a quantity.

Indecisive decision condition: The preference between two values on a utility scale is not discernible because ranges of uncertainty overlap in the case of uniform distributions of uncertainty or are above a certain error level for a normal distribution of uncertainty.

Independence: A property shared by two or more entities when the performance of any one or any group has no effect on the performance of any other one or any other group.

Indifference curves: Representation of a functional relationship of value expressed as equivalent quantities of an alternative that are identical in value.

Individual risk evaluation: The complex process, conscious or unconscious, whereby an individual accepts a given risk.

Inequitable risk: A risk agent is exposed to a risk and receives no direct benefits from such exposure, or the knowledge of the risk is purposely withheld from him.

Interdependence: A property shared by two or more entities whenever the performance of any one affects the performance of some or all the rest.

Intergenerational effects: Effects that are received by a generational group who are not party to a decision; usually the group is a descending generation of the decision makers.

Intrinsic parameter: A variable whose measurement is based on the value system of an individual and his perception of these values.

Learning curve: A graphic representation indicating the amount of acquired knowledge or skill versus time that is convertible to productive effort.

Level systemic control: Risks of a system do not increase over time any faster than the system's rate of expansion.

Loss function: A function used in decision theory for evaluating the losses incurred when certain decisions are made under uncertainty. If the loss function is independent of the decision value used, it is frequently called a cost function.

Measurable: *a.* Capable of being sensed, that which is sensed being convertible to an indication; the indication can be logical, axiological, numerical, or probabilistic. If probabilistic, it is empirical and subjective. *b.* Comparable to some unit designated as standard.

Measured risk level: The historic, measured, or modeled risk associated with a given activity.

Measurement uncertainty: The absence of information about the specific value of a measurable variable.

Methodology: An open system of procedures.

Model: An abstraction of reality that is always an approximation to reality.

Monitoring hazards: A recurrent process of observation, recording, and analysis of products, processes, phenomena, or persons for hazardous events or consequences.

Morphologic analysis: Study of the structure of form and features, following a definite behavioral approach and specific methodology.

Most likely value: Assignment of the value with the highest probability of occurrence to a stochastic process in contemplating alternative actions.

Motivation: The entire complex of factors whose consequence is specific behavior of a living entity.

Negative systemic control: The absence of a systemic control concept and/or a system whose risk behavior is characterized by an increase in risk over time.

Nominal scale (taxonomy): A classification of items that can be distinguished from one another by one or more properties.

Normative: An axiological term representing that which ought to be, within reasonable circumstances; hence representing on the one hand that which is thought desirable, and on the other that which is thought attainable; normally less than ideal.

Objective function: A specified mathematical relationship between a dependent variable (e.g., overall measure of benefits) and a set of independent variables (e.g., individual benefit measures and their relative weights). In choosing among alternatives, the decision maker typically seeks to maximize the (dependent variable of the) objective function.

Opinion survey/sampling: Any procedure for obtaining by oral or written interrogation or both the views of any portion of the affected population regarding benefit levels expected, their utility, and/or relative importance. Typically, scientific sampling procedures would be used to maximize (for a given level of effort) the accuracy and precision of the results obtained.

Opportunity cost: The value to society of the next best alternative use of a resource. This is the true economic cost to society of using a resource for a specific purpose or in a specific project.

Optimization: The process of finding a set of system parameters that maximizes the attainment of system goals and objectives.

Ordinal scale (rank scale): An ordering (ranking) of items by the degree to which they satisfy some criterion.

Outcome: The final result of an activity initiated by a causative event.

Paradigm: A structured set of concepts, definitions, classifications, axioms, and assumptions used in providing a conceptual framework for studying a given problem.

Parametric variation: A technique for sensitivity analysis of any given model in which the values of parameters that are input to the model's calculation are systematically varied to permit observation of how such variation affects the model's output (especially ranking of alternatives).

Pareto optimization: Optimization using a criterion that each person's needs be met as much as possible without diminishing the degree of achievement of any other person.

Pareto optimality: An ideal state in the sense that no further distribution of economic activity will improve one individual's welfare without decreasing the welfare of another individual.

Positive systemic control: Risk of a system decreases over time on either an absolute or a relative basis (i.e., a learning curve for risks can be shown).

Precision: The exactness with which a quantity is stated, that is, the number of units into which a measurement scale of that quantity may be meaningfully divided. The number of significant digits is a measure of precision.

Predictive modeling: Use of any mathematic model that estimates or predicts the value of a dependent variable in terms of component factors specified as independent variables.

Preference: Assignment of rank to items by an agent when the criterion used is utility to the ranking agent.

Preferential ordering: The ranking of alternative outcomes in order of preference without any evaluation of degree of preference.

Probability: A numerical property attached to an activity or event whereby the likelihood of its future occurrence is expressed or clarified.

Probability distribution: The representation of a repeatable stochastic process by a function satisfying the axioms of probability theory.

Probability of occurrence: The probability that a particular event will occur, or will occur in a given interval.

Probability threshold: A probability of occurrence level for a risk below which a risk agent is no longer concerned with the risk and ignores it in practice (*Threshold of concern*).

Propensity for risk acceptance: An individual, subjective trait designating the degree of risk one is willing to subject himself to for a particular purpose.

Quantification: The assignment of a number to an entity or a method for determining a number to be assigned to an entity.

Range of values: Evaluation of an uncertain outcome by estimation of maxima and minima for the event.

Relevance trees: Graphic representations of successively simpler elements of the concept, serving to extend the concept.

Reliability: The probability that the system will perform its required functions under given conditions for a specified operating time.

Risk: The potential for realization of unwanted, negative consequences of an event.

Risk acceptance: Willingness of an individual, group, or society to accept a specific level of risk to obtain some gain or benefit.

Risk acceptance function: A subjective operator relating the levels of probability of occurrence and value of a consequence to a level of risk acceptance.

Risk acceptance level: The acceptable probability of occurrence of a specific consequence value to a given risk agent.

Risk acceptance utility function: The profile of the acceptability of the probability of occurrence for all consequences involved in a risk situation for a specific risk agent.

Risk agent: See *Valuing agent.*

Risk assessment: The total process of quantifying a risk and finding an acceptable level of that risk for an individual, group, or society. It involves both risk determination and risk evaluation.

Risk averse: Displaying a propensity against taking risks.

Risk aversion: The act of reducing risk.

Risk aversive: Acting in a manner to reduce risk.

Risk consequence: The impact to a risk agent of exposure to a risky event.

Risk conversion factor: A numerical weight allowing one type of risk to be compared to another type. See Chapter 15.

Risk determination: The process of identifying and estimating the magnitude of risk.

Risk estimation: The process of quantification of the probabilities and consequence values for an identified risk.

Risk evaluation: The complex process of developing acceptable levels of risk to individuals or society.

Risk evaluator: A person, group, or institution that seeks to interpret a valuing agent's risk for a particular purpose.

Risk identification: The observation and recognition of new risk parameters, or new relationships among existing risk parameters, or perception of a change in the magnitude of existing risk parameters.

Risk proportionality derating factor: Quantifying the degree to which risks become less acceptable as indirect benefits to the risk agent declines.

Risk proportionality factor: That portion of the total societal risk that society will accept for a new technology.

Risk reduction: The action of lowering the probability of occurrence and/ or the value of a risk consequence, thereby reducing the magnitude of the risk.

Risk reference: Some reference, absolute or relative, against which the acceptability of a similar risk may be measured or related; implies some overall value of risk to society.

Risk referent: A specific level of risk deemed acceptable by society or a risk evaluator for a specific risk; it is derived from a risk reference.

Risky shift: The tendency of certain groups to become more extreme or take riskier positions in their judgments than they would, acting as individuals.

Satisficing: The selection of a decision function that optimizes an individual's freedom from anxiety, as opposed to a function that optimizes overall organization goals.

Screening (hazards): A process of hazard identification whereby a standardized procedure is applied to classify products, processes, phenomena, or persons with respect to their hazard potential.

Sensitivity analysis: A method used to examine the operation of a system by measuring the deviation of its nominal behavior due to perturbations in the performance of its components from their nominal values.

Societal risk evaluation: The complex process, formal or informal, whereby society accepts risks imposed on it.

States of nature: A concept from decision theory. In decision making under uncertainty, the outcomes (numerical results) associated with each available alternative are considered to be predictable as a set of *n* discrete values depending on conditions beyond the decision maker's control and for which he has no useful estimates of the respective probabilities. The *n* sets of conditions under which each one of the outcomes is expected are termed "states of nature."

Stochastic system: A system whose behavior cannot be exactly predicted.

Structured value (structured value analysis): The resultant value of a particular value set evaluated for a particular data set. This value lies between zero and unity and allows many data sets to be ranked numerically in relation to one another.

Structured value analysis: A multistage procedure for assessing the value of an action, project alternative, and so on, incorporating individual techniques at each stage for computing from quantitative measures of individual components a single figure expressing the overall value. A multistage procedure for assessing the value of an action, project, alternative, and so on, by structuring the complete entity into component elements, to each of which a numeric measure of value (positive or negative) can be assigned. These are then coverted to a common utility scale. Each component is assigned a weight expressing its relative significance in determining overall value of the entity. A single figure of worth or value is then computed from measures and weights of all individual components. The procedure permits considerable flexibility in choice of techniques used to perform each necessary optimal step.

Subjective probabilities: The assignment of subjective weights to possible outcomes of an uncertain event where weights assigned satisfy axioms of probability theory.

Surrogate or proxy measures: The use of a related quantity as a proxy for an unknown or difficult-to-measure value. The relationship may be established by armchair analysis, correlation techniques, scientific studies, or other means.

System: *a.* A complex entity formed of many, often diverse, parts subject to a common plan or serving a common purpose.
b. A composite of equipment, skills, and techniques capable of performing and/or supporting an operation.

Systems science: The branch of organized knowledge dealing with systems and their properties.

Systemic control of risk: A plan and implementation thereof to control the overall risk of a technological system.

Taxonomy: The identification and definition of properties of elements of the universe; a disaggregation, as contrasted with systematics (which is an aggregation) and as contrasted with morphology (which encompasses both taxonomy and systematics).

Technology: The tangible products of the application of scientific knowledge.

Teleology: The fact or character attributed to nature or natural processes of being directed toward an end or shaped by a purpose.

Threshold: A discontinous change of state of a parameter as its measure increases. One condition exists below the discontinuity, and a different one above it.

Uncertainty: The absence of information; that which is unknown.

Universe: The totality under consideration, often separated into system and environment.

Utility: The satisfaction of a rational economic man's wants and desires, as expressed on a scale.

Utility function: A scale of preference (ordinal) or value (cardinal) to one or many decision makers.

Valuation: The act of mapping an ordinal scale onto an interval scale (i.e., assigning a numerical measure to each ranked item based on its relative distance from the end points of the interval scale . . . assigning an interval scale value to a risk consequence.

Value: A quality quantified on a scale expressing the satisfaction of man's intrinsic wants and desires.

Value function (structured value analysis): A function relating points on the parameter measurement scale to the value scale for a particular parameter. These functions may result from explicit information or may be arrived at through value judgment.

Value set (structured value analysis): A specific set of model parameters made up of terms and factors, expressed in particular measurement scales, value functions, and weights.

Valuing: The act of assigning a value to a risk consequence.

Valuing agent: A person or group of persons who evaluates directly the consequence of a risk to which he is subjected. A *risk agent.*

Weight (structured value analysis): The relative importance of terms in a model expressed as a decimal fraction; weights for a set of terms add to unity.

Weighting factor: A coefficient used to adjust variable accuracy to a subjective evaluation; these factors are usually determined through surveys, Delphi sessions, or other formats of expressing social priorities.

Bibliography

This alphabetic bibliography, which provides additional references not cited in chapters directly, has been classified by subject as each entry pertains to risk assessment. The classifications are identified by the following symbols, which appear at the end of each entry.

A	Anthropology
C	Cost-Benefit Analysis
D	Data Bases
E	Epistomology and Philosophy of Science
EC	Economic Analysis
G	General
M	Medicine and Health
PH	Philosophy—General
PL	Political Science
PS	Psychology
R	Risk Analysis and Assessment
SO	Social Sciences
ST	Statistics and Decision and Utility Theory
SY	Systems Analysis and Science
T	Technological and Engineering

Ackoff, Russell (Ed.). *Progress in Operations Research,* Vol. 1. New York: Wiley, 1961. **ST**

Ackoff, Russell. *Scientific Method: Optimizing Applied Research Decisions.* New York: Wiley, 1962. **ST**

Ackoff, R. L. "General System Theory and Systems Research: Contrasting Conceptions of Systems Science," *General Systems,* Vol. 8, 1963. **SY**

Ackoff, R. L. and S. S. Sengupta. "Systems Theory from an Operations Research Point-of-View," *IEEE Transactions on Systems Science and Cybernetics,* Vol. SSC-1, No. 1, November 1965. **SY**

Acton, J. P. "Evaluating Public Programs to Save Lives: The Case of Heart Attacks," Report R-950-RC. Santa Monica, Calif.: RAND Corporation, January 1973. **EC**

Adams, Ernest W. "Survey of Bernoullian Utility Theory," in *Mathematical Thinking in the Measurement of Behavior,* Herman Solomon (Ed.). New York: Free Press, 1960, pp. 153–268. **ST**

Alchian, Armen. "The Meaning of Utility Measurement," *American Economic Review,* March 1953, pp. 26–50. **ST**

Alfven, Hannes. "Fission Energy and Other Sources of Energy," *Bulletin of the Atomic Scientists,* Vol. 30, No. 1, January 1974, pp. 4–8. **T**

Angyal, Andrus. *Foundations for a Science of Personality.* Cambridge, Mass.: Harvard University Press, 1941. **PS**

Argyris, Chris. "Resistance to Rational Management Systems," *Innovation,* No. 10. The Technology Group, New York, 1970. **PS**

Armstrong, W. D. "A Note on the Theory of Consumers' Behavior," *Oxford Economic Papers,* Vol. 2, 1950, pp. 119–122. **C**

Arriel, M. and A. C. Williams. "The Value of Information and Stochastic Programming," *Operations Research,* Vol. 18, No. 5, September–October 1970, pp. 947–954. **ST**

Arrow, Kenneth J. "Utility and Expectation in Economic Behavior," in *Psychology, A Study of a Science,* Vol. 6. Sigmund Koch (Ed.). New York: McGraw-Hill, 1963, pp. 724–752. **EC**

Arrow, K. J. "The Economic Implications of Learning by Doing," *Review of Economic Studies,* Vol. 29, No. 3, 1961–1962, pp. 155–173. **EC**

Arrow, K. J. "Alternative Approaches to the Theory of Choice in Risk-Taking Situations," *Econometrica,* Vol. 19, No. 4, October 1951, pp. 404–437. **R**

Arrow, K. J. "Limited Knowledge and Economic Analysis," *American Economic Review,* Vol. 64, No. 1, March 1974, pp. 1–10. **EC**

Ashby, W. R. "The Set Theory of Mechanism and Homeostasis," in *General Systems,* Yearbook of the Society for General Systems Research, Ann Arbor, Mich., Vol. 9, 1964, pp. 83–97. **SY**

Ashby, W. R. "The Cybernetic Viewpoint," *IEEE Transactions on Systems Science and Cybernetics,* Vol. SSC-2, No. 1, August 1966. **SY**

Atzinger, Erwin M., Wilbert J. Brook, Michael R. Chernick, Brian Elsner, and Ward V. Foster. "Compendium on Risk Analysis Techniques." U.S. Army Materiel Systems Analysis Agency Special Publication No. 4, Aberdeen Proving Grounds, Maryland, July 1972. **R**

Baram, Michael S. "Technology Assessment and Social Control," *Science,* Vol. 180, No. 4085, May 4, 1973, pp. 465–473. **PL**

Barrett, G. V. and R. H. Franke. "Social, Economic, and Medical Explanations of Psychogenic Death Rate Differences: A Cross-National Analysis," Technical Report 32, Management Research Center of the College of Business Administration, University of Rochester, Rochester, N.Y., October 1969, AD 700, 286. **M**

Bartley, Robert L. "On the Limits of Rationality," *Wall Street Journal,* September 10, 1971, p. 8. **G**

Bauer, Raymond (Ed.). *Social Indicators.* Cambridge, Mass.: MIT Press, 1966. **SO**

Baumol, William J. *Business Behavior, Value and Growth.* New York: Macmillan, 1959. **EC**

Bayliss, W. M. *Principles of General Psychology,* 4th ed. New York: Longmans, Green, 1927. **PS**

Bergmann, Gustav. *Philosophy of Science.* Madison: University of Wisconsin Press, 1957. **PH**

Black, D. *The Theory of Committees and Elections.* New York: Cambridge University Press, 1968. **PL**

Blake, V. E. "A Prediction of the Hazards from the Random Impact of Meteorites in the Earth's Surface," SC-RR-68-388 *Aerospace Nuclear Safety,* Albuquerque, N.M.: Sandia, December 1968. **D**

Block, M. K. and R. C. Lind. "Wealth Equivalents, Risk Aversion, and Marginal Benefit from Increased Safety." Prepared for Office of Naval Research, Arlington, Va., by Naval Postgraduate School, Monterey, Calif. AD 781375, May 1974. **R**

Boffey, Philip M. "Radiation Standards: Are the Right People Making Decisions?" *Science,* Vol. 171, No. 3973, February 26, 1971, pp. 780–783. **PL**

Bohnert, Gerbert G. "The Logical Structure of the Utility Concept," in *Decision Processes,* R. L. Davis, C. H. Coombs, and R. M. Thrall (Eds.). New York: Wiley, 1954, pp. 221–230. **ST**

Boulding, Kenneth. "General Systems Theory—The Skeleton of a Science," *General Systems,* Vol. 1, 1956. **SY**

Brown, Rex V. *Research and the Credibility of Estimates.* Cambridge, Mass.: Harvard Business School, Division of Research, 1969. **PH**

Bruner, Jerome S., Jacqueline J. Goodman, and George A. Austin. *A Study of Thinking.* New York: Wiley, 1956. **PS**

Burton, J., R. W. Kates, and G. F. White. "The Human Ecology of Extreme Geographical Events." Natural Hazard Research, Working Paper No. 1. Toronto, Can.: University of Toronto, 1968. **SO**

Campbell, N. R. "Symposium: Measurement and Its Importance to Philosophy," *Aristotelian Society Supplement,* Vol. 17, 1938. **PH**

Castore, C. H. and J. K. Murnighan. "Decision Rule and Group Member Responses to Group Decisions." Paper presented at the Psychonomic Society Meetings, St. Louis, Mo., November 1972. **SO**

Castore, C. H. and J. K. Murnighan. "Member Financial Support of Group Decisions Reached Under Alternative Decision Rules." Purdue University, ONR Technical Report No. 6, July 1973. (Contract N00014-67-A-0226.) **SO**

Castore, C. H. and J. K. Murnighan. "Group Member Responsiveness to Group Decisions Under Alternative Social Decision Schemes." *Journal of Personality and Social Psychology,* 1974 (submitted). **SO**

Castore, C. H. and Sung, Y. H. "Individual Preferences, Decision Procedures and Support of Group Decisions in Triads." Purdue University, ONR Technical Report No. 8, September 1973. (Contract N00014-67-A-0226.) **SO**

Carrol, J. J. and S. A. Parco. "Social Organization in a Crisis Situation: The Taal Disaster." Manila: A Special Publication of the Philippine Sociological Society, Inc., 1966. **A**

Cetron, Marvin J. and Joel D. Goldhan. *The Science of Managing Organized Technology.* New York: Gordon & Breach, 1971. **T**

Cetron, Marvin J. "The Social Responsibility of Business or the Changing Role of Goals in Corporate and R & D Planning," in *Technology Assessment in a Dynamic Environment.* New York: Gordon & Breach, 1973, pp. 451–470. **SO**

Charnes, A. and W. W. Cooper. "Deterministic Equivalents for Optimizing and Satisficing Under Chance Constraints," *Operations Research,* Vol. 2, No. 1, January–February 1963, pp. 18–39. **SY**

Chiang, C. L. "A Stochastic Model of Competing Risks of Illness and Competing Risks of Death," in *Stochastic Models in Medicine and Biology,* J. Gurland (Ed.). Madison: University of Wisconsin Press, 1964, pp. 323–354. **ST**

Chiang, C. L. *Introduction of Stochastic Processes in Biostatistics.* New York: Wiley, 1968. **ST**

Chiang, C. L. and R. Cohen. "How to Measure Health: A Stochastic Model for an Index of Health," *International Journal of Epidemiology,* Vol. 2, No. 1, 1973, pp. 7–13. **M**

Chipman, John S. "The Foundation of Utility," *Econometrica,* Vol. 28, No. 2, April 1960, pp. 193–224. **ST**

Christensen, Ronald. "Decision Theory and Group Behavior." Unpublished paper presented to Harvard Law School, April 1962. **PS**

Churchman, W. C. *Prediction and Optimal Decision.* Englewood Cliffs, N.J.: Prentice-Hall, 1961. **ST**

Churchman, W. C. *The Systems Approach.* New York: Dell, 1968. **SY**

Churchman, W. C., E. L. Arnoff, and R. L. Ackoff. *Introduction to Operations Research.* New York: Wiley, 1959, Chapter 6. **SY**

Churchman, W. C. and R. L. Ackoff. "An Appropriate Measure of Value," *Journal of the Operations Research Society of America,* Vol. 2, No. 2, May 1954, pp. 172–180. **ST**

Churchman, W. C. and P. Ratoosh (Eds.). *Measurement: Definition and Theories.* New York: Wiley, 1959. **PH**

Clark, Burton R. "Organizational Adaptation and Precarious Values," *American Sociological Review,* Vol. 21, 1956, pp. 327–336. **SO**

Coase, R. H. "The Problem of Social Cost," *Journal of Law and Economics,* Vol. 3, October 1960, pp. 1–44. **EC**

Coates, J. "Calculating the Social Costs of Automobile Pollution—An Exercise." Presented at the Symposium on Risk vs. Benefit, Los Alamos, N.M., November 1971. **C**

Colantoni, Claude S. "The Use of Mathematical Structures in Describing Measuring and Reporting the State of a Firm Under Conditions of Uncertainty." Unpublished doctoral dissertation, Purdue University, 1969. **ST**

Commoner, Barry. *The Closing Circle.* New York: Knopf, 1971. **G**

Commoner, B. *Science and Survival.* New York: Viking Press, 1967. **G**

"Comparative Risk Assessment: A Report on a Workshop on Comparative Risk Assessment of Environmental Hazards in an International Context, 1975." Woods Hole, Mass.: Scientific Committee on Problems of the Environment (SCOPE), Midterm Project 7, March 31–April 4, 1975. **R**

Connolly, T. H. and A. Mazur. "The Risks of Benefit-Risk Analysis," in *Proceedings of the 6th Annual Health Physics Society Midyear Symposium,* Richmond, Wash., November 1970. **E**

Cook, T. J. and F. P. Scioli, Jr. "A Research Strategy for Analyzing the Impacts of Public Policy," *Administrative Science Quarterly,* September 1972, pp. 328–339. **PL**

Coombs, C. H., H. Raiffa, and R. M. Thrall. "Some Views on Mathematical Models and Measurement Theory," in *Decision Processes,* R. L. Davis, C. H. Coombs, and R. M. Thrall (Eds.). New York: Wiley, 1954, pp. 19–37. **ST**

Coombs, C. H. "Psychological Scaling Without a Unit of Measurement," *Psychological Review,* Vol. 7, No. 3, May 1950. **PS**

Coombs, C. H. "Social Choice and Strength of Preference," in *Decision Processes.* R. L. Davis, C. H. Coombs, and R. M. Thrall (Eds.). New York: Wiley, 1954. **PS**

Cooper, George R. "Decision Theory," in *System Engineering Handbook,* Robert E. Machol (Ed.). New York: McGraw-Hill, 1965, Chapter 24. **SY**

Cox, John L. "An Information Decision Theory Approach to the Allocation of Resources." Unpublished doctoral dissertation, Arizona State University, 1971. **ST**

Crow, J. F. "Radiation and Chemical Mutagens: A Problem in Risk Estimation," in *Perspectives on Benefit-Risk Decision Making."* Report of a Colloquium in Benefit-Risk Relationships for Decision Making Conducted by the Committee on Public Engineering Policy, National Academy of Engineering, April 26–27, 1971. Washington, D.C.: The Academy, 1972, pp. 56–58. **R**

Dadario, E. Q. "Academic Science and the Federal Government," *Science,* Vol. 162, December 13, 1968, pp. 1249–1251. **PL**

Dalkey, N. C. "Quality of Life," in *The "Quality of Life" Concept.* Washington, D.C.: Government Printing Office, 1972, pp. V-19–V-29. **SO**

David, H. A. and M. L. Moeschberger. "Life Tests Under Competing Causes of Failure and the Theory of Competing Risks," *Biometrics,* Vol. 27, No. 4, December 1971, pp. 909–933. **ST**

Davidson, Donald, S. Siegel, and P. Suppes. "Some Experiments and Related Theory in the Measurement of Utility and Subjective Probability," Report No. 4. Stanford, Calif.: Stanford Value Theory Project, 1955. **ST**

Debreu, G. "Representation of a Preference Ordering by a Numerical Function," in R. M. Thrall, C. H. Coombs, and R. L. Davis (Eds.). *Decision Processes.* New York: Wiley, 1954. **ST**

DeGroot, Morris H. "Some Comments on the Experimental Measurement of Utility," *Behavioral Science,* Vol. 8, No. 2, 1963, pp. 146–149. **ST**

De Jong, F. J. *Dimensional Analysis for Economists.* Amsterdam: North-Holland, 1967. **EC**

Diamond, P. "Economic Factors in Benefit-Risk Decision Making," in *Perspectives on Benefit-Risk Decision Making.* Report of a Colloquium on Benefit-Risk Relationships for Decision Making Conducted by the Committee on Public Engineering Policy, National Academy of Engineering, April 26–27, 1971. Washington, D.C.: The Academy, 1972, pp. 115–120. **EC**

Douglas, J. D. *The Technological Threat.* Englewood Cliffs, N.J.: Prentice-Hall, 1971. **PL**

Edwards, Ward. "The Theory of Decision-Making," *Psychological Bulletin,* Vol. 51, No. 4, July 1954, pp. 380–417. **PS**

Edwards, W. "Utility, Subjective Probability, Their Interaction and Variance Preferences," *Journal of Conflict Resolution,* Vol. 6, No. 1, 1962, pp. 42–51. **ST**

Ehrlich, Paul R. and Anne H. Ehrlich. *Population, Resources, Environment: Issues in Human Ecology.* San Francisco: Freeman, 1970. **SO**

Enke, Stephen and Richard A. Brown. "Economic Worth of Preventing Death at Different Ages in Developing Countries," *Journal of Biosocial Science,* Vol. 4, No. 3, 1972. **C**

Epstein, S. S. "Information Requirements for Determining the Benefit-Risk Spectrum," in *Perspectives on Benefit-Risk Decision Making.* Report of a Colloquium on Benefit-Risk Relationships for Decision Making Conducted by the Committee on Public Engineering Policy, National Academy of Engineering, April 26–27, 1971. Washington, D.C., The Academy, 1972, pp. 50–55. **C**

Federal Radiation Guide, Reports 1–7, Washington, D.C.: Federal Radiation Council, 1962. **T**

Farris, Donald R. and Andrew P. Sage. "Worth Assessment Methods for Urban Systems: A Comparison of Worth Assessment with Other Techniques for Determining Multi-

Attribute Preferences," In *Proceedings of the National Electronics Conference,* Chicago, October 1974. **ST**

Farris, Donald R. and Andrew P. Sage. "Worth Assessment in Large Scale Systems," *In Proceedings of the Milwaukee Symposium on Automatic Control,* Milwaukee, March 1974. **ST**

Festinger, L. *A Theory of Cognitive Dissonance.* Stanford, Calif.: Stanford University Press, 1957. **PS**

Fine, William T. "Mathematical Evaluations for Controlling Hazards." Naval Ordinance Laboratory Technical Report NOLTR 71-31, White Oak, Md., March 8, 1971. **ST**

Fishburn, Peter C. *Decision and Value Theory.* New York: Wiley, 1964. **ST**

Fishburn, P. C. "On the Prospects of a Useful Unified Theory of Value for Engineering." *IEEE Transactions on Systems Science and Cybernetics,* Vol. SSC-2, August 1966, pp. 27–35. **PH**

Fishburn, P. C. "A Note on Recent Developments in Additive Utility Theories for Multiple Factor Situations," *Operations Research,* Vol. 14, No. 6, November–December 1966, pp. 1143–1148. **ST**

Fishburn, P. C. "Methods of Estimating Additive Utilities," *Management Science,* Vol. 13, No. 7, March 1967, pp. 435–453. **ST**

Fishburn, P. C. "Utility Theory," *Management Science,* Vol. 14, January 1968, pp. 335–378. **ST**

Fishburn, P. C. "Utility Theory with Inexact Preferences and Degrees of Preference," *Synthese,* Vol. 21, 1970, pp. 204–221. **ST**

Fishburn, P. C. *Utility Theory for Decision Making.* New York: Wiley, 1970. **ST**

Fleming, M. "A Cardinal Concept of Welfare," *Quarterly Journal of Economics,* Vol. 66, No. 3, August 1952. **EC**

Foster, R. B. and F. P. Hoeben. "Cost-Effectiveness Analysis for Strategic Decisions," *Operations Research,* Vol. 5, No. 4, November 1955. **C**

Gardiner, Peter C. and Ward Edwards. *Public Values: Multi-Attribute Utility Measurement for Social Decision Making.* Los Angeles: University of Southern California, Social Science Research Institute Report 75-5, 1975. **SY**

Gardenier, John S. "Concepts for Analysis of Massive Spill Accident Risk in Maritime Bulk Liquid Transport." U.S. Coast Guard Office of Research and Development Report No. 723111, Washington, D.C., December 1971. **T**

Giddings, J. Calvin. "World Population, Human Disaster and Nuclear Holocaust," *Bulletin of the Atomic Scientists,* Vol. 29, No. 7, September 1973, pp. 24–50. **G**

Gilfillan, S. C. *The Sociology of Invention.* Cambridge, Mass.: MIT Press, 1963. **SO**

Gilson, Charlotte R. "Individual Differences in Risk Taking." Yale University, Department of Psychology Technical Report No. 13, New Haven, Conn., June 1968. **PS**

Golant, Steven, and Ian Burton. *Avoidance-Response to the Risk Environment.* University of Chicago, Department of Geography; Clark University, Graduate School of Geography; University of Toronto, Department of Geography; Working Paper No. 6, 1969. **R**

Greenburg, Daniel S. *The Politics of Pure Science.* New York: New American Library, 1967. **PL**

Gruenstein, Peter and Richard H. Sandler. "Power From Fission: Potential for Catastrophe," *The Progressive,* Vol. 37, No. 11, November 1973, pp. 36–41. **G**

Gupta, R. B. and G. Rama Rao. "Effect of Elimination of Different Causes of Death on Expectation of Life—Bombay, 1960–1961," *Indian Journal of Medical Research,* Vol. 61, No. 6, June 1973, pp. 950–961. **ST**

Haaland, Gordon A., Dean G. Pruitt, Richard St. Jean, and Allan I. Teger. "A Re-examination of the Familiarization Hypothesis in Group Risk Taking." State University of New York Technical Report No. 9, New York, New York, February 17, 1969. **PS**

Hammond, R. J. *Benefit-Cost Analysis and Water Pollution.* Stanford, Calif.: Stanford University Press, 1958. **C**

Harsanyi, J. C. "Cardinal Welfare, Individualistic Ethics, and Interpersonal Comparison of Utility," *Journal of Political Economy,* Vol. 63, No. 4, August 1955. **ST**

Hausner, M. "Multi-Dimensional Utilities," in *Decision Processes,* R. L. Davis, C. H. Coombs, and R. M. Thrall (Eds.). New York: Wiley, 1954, pp. 167–180. **ST**

Hayelden, J. E. "The Value of Human Life," *Public Administration,* Vol. 46, No. 427, 1968. **E**

Heinrich, H. W. *Industrial Accident Prevention,* 3rd ed. New York: McGraw-Hill, 1950, pp. 332–334. **R**

Henderson, W. Paul. "Pyrotechnic Hazard Evaluation and Risk Concepts." Paper presented at the Conference on Hazard Evaluation and Risk Analysis, Houston, August 18–19, 1971. **R**

Hertz, D. B. "Risk Analysis in Capital Investment," *Harvard Business Review,* Vol. 42, 1964. **EC**

Hespos, R. F. and P. A. Strassman. "Stochastic Decision Trees for Analysis of Investment Decisions," *Management Science,* Vol. 11, No. 10, August 1965. **EC**

Higbee, Kenneth L. and Siegfried Streufert. "Group Risk Taking and Attribution of Causality." Purdue University Technical Report No. 21, Lafayette, Ind, April 1969. **R**

Hirshleifer, J., T. Bergstrom, and E. Rappoport. "Applying Cost-Benefit Concepts to Projects Which Alter Human Mortality," UCLA-ENG-7478. Los Angeles: UCLA School of Engineering and Applied Science, November 1974. **C**

Hoel, D. "A Representation of Mortality Data by Competing Risks," *Biometrics,* Vol. 28, 1972, pp. 475–488. **R**

Houthakker, H. S. "Revealed Preference and the Utility Function," *Economica* (new series), Vol. 7, May 1950, pp. 159–174. **ST**

Howard, R. A. "Information Value Theory." *IEEE Transactions on Systems Science and Cybernetics,* Vol. SSC-2, No. 1, August 1966, pp. 22–26. **ST**

Howard, R. A. "Value of Information Lotteries." *IEEE Transactions on Systems Science and Cybernetics,* Vol. SSC-3, June 1967, pp. 54–60. **ST**

Howard, R. A. "Baynesian Decision Models for System Engineering," *IEEE Transactions on Systems Science and Cybernetics,* Vol. SSC-4, No. 3, September 1968, pp. 211–219. **SY**

Howard, R. A. "The Foundations of Decisions Analysis," *IEEE Transactions on Systems Science and Cybernetics,* Vol. SSC-4, No. 3, September 1968. **ST**

Howard, R. A. "Social Decision Analysis," *Proceedings of the IEEE* (*Special Issue on Social Systems Engineering*), Vol. 63, March 1975, pp. 359–371. **ST**

Huebner, S. S. and Kenneth Black. *Life Insurance,* 5th ed. New York: Appleton-Century-Crofts, 1959. **ST**

Huebner, S. S. *The Economics of Life Insurance,* 3rd ed. New York: Appleton-Century-Crofts, 1959. **EC**

Insurance Information Institute. *Insurance Facts,* 1974 ed. New York: The Institute. **D**

Jeffrey, R. C. *Theory of Probability,* 3rd ed. Oxford: Clarendon Press, 1961. **ST**

Jeffrey, R. C. *The Logic of Decision.* New York: McGraw-Hill, 1965. **ST**

Jensen, Neils Erik. "An Introduction to Bernoullian Utility Theory," *Swedish Journal of Economics,* Vol. 69, Nos. 3 and 4, 1967. **ST**

Jessen, Peter J. "Defining 'Quality of Life' Measures—The State of the Art," in *The "Quality of Life" Concept.* Washington, D.C.: Government Printing Office, pp. I-1–I-15. **SO**

Johnson, Erik. *Studies in Multi-Objective Decision Models.* Lund, Sweden: Economic Research Center. 1968. **ST**

Johnson, R. A., F. E. Cast, and J. E. Rosenzweig. *The Theory and Management of Systems.* New York: McGraw-Hill, 1967. **SY**

Jones-Lee, M. "The Value of Changes in the Probability of Death or Injury," *Journal of Political Economy,* July–August 1974, pp. 835–849. **E**

Kaplan, Abraham. *The Conduct of Inquiry.* San Francisco: Chandler, 1966. **PL**

Keeler, Emmett and Richard Zeckhauser. *Another Type of Risk Aversion,* RM-5996-PR. Santa Monica, Calif.: The RAND Corporation, May 1969. **R**

R. L. Keeney. "Utility Independence and Preferences for Multiattributed Consequences," *Operations Research,* Vol. 19, 1971. **ST**

R. L. Keeney. "Utility Function for Multiattributed Consequences," *Management Science,* Vol. 18, No. 5, January 1972. **ST**

Kirkwood, C. W. "Decision Analysis Incorporating Preferences of Groups." Operations Research Center, Massachusetts Institute of Technology, Technical Report No. 74, June 1972. **ST**

Klarman, H. E. *The Economics of Health.* New York: Columbia University Press, 1965. **EC**

Klir, G. J. *An Approach to General Systems Theory.* New York: Van Nostrand Reinhold, 1969. **SY**

Kroeber, A. L. *Anthropology: Culture Patterns and Processes.* New York: Harbinger, 1963. **A**

Krouse, Clement G. "Complete Objectives, Decentralization and the Decision Process of the Organization." *Administrative Science Quarterly,* Vol. 17, No. 4. Worchester, Mass.: McHarmon Press, December 1972, pp. 544–554. **SY**

Kuhn, Thomas A. *The Structure of Scientific Revolutions.* Chicago: University of Chicago Press, 1962. **PH**

Lane, J. Michael, J. Donald Miller, and John M. Neff. "Smallpox and Smallpox Vaccination Policy," *Annual Review of Medicine,* Vol. 22, 1971, pp. 251–272. **M**

La Porte, T. and D. Metlay. "Technology Observed: Attitudes of a Wary Public," *Science,* Vol. 188, 1975, pp. 121–127. **PL**

Lave, L. and E. P. Seskin. "Air Pollution and Human Health," *Science,* Vol. 169, No. 3947, August 21, 1970, pp. 723–733. **M**

Lave, L. B. "Air Pollution Damage: Some Difficulties in Estimating the Value of Abatement," in *Environmental Quality Analysis—Theory and Method in the Social Sciences,* A. V. Kneese and B. T. Bower (Eds.). Baltimore: Johns Hopkins Press, 1972, pp. 213–242. **M**

Lave, L. B. "Risk, Safety, and the Role of Government," in *Perspectives on Benefit-Risk Decision Making.* Report of a Colloquium on Benefit-Risk Relationships for Decision Making Conducted by the Committee on Public Engineering Policy, National Academy of Engineering, April 26–27, 1971. Washington, D.C.: The Academy, 1972, pp. 96–108. **R**

Lenox, Hamilton D. "Risk Assessment." Unpublished thesis, Air Force Institute of Technology, June 1973. **R**

Lieblich, Amia. "The Effects of Stress on Risk Taking," *Journal of Psychonomic Science,* Vol. 10, No. 8, 1968, pp. 303–304. **PS**

Lindblom, C. E. *The Policy Making Process.* Englewood Cliffs, N.J.: Prentice-Hall, 1968. **PL**

Marschak, Jacob. "Actual Versus Consistent Decision Behavior," *Behavioral Science,* Vol. 9, No. 2, April 1964, pp. 103–110. **PS**

Marwick, M. G. "Witchcraft and the Epistemology of Science," *Science and Public Policy,* 1974, pp. 335–341. **PH**

Mazur, A. "Opposition to Technological Innovation," *Minerva,* Vol. 13, 1975, pp. 58–81. **SO**

Mead, M. (Ed.). *Cultural Patterns and Technological Change.* New York: New American Library, 1955. **A**

Melinek, S. J. "A Method of Evaluating Human Life for Economic Purposes." Fire Research Note No. 950, Fire Research Station, Boreham Wood, Hertshire, England, November 1972. **EC**

Mesarovic, M. D. "Foundations for a General Systems Theory," in *Views on General Systems Theory, Proceedings of the Second Systems Symposium.* New York: Wiley, 1964, Chapter 1. **SY**

Migdal, J. S. "Why Change? Toward a New Theory of Change Among Individuals in the Process of Modernization," *World Politics,* Vol. 26, 1974, pp. 189–206. **SO**

Mihalasky, John. "Decision Risk Analysis: Problems in Practice." Paper presented at the Conference on Hazard Evaluation and Risk Analysis, Houston, August 18–19, 1971. **ST**

Mishan, E. J. *Cost-Benefit Analysis.* London: Allen & Urwin, 1971. **C**

Mitroff, Ian I. and Murray Turoff. "The Whys Behind the Hows," *IEEE Spectrum,* March 1973, pp. 62–71. **PH**

Mitchell, Joyce M. and William C. Mitchell. *Political Analysis and Public Policy.* Skokie, Ill.: Rand-McNally, 1969. **PL**

Mumford, L. *The Myth of the Machine: Techniques and Human Development.* London: Secker & Warburg, 1967. **SO**

Murphy, A. H. and R. L. Winkler. "Subjective Probability Forecasting in the Real World: Some Experimental Results," Research Report RR-73-16. Laxenburg, Austria: International Institute for Applied Systems Analysis, 1973. **PS**

Mushkin, S. J. (Ed.). *Public Prices for Public Goods.* Washington, D.C.: The Urban Institute, 1972. **EC**

Nagel, Ernest. *The Structure of Science.* New York: Harcourt Brace Jovanovich, 1961. **PH**

National Academy of Engineering. *Perspectives on Benefit-Risk Decision Making.* Report of a Colloquium on Benefit-Risk Relationships for Decision Making Conducted by the Committee on Public Engineering Policy, April 26–27, 1971. Washington, D.C.: The Academy, 1972. **C**

National Bureau of Standards. *Statistical Theory of Extreme Values and Some Practical Applications* (Applied Mathematics Series 33). Washington, D.C.: The Bureau 1954. **ST**

Nelson, W. R. (Ed.) *The Politics of Science: Readings in Science, Technology and Government.* New York: Oxford University Press, 1968. **PL**

Newell, Allen and Herbert A. Simon. *Human Problem Solving.* Englewood Cliffs, N.J.: Prentice-Hall, 1972. **SY**

Nicosia, Francesco M. *Consumer Decision Processes.* Englewood Cliffs, N.J.: Prentice-Hall, 1968. **EC**

Otway, Harry J. and Robert C. Erdmann. "Reactor Siting and Design from a Risk Standpoint," *Nuclear Engineering and Design,* Vol. 13, 1970. **T**

Otway, H. J., Philip D. Pahner, and Joanne Linnerooth. *Social Values in Risk Acceptance,* RM-75-54. Laxenburg, Austria: International Institute for Applied Systems Analysis, 1975. **R**

Otway, H. J., R. Maderthaner, and G. Galtman. "Avoidance Response to the Risk Environment: A Cross-Cultural Comparison," RM-75-14. Laxenburg, Austria: International Institute for Applied Systems Analysis, 1975. **SO**

Pratt, John W., Howard Raiffa, and Robert Schlaifer. "The Foundations of Decisions Under Uncertainty: An Elementary Exposition," *Journal of the American Statistical Association,* 1964, pp. 353–375. **ST**

Price, Derek J. de Solla. *Little Science, Big Science.* New York: Columbia University Press, 1963. **E**

Price, Don K. *Government and Science: Their Dynamic Relation in American Democracy.* New York: New York University Press, 1954. **PL**

Price, D. K. *The Scientific Estate.* Cambridge, Mass.: Belknap Press of Harvard University Press, 1965. **PL**

Pruitt, D. C. "Choice Shifts in Group Discussion: An Introductory Review," *Journal of Personality and Social Psychology,* Vol. 20, 1971, pp. 339–360. **SO**

Pruzan, P. Mark and J. T. Ross Jackson. "On the Development of Utility Spaces for Multi-Goal Systems," *Erhvervsokonomisk Tidsskrift* (Copenhagen), No. 4, 1963, pp. 257–274. **ST**

Quade, E. S. "A History of Cost-Effectiveness." Paper Presented at IFORS (International Federation of Operations Research Societies) International Cost-Effectiveness Conference, Washington, D.C., April 12–15, 1971. **EC**

Quade, E. S. "The Systems Approach and Public Policy." Santa Monica, Calif.: The RAND Corporation. **SY**

Raiffa, Howard. *Decision Analysis.* Reading, Mass.: Addison-Wesley, 1968. **SY**

Raiffa, H. *Decision Analysis: Introductory Lectures on Choices Under Uncertainty.* Reading; Mass.: Addison-Wesley, 1968. **SY**

Raiffa, H. and R. Schlaifer. *Applied Statistical Decision Theory.* Boston: Graduate School of Business, Harvard University, 1961. **ST**

Ramsey, Frank P. "Truth and Probability," in *The Foundations of Mathematics and Other Logical Essays.* New York: Harcourt Brace Jovanovich 1931. **PH**

Rapoport, Anatol. "Some Comments on Accident Research," in *Behavioral Approaches to Accident Research.* New York: Association for the Aid of Crippled Children, 1961. **PS**

Rescher, Nicholas. *Introduction to Value Theory.* Englewood Cliffs, N.J.: Prentice-Hall, 1969. **PH**

Roby, Thornton B. "Utility and Futurity," *Behavioral Science,* Vol. 17, No. 2, April 1962, pp. 194–210. **PS**

Rosenhan, D. L. "On Being Sane in Insane Places," *Science*, Vol. 179, January 1973, pp. 250–258. **PS**

Rothenberg, J. F. *The Measurement of Social Welfare*. Englewood Cliffs, N.J.: Prentice-Hall, 1961. **EC**

Rubenstein, Albert H. "Studies of Project Selection Behavior in Industry," in *Operations Research in Research and Development*, B. V. Dean (Ed.). New York: Wiley, 1963. **PS**

Sagan, L. A. "Health Costs Associated with the Mining, Transport, and Combustion of Coal in the Steam-Electric Industry," *Nature*, Vol. 250, July 12, 1974. **EC**

Sather, H. N. "Biostatistical Aspects of Risk-Benefit: The Use of Competing Risk Analysis," UCLA-ENG-7477. Prepared for the National Science Foundation by University of California School of Engineering and Applied Science, Los Angeles, September 1974. **R**

Savage, Leonard. "The Theory of Statistical Decisions." *Journal of the American Statistical Association*, Vol. 46, 1951, pp. 55–62. **ST**

Schlaifer, Robert. *Probability and Statistics for Business Decisions*. New York: McGraw-Hill, 1959. **ST**

Selman, Jerome. "Decision Risk Analysis: Risk Theory." Paper presented at the Conference on Hazard Evaluation and Risk Analysis, Houston, August 18–19, 1971. **ST**

Shelley, Maynard W. and Glenn L. Bryan (Eds.). *Human Judgments and Optimality*. New York: Wiley, 1964. **PS**

Shubik, Martin. "Behavioristic or Normative Decision Criteria." *Proceedings of the Third International Conference on Operational Research*. Oslo, Norway: Dunod, 1963. **ST**

Simmons, D. C. "Efik Divination, Ordeals and Omens," in *Cultural and Social Anthropology*, Peter B. Hammond (Ed.). New York: Macmillan, 1969, pp. 330–333. **A**

Simmons, John A. "The Risk of Catastrophic Spills of Toxic Chemicals," UCLA-ENG-7425. Los Angeles: UCLA School of Engineering and Applied Sciences, May 1974. **T**

Simmons, J. A., R. C. Erdmann, and B. N. Naft. "Risk Assessment of Large Spills of Toxic Materials." *Proceedings of 1974 Conference on Control of Hazardous Material Spills*, cosponsored by American Institute of Chemical Engineers and the Environmental Protection Agency, San Francisco, August 1974. **T**

Simon, Herbert A. "Some Strategic Considerations in the Construction of Social Science Models," in *Mathematical Thinking in the Social Sciences*, Paul L. Lazarsfeld (Ed.). New York: Free Press, 1954. **SO**

Simon, H. A. *Models of Man*. New York: Bailey, 1957. **EC**

Simon, H. A. "The Architecture of Complexity," *Proceedings of the American Philosophical Society* Vol. 106, December 1962, pp. 467–482. **PH**

Simon, H. A. *The Sciences of the Artificial*. Cambridge, Mass.: MIT Press, 1969. **SY**

Smith, Nicholas M., Jr. "A Calculus of Ethics: A Theory of the Structure of Value," *Behavioral Science*, Vol. 1, Nos. 1 & 2, 1956. **PH**

Smith, Nicholas M., Jr., Stanley S. Walters, Franklin C. Brooks, and David H. Blackwell. "The Theory of Value and the Science of Decision: A Summary," *Journal of Operational Research*, May 1953. **PH**

Soelberg, Peer O. "Unprogrammed Decision Making," *Industrial Management Review*, Spring 1967. **ST**

Spiegelman, M. *Introduction to Demography*. Cambridge, Mass.: Harvard University Press, 1968. **G**

Starr, Chauncey. "Social Benefit vs. Technological Risk," *Science,* Vol. 165, 1969, pp. 1232–1238. **R**

Starr, C. "General Philosophy of Risk-Benefit Analysis." Presented at Electric Power Research Institute (EPRI)/Stanford IES Seminar, Stanford, Calif., September 30, 1974. **R**

Starr, C. "Benefit-Cost Relationships to Socio-Technical Systems," in *Environmental Aspects of Nuclear Power Stations,* IAEA SM-146/47. Vienna: International Atomic Energy Agency, 1971, p. 900. **T**

Stevens, S. S. "Measurement and Man," *Science,* Vol. 127, No. 3295, February 1938, pp. 383–389. **PH**

Stevens, S. S. "Mathematics, Measurement, and Psychophysics," in *Handbook of Experimental Psychology.* S. S. Stevens (Ed.). New York: Wiley, 1951. **PH**

Stevens, S. S. "Measurements, Statistics, and the Schemapiric View," *Science,* Vol. 161, No. 3844, August 30, 1968, pp. 849–856. **PH**

Stigler, G. J. "The Development of Utility Theory," *Journal of Political Economy,* Vol. 58, 1971, pp. 307–327, 373–396. **ST**

Storer, John B. "Late Effects: Extrapolation to Low Dose Rate Exposures," *Health Physics,* Vol. 17, No. 1, July 1969, pp. 3–9. **M**

Suppes, Patrick and Karol Walsh. "A Non-Linear Model for the Experimental Measurement of Utility," *Behavioral Science,* Vol. 4, No. 3, July 1959, pp. 204–211. **ST**

Swalm, Ralph O. "Utility Theory Insights into Risk Taking," *Harvard Business Review,* November–December 1966, pp. 123–136. **ST**

Tamerin, J. S. and H. L. P. Resnik. "Risk Taking by Individual Option-Case Study," in *Perspectives on Benefit-Risk Decision Making.* Report of a Colloquium on Benefit-Risk Relationships for Decision-Making Conducted by the Committee on Public Engineering Policy, National Academy of Engineering, April 26–27, 1971. Washington, D.C.: The Academy, 1972, pp. 73–84. **R**

Taviss, I. "A Survey of Popular Attitudes Toward Technology," *Technology and Culture,* Vol. 13, No. 4, 1972, pp. 606–621. **A**

Teich, Albert H. (Ed.) *Technology and Man's Future.* New York: St. Martin's, 1972. **G**

Thaler, R. and S. Rosen. "The Value of Saving a Life: Evidence from the Labor Market." Paper presented at the National Bureau of Economic Research Conference on Income and Wealth, Washington, D.C., November 30, 1973, published by the Department of Economics, University of Rochester, Rochester, N.Y. **EC**

Torgerson, Warren S. *Theory and Methods of Scaling.* New York: Wiley, 1960. **PH**

Torrance, E. P. and R. C. Ziller. "Risk and Life Experience: Development of a Scale for Measuring Risk-Taking Tendencies," Research Report AFPTRC-TN-57-23, ASTIA Document No. 098926. Randolph Air Force Base, Texas: Air Force Personnel and Training Center, pp. 5–7. Feb. 1957. **PS**

Tribus, Myron. *Rational Descriptions, Decisions, and Designs.* New York: Pergamon Press, 1970. **ST**

U.S. Atomic Energy Commission. "Theoretical Possibilities and Consequences of Major Accidents in Large Nuclear Power Plants (WASH-740)." Report to the Joint Committee on Atomic Energy, Congress of the United States, March 1957. **T**

U.S. Atomic Energy Commission. Proposed Appendix I to AEC Regulation 10CFR50. **T**

U.S. Atomic Energy Commission. "1972 Atomic Energy Programs: Operating and Developmental Functions." Washington, D.C.: Government Printing Office, 1973. **T**

U.S. Department of Health, Education, and Welfare. "Vaccination Against Smallpox in the United States: A Reevaluation of the Risks and Benefits." Atlanta: Public Health Service, Center for Disease Control, revised February 1972. **M**

U.S. Department of Health, Education, and Welfare. "The Health Consequences of Smoking," January 1974. **M**

U.S. Department of the Interior. "Coal-Mine Fatalities in 1970." Bureau of Mines, January 10, 1971. **D**

U.S. Department of the Interior. "Coal-Mine Fatalities in 1972." Bureau of Mines, January 14, 1973. **D**

U.S. Department of the Interior. "Coal-Mine Fatalities in 1973." Bureau of Mines, January 1974. **D**

U.S. Department of the Interior, Bureau of Mines. "Fuels," in *Minerals Yearbook, 1953–1971*. Washington, D.C.: Government Printing Office. **D**

U.S. Department of Labor. *Occupational Injuries and Illnesses by Industry, 1972*. Bulletin 1830, Bureau of Labor Statistics, 1974. **D**

U.S. Food, Drug, and Cosmetic Act, Section 409c(3)A. **G**

U.S. House of Representatives, Committee on Science and Astronautics. *Technology: Processes of Assessment and Choice*. Committee Report. Washington, D.C.: Government Printing Office, 1969. **PL**

U.S. Water Quality Act of 1972—Public Law 92-500. **G**

Van der Meer, H. D. "Decision Making: The Influence of Probability Preference, Variance Preferences, and Expected Value on Strategy in Gambling," *Acta Psychologica*, Vol. 21, 1963, pp. 231–259. **ST**

von Hippel, Frank and Joel Primack. "Public Interest Science," *Science*, Vol. 177, 1972. **E**

Wald, Abraham. "Contributions to the Theory of Statistical Estimation and Testing Hypotheses," *Annals of Mathematics and Statistics*, Vol. 10, 1939, pp. 299–326. **ST**

Webre, A. L. and P. H. Liss. *The Age of Cataclysm*. New York: Putnam, 1974. **R**

Webster's Third New International Dictionary, Unabridged. Springfield, Mass.: Merriman, 1971. **G**

Weinberg, Alvin M. "Science and Trans-Science," *Minerva*, Vol. 10, 1972. **E**

Weinberg, A. M. "Social Institutions and Nuclear Energy," *Science*, Vol. 177, No. 4043, July 7, 1972, pp. 27–34. **SO**

Weisbrod, B. A. "Income Redistribution Effects and Benefit-Cost Analysis," in *Problems in Public Expenditure Analysis*, S. B. Chase, Jr. (Ed.). Washington, D.C.: The Brookings Institution, 1968, pp. 176–222. **EC**

Wiggins, J. H. "Earthquake Safety in the City of Long Beach Based on the Concept of Balanced Risk," in *Perspectives on Benefit-Risk Decision Making*. Report of a Colloquium on Benefit-Risk Relationships for Decision Making conducted by the Committee on Public Engineering Policy, National Academy of Engineering, April 26–27, 1971. Washington, D.C.: The Academy, 1972, pp. 87–95. **R**

Williams, J. D. *The Compleat Strategyst*. New York: McGraw-Hill, 1954. **ST**

Wilson, R. "Tax the Integrated Pollution Exposure," *Science*, Vol. 178, October 1972. **EC**

World Health Organization. *Health Hazards of the Human Environment.* Geneva: 1972. **D**

Wyler, Allen R., Minoru Masuda, and Thomas H. Holmes. "Seriousness of Illness Rating Scale," *Journal of Psychosomatic Research,* Vol. 11, 1968. **PS**

Zadeh, Lotfi A. "Outline of a New Approach to the Analysis of Complex Systems and Decision Processes," *IEEE Transactions on Systems, Man, and Cybernetics,* Vol. SMC-3, No. 1, January 1973. **SY**

Zamora, Ramon. "Decision Analysis Software," *IEEE Transactions on Systems, Man, and Cybernetics,* Vol. SMC-2, No. 2, April 1972. **ST**

Zwicky, F. and A. G. Wilson (Eds.). *New Methods of Thought and Procedure.* New York: Springer, 1967, pp. 273–295. **SY**

Index